AF248700

PHYSICAL PRINCIPLES OF INFRARED IRRADIATION OF FOODSTUFFS

PHYSICAL PRINCIPLES OF INFRARED IRRADIATION OF FOODSTUFFS

Revised, Augmented and Updated Edition

S. G. Il'yasov and V. V. Krasnikov

Moscow Institute of the Food Processing Industry
Moscow, USSR

Translated and Edited by

A. P. Kotlobye

Moscow

⬤ **HEMISPHERE PUBLISHING CORPORATION**
A member of the Taylor & Francis Group

New York   Washington   Philadelphia   London

Originally published as Fizicheskiye osnovy infrakrasnogo oblucheniya pischevykh produktov by Pishchevaya Promyshlennost in 1978.

PHYSICAL PRINCIPLES OF INFRARED IRRADIATION OF FOODSTUFFS

1 2 3 4 5 6 7 8 9 0   E B E B   9 8 7 6 5 4 3 2 1 0

Cover design by Debra Eubanks Riffe.
A CIP catalog record for this book is available from the British Library.

Library of Congress Cataloging-in-Publication Data

Il'iasov, S. G. (Safo Garifullovich)
   [Fizicheskie osnovy infrakrasnogo oblucheniia pischevykh produktov. English]
  Physical principles of infrared irradiation of foodstuffs/S. G. Il'yasov and V. V. Krasnikov; translated and edited by A. P. Kotlobye.—Rev., augm., and updated ed.
    p. cm.
  Translation of: Fizicheskie osnovy infrakrasnogo oblucheniia pischevykh produktov.
  1. Radiation preservation of food. 2. Infrared radiation—Industrial applications. I. Krasnikov, V. V. (Valerii Vladimirovich) II. Title.
TP371.8.I5613   1990
664'.0288—dc20   90-41433
                                                        CIP

ISBN 0-89116-958-X

CONTENTS

The purpose of this book is to explain the physical principles of infrared irradiation and how they should be applied in the thermal processing of foodstuffs.

The phenomena of the propagation and attenuation (both of absorption and scattering) of infrared radiation in foodstuffs are considered, and analytical and experimental methods for studying the optical properties of irradiated substances are described. These methods are developed on the basis of research into the theoretical aspects of both heat and mass transfer in capillary-porous colloidal systems and radiation energy transfer in turbid media.

To carry out the thermal processing of foodstuffs, we need to know the principles governing the transfer of radiant energy in multilayered systems. These principles, as worked out by the present authors, are set forth in this book, along with the principle of mutual effect of transfer of radiant energy and heat and mass transfer during infrared irradiation. On the basis of these principles control systems can be designed for various technological processes in the food industry, such as: desiccation, baking, frying, blanching, curing, and pasteurization. These principles can also be used in a quantitative evaluation of heat and mass transfer inside different layers of the foodstuffs subjected to infrared irradiation; they make it possible to formulate the basic equations for heat and mass transfer and to determine the boundary conditions for processing water-containing substances in the presence of mobile boundaries of phase transformations.

The theoretical equations developed here enable us to get a clear picture of the radiation field in foodstuffs subjected to infrared irradiation and to calculate the distribution of the absorbed energy with respect to the thickness of the layers and the spectral composition of the infrared radiation source. We can then assess the efficiency of a thermal process and work out the optimal configurations of the articles to be irradiated. With the aid of these equations we can also forecast the peculiarities of the heat and mass transfer processes in the foodstuffs being treated and thus improve technological processes in general.

Methods and spectral equipment have been developed for the simultaneous and systematic measurements of the thermal radiational characteristics of scattering materials under different conditions of irradiation. Thus, we can determine the optical properties of the foodstuffs being irradiated by taking into account the following factors: the thermolability of the foodstuffs, the continuity and irreversibility of the biochemical processes occurring in the foodstuffs, the intensive scattering and spreading of the radiant flux, the commensurability of large optical inhomogeneities and the dimensions of the cross section of the radiant flux, and the dependence of the thermal radiational characteristics on the technological parameters of the irradiation process and equipment used.

The spectral measuring equipment mentioned above has already been used for a number of years not only at food processing plants, but also at research institutes, in the textile, chemical and pharmaceutical industries, in peat processing, at paint and varnish plants, and in some branches of light industry.

On the basis of our investigations into the optical properties of foodstuffs being irradiated, methods of quality control have been developed. The quality indicators include: the degree of thermal processing of meat, baked products, and cereals; the translucency of wheat grains; the color and whiteness of starch, flour, and sugar; the taste, aroma and degree of curing of tobacco leaves, and the moisture content of foodstuffs.

The principles of the transfer of energy of integral radiation in selectively absorbing and scattering substances are defined in this book, and based on them, more intensive technological processes have been developed, along with kinetic methods for calculating the engineering parameters of thermal radiation installations. For example, a way has been found for making modified starches and dextrin by treating starch with short-wave infrared radiation. With this method the efficiency of dextrinization increases eight to ten times as compared with the usual thermal conduction method used in industry. The new, intensive process is also more economical; it requires less energy and improves output quality.

The book contains a complete review of the work done in the field of thermal processing of foodstuffs in the Soviet Union and abroad. The list of references covers both theoretical and applied research and development in this field and pertinent sources of patent information.

The book is intended for researchers and engineers specializing in food technology. It can also be used as a university textbook.

SYMBOLS AND DESIGNATIONS

A	absorptance
B	radiance
C_1, C_2	Duntley's parameters
D	sensitivity of the detector
E	radiation flux
E	irradiance
F	radiant power
I	radiant intensity
N	detector signal
P	power
$P(\gamma)$	degree of polarization
R	reflectance
R_∞	reflectivity of a layer with an infinite optical thickness
R	radiant emittance
T	transmittance
T_e	temperature of a radiant body
W_e	radiant energy
W	whiteness
C_λ	velocity of radiation in a medium
C	specific heat capacity
f	forward scattering coefficient
j	flux per unit area
K_λ	absorption coefficient
m	coefficient describing the spatial distribution of incident radiation
$\vec{n}$	normal vector
n_λ	index of refraction
$\vec{q}, q$	net radiation flux vector and net radiation flux, respectively
$r(\Theta)$	indicatrix of reflectance
s	backward scattering coefficient
$t(\Theta)$	indicatrix of transmittance
w	energy absorbed per unit time by volume element
β	absorption criterion
γ	angle of scattering
σ	coefficient describing the spatial distribution of scattering radiation
ϵ	extinction coefficient
$\epsilon(T_e)$	emissivity
Θ	angle of incidence
X_λ	absorption index
λ_t	coefficient of thermal conductivity
Λ	scattering criterion
μ	cosine of incidence angle
ρ_λ	boundary reflection coefficient
$\rho_\lambda(\Theta)$	coefficient of radiance
ρ	density
σ	scattering coefficient

τ	time
ζ	azimuth angle
$X(\gamma)$	indicatrix of scattering
λ	spectral
a	blackbody
*	dimensionless value
'	directional irradiation

FOODSTUFFS AS RADIATION-ABSORBING AND RADIATION-SCATTERING OBJECTS

The transfer of thermal radiation energy in foodstuffs takes place through the absorption and scattering of radiation. Foodstuffs consist mainly of moisture, proteins, carbohydrates, and fats. By their physico-chemical properties and structure they are classified as capillary-porous colloidal systems. It has been experimentally shown [1,6,7,10,39] that in a study of the transfer of energy of thermal radiation in foodstuffs, in various capillary-porous systems, as well as in pigments and powdery, fibrous, cellular and other optically heterogeneous substances, all these can be regarded as energy-absorbing and energy-scattering polydispersed media having a forward scattering indicatrix with a diffraction parameter within 0.1-1000.

1.1 Absorption and scattering of thermal radiation in foodstuffs

When a substance is exposed to electromagnetic radiation its temperature, moisture content, structure and other properties undergo changes. During this process the main characteristics of radiation also change; these include the volume, angular and surface densities of the radiant energy and the polarization characteristics of the radiation [33,39].

By the energy of thermal radiation (radiant energy) is meant the energy transferred by electromagnetic radiation, which is emitted by all bodies whose temperature lies above absolute zero. It is usually considered that thermal radiation falls mainly in the infrared spectral region ranging from 0.75 to 1000 μm. This region is subdivided into a near-infrared region (from 0.75 to 1.50 μm), a middle-infrared region (from 1.50 to 15.0-50.0 μm) and a far-infrared region (from 50.0 to 350-1000 μm).

By the energy of integrated radiation, W_e (joule), is meant the amount of energy radiated (or transmitted) by a medium and falling in the spectral range $\Delta\lambda=\lambda_2-\lambda_1$. The total energy of integrated radiation is the sum of the energy radiated (or transmitted) by a medium in the range of wavelengths from 0 to ∞.

Radiant flux (radiant power) F (watt) is described as follows:

$$F=dW_e/d\tau=\int_{\lambda_1}^{\lambda_2}(dW_\lambda/d\tau)d\lambda=\int_{\lambda_1}^{\lambda_2}F_\lambda d\lambda ; \qquad (1.1)$$

where, F_λ - spectral (monochromatic) radiant flux (watt/μm).

Radiant intensity I (watt/sr) is introduced to describe the angular density of the radiant energy:

$$I=dF/d\omega=\int_0^\infty(dF_\lambda/d\omega)d\lambda=\int_0^\infty I_\lambda d\lambda; \qquad (1.2)$$

here, I_λ - the spectral density of radiant intensity (watt/sr·μm).

The elementary solid angle, according to the polar reference system (Fig.1.1a), is determined with the aid of the azimuth angle $d\phi(0\leq\theta\leq\pi/2)$:

$$d\omega=\sin\theta d\theta d\phi. \qquad (1.3)$$

Radiant emittance (irradiance) (watt/m^2) is numerically equal to the radiant energy emitted in all directions from a unit surface area of a body:

$$R=dF/ds=\int_0^\infty(dF_\lambda/ds)d\lambda=\int_0^\infty R_\lambda d\lambda ; \qquad (1.4)$$

where, R_λ - the spectral density of radiant energy (watt/m^2·μm), which for a blackbody is determined according to the Planck radiation equation:

$$R_\lambda(T)=(2\pi hc^2/\lambda^5)[\exp(hc/\lambda kT)-1]^{-1}. \qquad (1.5)$$

The intensity of radiation, also known as radiance, $B(\text{watt/m}^2\cdot\text{sr})$, determines the specific flux density of thermal radiant energy within a solid angle $d\omega$ in a given direction. This direction is determined by angle θ formed with the surface normal $\vec{n}$ (Fig.1.1b) in the vicinity of the point under consideration:

$$B=(1/\cos\theta)[dI(\theta)/ds]=\int_0^\infty B_\lambda d\lambda; \qquad (1.6)$$

where, B_λ – the spectral intensity of radiation $(\text{watt/m}^2\cdot\text{sr}\cdot\mu m)$. The mag-

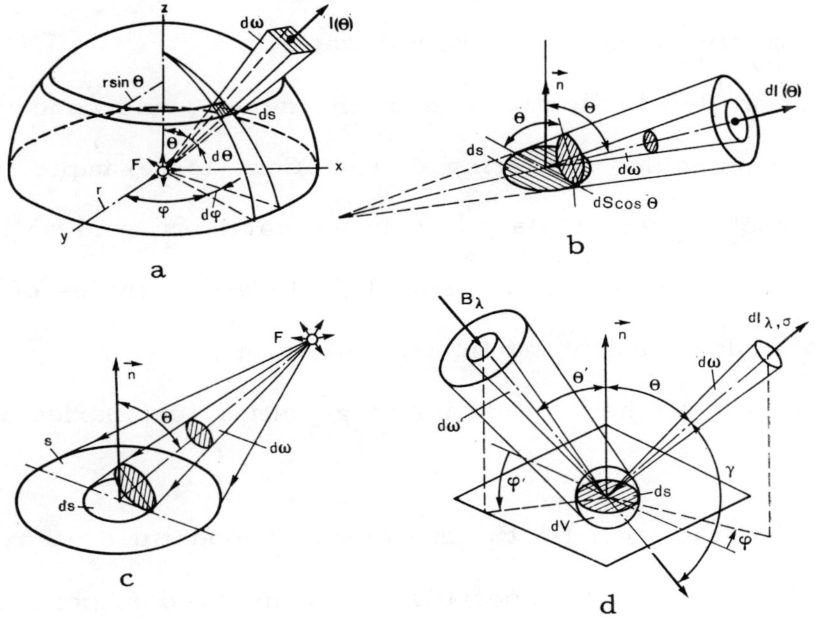

Figure 1.1 Diagrams for determining the energy parameters [39,44]:
    a – radiant intensity;
    b – radiance;
    c – radiant emittance;
    d – scattering indicatrix.

nitude of the spectral specific density of the radiation flux, i.e., the spectral intensity B_λ emitted by the surface, is defined for a blackbody with the aid of Planck's function. Here the spectral intensity of radiation is related to the volume density of radiant energy U_λ and R_λ as follows:

$$B_\lambda=(c_\lambda/4\pi)U_\lambda=(1/\pi)R_\lambda. \qquad (1.7)$$

The flux density of radiant energy transmitted per unit time across a unit surface area in the vicinity of the point under consideration inside a layer of the medium or incident upon a unit surface area of the layer (Fig.1b) is called irradiance E (watt/m^2):

$$E = \int_{(\omega)} B\cos\theta \, d\omega = dF/ds = \int_0^\infty E_\lambda d\lambda; \tag{1.8}$$

where, E_λ – spectral irradiance (watt/m$^2 \cdot \mu$m).

The absorption and scattering of thermal radiation in foodstuffs is determined mainly by the following four processes:

- resonance absorption of radiation by the molecules of a dry substance and by the molecules of structural water and bound water;

- scattering of radiation due to fluctuations in the density and concentration of a substance; and molecular scattering of radiation (for example, scattering by the molecules of proteins, starch, polymer particles, and so on);

- scattering of radiation by suspended colloidal particles, granules of starch, plant cells, microfibrils, pigment particles, and so on;

- scattering of radiation by other optical inhomogeneities (capillaries and pores).

Resonance absorption of radiation by the molecules of foodstuffs occurs in the ultraviolet and visible region and especially in the infrared region. The frequencies of absorption correspond to the quantum transitions. These frequencies arise owing to the quantized changes in the total energy of a molecule – the electron energy and the energy due to mutual atomic vibrations within the molecule and to molecular rotation.

A comparison of the absorption spectra of a dielectric substance and those of foodstuffs and other radiation-scattering substances (Fig.1.2) [107] shows that the optical properties of most foodstuffs are similar to those of wood, paper, and other substances on which radiation is scattered. This is explained by their common cellular structure.

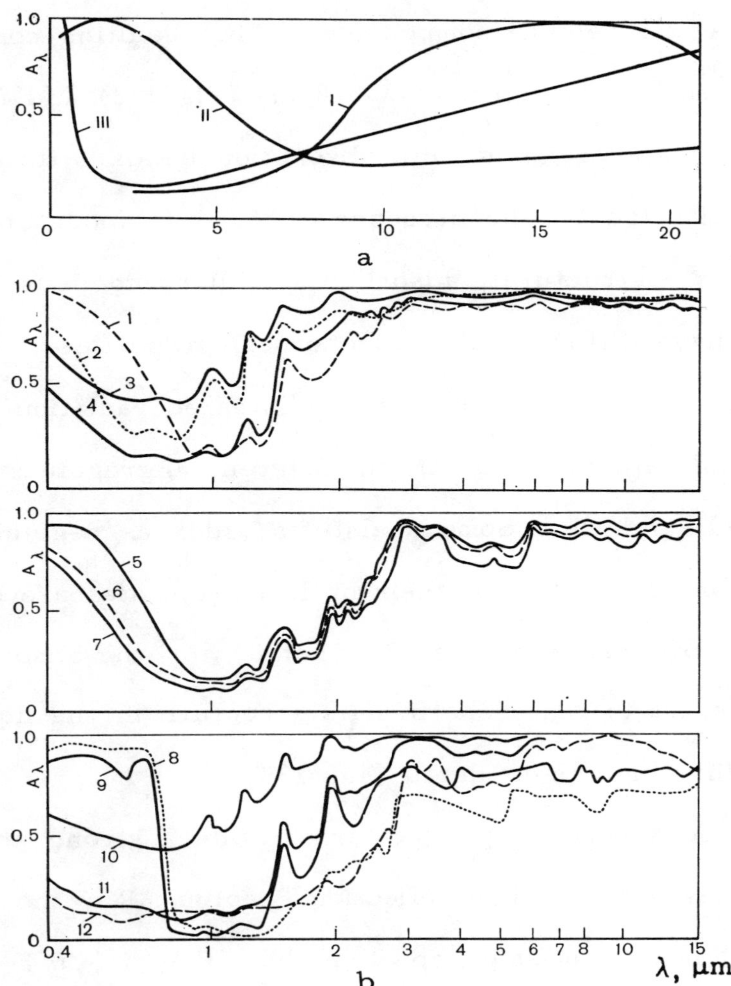

Figure 1.2 Characteristic dependences of absorptance [107]
    a - for solids:
        I - dielectric;
       II - semiconductor;
      III - metal;
    b - for foodstuffs and other radiation-scattering substances:
        1 - dried apple, ℓ(undried)=10 mm;
        2 - apple, ℓ=10 mm, W^d=86.6%;
        3 - potato, ℓ=10 mm, W^d=74.5%;
        4 - dried potato, ℓ(undried)=10 mm;
        5 - redwood, ℓ=10 mm;
        6 - beech, ℓ=10 mm;
        7 - pine wood, ℓ(air-dried state)=10 mm;
        8 - dried tea leaves
        9 - fresh green tea leaves;
       10 - potato starch, W^d=76.5%;
       11 - potato starch, W^d=11.8%;
       12 - enamel VL-55.

Spectral investigations of foodstuffs subjected to infrared irradiation show that the amount of radiation absorbed by a sample is the sum total of the radiation absorbed by all its cellular components. The resulting continuous absorption spectrum lies in the region at $\lambda > 3.0$ μm (Fig.1.2) and is formed by the overlapping of spectral lines and absorption bands. Its overall appearance is strongly affected by the presence of water in foodstuffs. And different types of water - structural, adsorbed, capillary-condensed, and free water - in foodstuffs have different physico-chemical properties.

In the technological processing of foodstuffs by infrared radiation the water contained in pores and capillaries can be in different aggregate states: in the form of vapor, liquid, and ice. Some foodstuffs, such as vegetables and fruits, consist mostly of water, and so their optical properties are determined mainly by the optical properties of water. That is why the absorption spectra of water are of interest to researchers. Its structure in the liquid and solid states has been investigated in detail [130,138].

Throughout the infrared spectral region water exhibits a strong absorption and weak scattering of radiation. The reflection spectrum R_λ, the absorption spectrum k_λ, and the transmission spectrum T_λ (Figs.1.3,1.4) of water are characterized by a series of strong absorption bands. They correspond to the fundamental stretching vibrations: symmetrical vibrations (3420 cm^{-1}) and asymmetrical vibrations (3490 cm^{-1}) in the vicinity of 2.99 μm, their overtones in the vicinity of 0.75, 0.85, 0.98, 1.118, 1.44, and 1.93 μm, deformational vibrations (1640 cm^{-1}) and librational vibrations (700 cm^{-1}), and multiplet vibrations (2130 cm^{-1}) in the vicinity of 6.12, 15.0 and 4.74 μm.

It has been shown experimentally (Fig.1.2) that the presence of even trace amounts of structural or bound water in the substance results in the appearance of intense, wide absorption bands in the vicinity of wavelengths

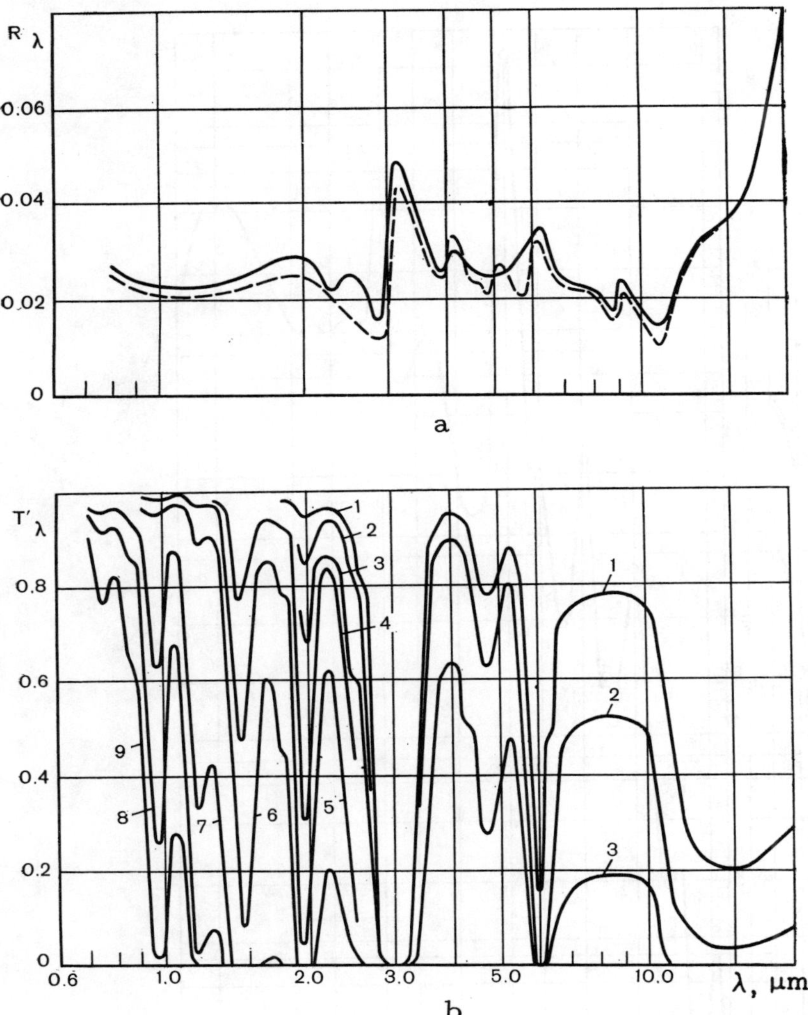

Figure 1.3 Spectral reflectance (a) and transmittance (b) of distilled water as a function of the wavelength and the thickness of a layer (in mm) (dashed line - water from the Black Sea and NaCl solution) [4,60,130,183]:

1 - 0.005;	4 - 0.1;	7 - 10.0;
2 - 0.01;	5 - 0.3;	8 - 30.0;
3 - 0.03;	6 - 1.0;	9 - 100.0.

corresponding to the fundamental normal oscillatory modes. The less intense absorption bands lie in the vicinity of wavelengths corresponding to the over-tones and their combinations. Thus, the spectra of thoroughly dried samples of

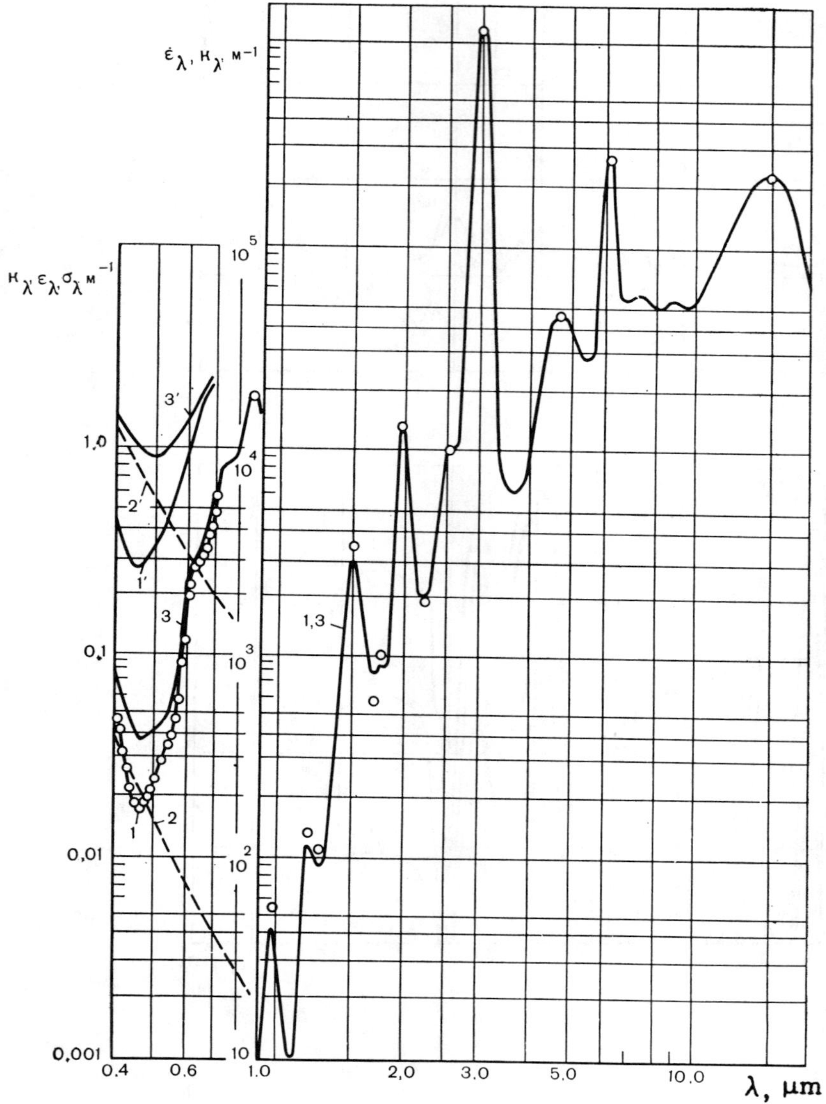

Figure 1.4 Coefficients of absorption, scattering, and extinction as a function of the wavelength in the visible and infrared regions [130]:
    1-3 – distilled water;
    1'-3' – sea water (solid line [64], dotted line [179];
      1' – k_λ; 2,2' – σ_λ; 3,3' – ε_λ.

foodstuffs and other substances also show characteristic absorptions in the vicinity of 0.98, 1.18, 1.44, 2.11, 2.32, 2.52, 2.92, 5.82, and 6.12 μm (Fig. 1.2). This indicates that the absorption bands are due to the presence of hydroxyl groups held through hydrogen bonding with the molecules of the substances being examined.

When the samples are moistened additional absorption bands appear near 1.93 and 4.74 µm (Fig.1.2). This indicates that absorption bands are characteristic of the vibrations of the adsorbed water molecules only.

Molecular scattering of radiation can be observed only in an optically heterogeneous medium. Such a medium is characterized by gradients in density, temperature, and concentration; it exhibits anisotropic properties and is structurally heterogeneous. Therefore, its refractive index depends on the coordinates and on time. Optical microscopic inhomogeneities can appear in a substance owing to fluctuations in very small volumes. Such fluctuations can be caused by thermally induced random movement of atoms or molecules and other particles that make up the substance. Solutions of proteins, polymers and other similar substances exhibit a strong scattering of radiation in the visible and infrared spectral regions. Starch molecules exhibit an even more intense scattering of radiation.

The attenuation (both of absorption and scattering) of radiation by suspended colloidal particles (e.g., starch grains, plant cells, pigment molecules) is analogous to the attenuation by aerosols [33,69], but far more complex in nature.

An electromagnetic wave (whether from a light or a radiation source), when passing through a medium, undergoes diffraction from the surface of a particle whose refractive index differs from that of the medium. As a result we have a phenomenon known as scattering. The scattering of radiation by particles also consists in a combination of the effects of reflection, refraction and secondary radiation by the particles. The scattering and attenuation of radiation thus depend on the size and the form of the particles, on their microstructure and optical properties, and on the wavelength of the radiation.

Colloidal particles, starch grains, plant cells, etc., of which most food-

stuffs are composed, differ in shape and dimension and are optically anisotropic. Moreover, they undergo changes when they absorb moisture [39,61,131].

Milk, an oil-in-water emulsion, contains many particles some of which are
in a dissolved and others in a colloidal state. Thus, casein is present in milk
as a phosphorotein - calcium complex in a colloidal dispersion, with the average
size of the particles being about 100 μm.

The cells of tea and tobacco leaves, etc. are in the form of irregular
polygons 10-40 μm in size on the average, those of green algae (of the order
chlorococcales) - 5-15 μm, and those of erythrocytes - approximately 8 μm.

Because of the diverse properties of the particles it is difficult to determine the spatial distribution of the radiation and derive sufficiently simple
equations for calculating its scattering and attenuation by a polydispersed medium. In a general case the spatial distribution of scattered radiation is a function of the size, the refraction index (n_λ), and the absorption coefficient (χ_λ)
both of the particles and the medium, and is determined by the indicatrix of
scattering.

So long as the linear dimensions of the particles are 2r, which are much
smaller than the radiation wavelength, the secondary radiation waves can be
regarded as being emitted by the electric dipoles, and the character of the
scattering will be analogous to that of molecular scattering. When the dimensions of the particles are commensurable with the incident wavelengths, the
induced vibrations are of a complex nature since the electromagnetic field
varies within each particle.

For a majority of cellular particles the diffraction parameter ρ, which
characterizes their relative dimensions, fluctuates in the infrared spectral region within $0.1 \leq \rho \leq 1000$ [21,64,67,175]. Therefore, the shape of the indicatrix of scattering varies with the wavelength of the radiation. As the dimension of the particles increases $(\rho > 1)$ the scattering indicatrix becomes

elongated in the direction of the incident radiation (the Mie effect [175,64].

For example, for water droplets (n=1.33) the indicatrix of scattering is strongly elongated in a forward direction (Fig. 1.5a) when the diffraction parameters are 4, 8, 15, and 30. It has been shown experimentally that the indicatrix of scattering for milk is elongated in a forward direction and that its shape is typical for the indicatrix of scattering of seawater [33,64].

For different types of starch   the diffraction parameter at $\lambda=1.0$ μm varies from 3.4 to 500. According to a theoretical analysis [64,67] the ratio of the fraction of energy flux which is scattered forward to that which is scattered backward exceeds 10^2-10^3 by an order of magnitude; i.e., the scattering indicatrix is strongly elongated in a forward direction.

When moistened, starch particles and plant cells absorb water gradually, and their structure undergoes changes. As a result the polydispersed particles acquire an irregular form, and their radial refractive index becomes non-uniform (for water n=1.33 and for plant cells n=1.3-1.42 at $\lambda=0.4$-1.4 μm; cf. Fig. 2.11).

The scattering function $K(\rho)$ is dependent in a complex manner on the diffraction parameter ρ, on the complex refractive index of the particles $m_\lambda = n_\lambda - i\chi_\lambda$, and on the wavelength of the radiation. This dependence of $K(\rho)$ in the case of spherical water droplets is shown in Fig. 1.5b (in the visible spectrum – curve 1, and in the infrared spectral region – curves 2,3, and 4). The oscillating character of curve 1 is explained by the interference phenomena of the light waves diffracted from the particles: the scattered field represents the superposition of the electromagnetic fields of dipoles, quadrupoles, etc. [93]. The position of the main diffraction maxima depends on the complex refractive index $m_\lambda = n_\lambda - i\chi_\lambda$. The weak oscillations are due to the fine structural interference pattern. With an increase in the absorption index χ_λ the $K(\rho)$ function levels off and the oscillations of the curve become unnoticeable.

The $K(\rho)$ function has been calculated for heterogeneous and two-layer systems according to a very simple theoretical model for a spherical particle; the refractive index is found to vary from the center to the edge of the particle from 1.33 to 1.5. It has been established that the $K(\rho)$ function for heterogeneous particles differs considerably from that for particles of water (n=1.33) and homogeneous particles (n=1.5) [93].

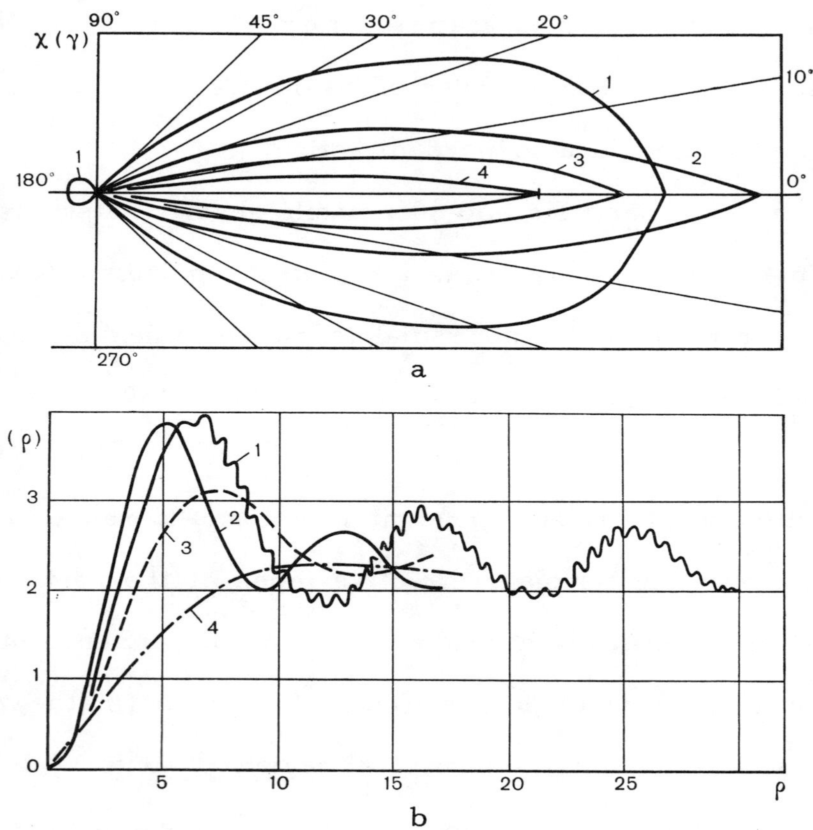

Figure 1.5 Scattering indicatrix $\chi(\gamma)$ for water droplets [64]:
a: 1 - ρ=4;   b: 1 - λ=0.4-1.0 μm; n=1.33;
    2 - ρ=8;   2 - λ=3.45 μm; m=1.42 - i0.014;
    3 - ρ=15;   3 - λ=8.2 μm; m=2.28 - i0.051;
    4 - ρ=30;   4 - λ=11.0 μm; m=1.14 - i0.114.

In foodstuffs any layer consists of numerous particles; when subjected to infrared irradiation each particle is irradiated not only by the initial radiant flux but also by the radiation scattered from other particles. In other words, there is secondary and multiple scattering. It should be borne in mind that radiation which is scattered in all directions except the direction of its incident propagation is always partially polarized [33,64,67,117].

The scattering of radiation on other optical heterogeneities in foodstuffs is a more complex phenomenon than the usual scattering of radiation by particles. The former results from the combined effects of scattering, reflection, refraction, and secondary radiation emitted both at the interface between the sample and the water vapor-air medium and by the particles.

Here not only colloidal particles and fluctuations in density but also pores and capillaries act as the scattering centers (optical inhomogeneities). The pores and capillaries are located at random throughout the irradiated object. And they not only scatter radiation, but also divert its direction at the interface between two media – a water vapor-air medium inside the pores and capillaries, and water on their walls, as well as at the interface formed by water and the walls of the pores or capillaries.

The refractive indices for various components of plant cells differ very little from one another, and so the scattering of radiation in them occurs mainly at the interface between the cell and the binding medium. It can be shown that for a majority of foodstuffs, within a layer of a sample that is more than 0.1 mm thick, there can be multiple scattering, with the multiplicity factor being greater than two [39].

An investigation of the basic phenomena related to the absorption and scattering of radiation in foodstuffs shows how complex the problem is of explaining theoretically the attenuation of thermal radiation in such substances. The fact that absorption and scattering depend on numerous factors and that these factors influence one another makes a theoretical analysis of the subject a difficult task. Therefore, at this stage more attention is being paid to experimental research into the process of attenuation of radiation in foodstuffs.

1.2   Polarization of scattered radiation

To understand the propagation of electromagnetic radiation in foodstuffs it is necessary to know the basic correlations between the parameters of the particles and their light-scattering and polarization properties [116,211]. It has been shown [193] that when natural light is reflected from light-scattering particles radiation is most strongly polarized in those spectral regions where absorption is the most intense; in other words, the polarization of light is selective in nature. Our research [39,40,44] into the reflection spectra of bread crust, meat and other foods subjected to plane-polarized radiation confirms this conclusion. For the perpendicular incident plane $\vec{E}_s$ and the parallel incident plane $\vec{E}_p$ of the component of the light vector $\vec{E}$, the reflection spectra differ most markedly in the region of the absorption bands.

Scattered radiation is always partially polarized. The properties of a particle subjected to scattered radiation are determined by two factors: the indicatrix of scattering $\chi(\gamma)$ and the angular function of the degree of polarization $P(\gamma)$. Both of these factors depend on the diffraction parameter ρ, the index of refraction n_λ, and the index of absorption χ_λ. The degree of polarization $P(\gamma)$ is defined as the ratio of the difference in extreme values to the sum of the extreme values of the amplitude of the electric intensity vector of radiation squared $|\vec{E}|^2$ in the plane that is perpendicular to the direction of the radiation.

By measuring the intensity of scattered radiation B_λ, which numerically equals $|\vec{E}|^2$, we can calculate the degree of polarization from a known equation [33,67]:

$$P_\lambda(\gamma)=(B_{max\lambda} - B_{min\lambda})/(B_{max\lambda} + B_{min\ \lambda}). \tag{1.9}$$

In Fig.1.6 are shown the polar diagrams of scattered radiation for different forms of the scattering indicatrix $\chi(\gamma)$. A simple symmetrical form is characteristic of spherical nonabsorbing particles (Fig.1.6a). In this case at $\rho \ll 1$ (Rayleigh's scattering) the radiation is completely polarized at a $90°$ angle in the direction of the incident flux. At $\alpha = 90°$ the component vector $\vec{E}_p$ (dashed line) equals zero, and the radiation scattered in this direction has the vibration of the electric vector only in the plane that is perpendicular to the plane of the drawing, $\vec{E}_s$ (solid line); in other words, the radiation is completely polarized ($P=1$). In all other directions the degree of polarization is less than one ($P<1$). In this case the scattered radiation is partially polarized since both components, $\vec{E}_p$ and $\vec{E}_s$, differ in their magnitude except for the direction in which radiation propagates. In the directions $\theta = 0°$ and $\theta = 180°$ the magnitude of $\vec{E}_p$ and that of $\vec{E}_s$ coincide, and the degree of polarization equals zero.

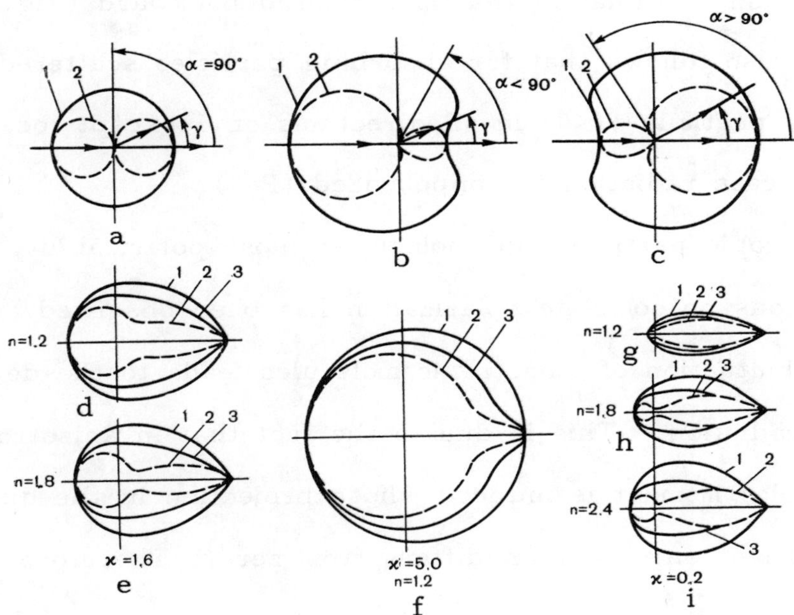

Figure 1.6 Main forms of scattering indicatrixes [92]:
    a – Rayleigh's;   d,e,f,g,h,i – for large particles at different n and χ;
    b – backward;   1,2 – of polarized light;
    c – forward;   3 – of natural light;

$$1 - \left|\vec{E}\right|^2; \qquad 2 - \left|\vec{E}_p\right|^2; \qquad 3 - \left|\vec{E}\right|^2.$$

In the case of most foodstuffs and other polydispersed systems, $0.1 \leq \rho \leq$

1000, radiation is scattered both by small and large energy-absorbing particles. This has been confirmed by experimental data [39,44]. The indicatrix of scattering for a volume element of large particles is strongly elongated forward in the direction of the propagation of incident radiation.

In Fig.1.6d-i are shown the indicatrixes of scattering, calculated by the method of geometrical optics [92,93], for large particles having the following characteristics: $\rho \geq 100$, n=1.2, 1.8, 2.4, χ=0.2, 1.6, 5.0. With an increase in the absorption index from 0.2 to 5.0 at a constant index of refraction (n= 1.2) the form of the scattering indicatrix changes and assumes a backward direction. The fraction of radiation that is scattered backward increases with an increase in the absorption of particles (cf., Fig.1.6g,d,f). An increase in the refractive index at a constant absorption index (χ=0.2) likewise leads to an increase in the fraction of radiation that is scattered backward (Fig.1.6g, h,i). From Fig.1.6d-i it also follows that for absorbing particles scattered radiation is polarized only partially (P<0) in all directions of θ except for θ=0° and θ=180° in which case radiation is nonpolarized (P=0).

In the case of anisotropic particles and molecules whose polarizability is not the same in all directions no complete polarization has been observed at any angle α [67]. The fluctuation of anisotropic molecules leads to the depolarization of scattered radiation. This is due to the fact that in anisotropic molecules an electric (dipole) moment is induced whose projection in the direction of the propagation of incident radiation differs from zero. Therefore, even in a direction that is perpendicular to the propagation of radiation there is no complete polarization. This phenomenon is known as the depolarization of scattered radiation [67].

Such depolarization by the leaves of plants and by other light-scattering substances has been described [95,116]. We are at present investigating the relationship between the polarization characteristics of foodstuffs and their quality.

A very simple method for estimating the energy and polarization characteristics of scattered radiation with the aid of four specific parameters was proposed by Stokes in 1852. The method is widely used to this day [33,69, 195].

Approximate methods for investigating this phenomenon usually disregard the polarization characteristics of radiation. In this case the absorption and scattering of radiation are accounted for with the aid of experimentally determined optical properties.

1.3   The absorbing and scattering properties of foodstuffs

To understand and analytically describe the absorption and scattering of radiation in a substance it is necessary to know the optical properties of this substance and the principles governing its complex microstructure. For this purpose it is enough to know some of the optical properties which can be determined experimentally. Depending on the type of the problem set (spatial - three-dimensional or plane - unidimensional) in investigating the transfer of energy and radiation, these optical properties can be conditionally referred to a volume element dV or to a layer having dx thickness. Such properties will represent the total effect of all the above-mentioned factors which determine the absorption and scattering of radiation and will furthermore take into account the mutual effect of these factors.

Thus, if we know such optical properties of a specific substance, we can formulate and numerically solve the problem regarding the propagation of radiation and the attenuation of its energy (both for absorption and scattering) in this substance.

The condtionality of the linear dimensions of the volume element and the layer element is explained by the fact that their dimensions can be smaller than those of the optical inhomogeneities (such as large pores in bread). There-fore, the optical properties and the established regularities for the attenua-tion of radiation become meaningless if the thickness of the layer or the volume of the substance is less than the dimensions of individual optical inhomogene-ities.

Let us assume that the volume element dV or a layer having dx thick-ness inside an absorbing and scattering medium is exposed to a radiation flux having spectral radiance B_λ within a solid angle $d\omega'$ in the direction θ' (Fig. 1.1d). The incident radiant power F will partially decrease owing to the ab-sorption of radiation inside this volume element or the layer element and to the scattering of radiation in all directions by either of these elements.

As a result of multiple scattering of radiation either the volume element or the layer element will be irradiated from all directions by scattered radia-tion even if the original incident flux was unidirectional in nature. In this case the irradiation conditions for the volume or the layer element from the half-space $-\pi/2<\theta'<\pi/2$ and from the half-space $\pi/2\theta'<-\pi/2$ will be different because of the difference in the spatial distribution of the radiation energy.

It has been shown [33,39] that in those cases where the polarizational effects can be disregarded in describing the propagation and attenuation of radiation in turbid media it is sufficient to know the magnitude of the absorp-tion coefficient k_λ, the scattering coefficient σ_λ, and the indicatrix of scat-tering $\chi_\lambda(\gamma)$. Here, the indicatrix of scattering can be used in estimating multiple scattering in strongly scattering media.

However, because of the complex nature of scattering and absorption for most foodstuffs it is practically impossible to determine these character-istics, in particular, the indicatrix of scattering $\chi_\lambda(\gamma)$. To overcome this difficulty the main characteristics are replaced by auxiliary averaged character-istics which are determined experimentally. The latter represent the overall information on the spatial distribution of radiation inside the volume or the layer element and on the optical properties of the substance.

When a sample is exposed either to unidirectional or to incident multi-directional diffuse radiation the scattering coefficient σ_λ can be conditionally represented as the sum of two semispherical coefficients averaged with respect to the angle - the coefficient of backward scattering s'_λ and the coefficient of forward scattering f'_λ:

$$\sigma_\lambda = s'_\lambda + f'_\lambda; \tag{1.10}$$

where, $s'_\lambda = \sigma_\lambda \delta_s$, $f'_\lambda = \sigma_\lambda \delta'_f$.

The magnitude of s'_λ and f'_λ, unlike that of σ_λ, depends not only on the optical properties of a substance but also on the form of the scattering indicatrix which is taken into account by the coefficients δ'_s and δ'_f. To assess the significance of these coefficients we shall determine the scattered and absorbed fractions of the radiation flux incident upon a volume element or a layer element from a half-space ($\omega \leq 2\pi$) in a general case of arbitrary energy distribution:

$$dF_{\lambda k} = k_\lambda \, dV d\lambda \int_{(\omega')} B_\lambda(\omega') d\omega'. \tag{1.11}$$

After multiplying Eq.1.11 by a fraction whose numerator and denominator contain the term $\int_{(\omega')} B_\lambda(\omega')\cos\theta \, d\omega'$, and by taking into account Eq.1.8, we obtain:

$$dF_{\lambda k} = k_\lambda m_\lambda E_\lambda \, dV d\lambda = \bar{k}_\lambda E_\lambda \, dV d\lambda. \tag{1.12}$$

Here, $m_\lambda = [\int_{(\omega')} B_\lambda(\omega') d\omega'] / [\int_{(\omega')} B_\lambda(\omega')\cos\theta \, d\omega'] -$ (1.13)

the proportionality factor, known as the coefficient of spectral distribution of incident radiation; it is equal to the reciprocal of the mean cosine ($\mu = \cos\theta$) of the angle formed by the incident flux ($m_\lambda = \mu^{-1}$). In the case where all the rays are in normal incidence with respect to the surface of the volume or the layer element, $m_\lambda = 1$; and in the case of a hemispherical, perfectly diffuse incident flux, $m_\lambda = 2$. When μ approaches zero (i.e., when all the rays are incident upon the surface of the volume element or the layer element at angle θ close to $90°$), $m_\lambda \to \infty$.

The fraction of an incident flux scattered in all directions within a solid angle $\omega < 2\pi$ can be described as follows, with account taken of Eqs. 1.12 and 1.13:

$$dF_{\lambda\sigma} = dVd\lambda \int_{(\omega')} B_\lambda(\omega')(\sigma_\lambda/4\pi) \int_{(4\pi)} \chi_\lambda(\gamma)d\omega. \qquad (1.14)$$

By taking into consideration Eqs. 1.8, 1.12-1.14 and the normalization condition for the indicatrix of scattering $\chi_\lambda(\gamma)$ we can transform Eq. 1.14 into:

$$dF_{\lambda\sigma} = \sigma_\lambda m_\lambda E_\lambda dVd\lambda . \qquad (1.15)$$

For many practical cases it is convenient, when solving the problem of radiation energy transfer, to represent the term $dF_{\lambda\sigma}$ in the form of the sum of two addenda: backward scattered radiation $dF_{\lambda\sigma}$ (into the half-space $2\pi, n_+$; i.e., into a half-space with a positive direction of the normal $\vec{n}$ with respect to the irradiated surface of a volume element or a layer element; cf., Fig. 1.1d); and forward scattered radiation $dF_{\lambda f}$ (into the half-space $2\pi, n_-$):

$$dF_{\lambda\sigma} = dF_{\lambda s} + dF_{\lambda f}. \qquad (1.16)$$

On the basis of Eq. 1.14 the terms $dF_{\lambda s}$ and $dF_{\lambda f}$ can be defined as follows:

$$dF_{\lambda s} = dVd\lambda \int_{(\omega' \leqq 2\pi, n_+)} B_\lambda(\omega')d\omega'(\sigma_\lambda/4\pi) \int_{(2\pi, n_+)} \chi_\lambda(\gamma)d\omega; \qquad (1.17)$$

$$dF_{\lambda f} = dVd\lambda \int_{(\omega' \leqq 2\pi, n_+)} B_\lambda(\omega')d\omega'(\sigma_\lambda/4\pi) \int_{(2\pi, n_-)} \chi_\lambda(\gamma)d\omega. \qquad (1.18)$$

By using Eq. 1.13 we can rewrite the last two equations in the following form:

$$dF_{\lambda s}=\sigma_\lambda \delta_s m_\lambda E_\lambda dVd\lambda = s_\lambda E_\lambda dVd\lambda; \qquad (1.19)$$

$$dF_{\lambda f}=\sigma_\lambda \delta_f m_\lambda E_\lambda dVd\lambda = f_\lambda E_\lambda dVd\lambda. \qquad (1.20)$$

Here, σ_s and δ_f - coefficients which are numerically equal to the fraction of the radiation that is scattered backward and forward, respectively, by a volume or a layer element having an indicatrix of scattering $\chi_\lambda(\gamma)$ when irradiated within a solid angle $\omega' \leqq 2\pi, n_+$ with an angular radiance distribution $B_\lambda(\omega')$. The coefficients σ_s and δ_f can be determined from the equations:

$$\delta_s = [\int_{(\omega'\leqq 2\pi, n_+)} B_\lambda(\omega')d\omega'(1/4\pi) \int_{(2\pi, n_+)} \chi_\lambda(\gamma)d\omega] / \int_{(\omega'\leqq 2\pi, n_+)} B_\lambda(\omega')d\omega'; \qquad (1.21)$$

$$\delta_f = [\int_{(\omega'\leqq 2\pi, n_+)} B_\lambda(\omega')d\omega'(1/4\pi) \int_{(2\pi, n_-)} \chi_\lambda(\gamma)d\omega] / \int_{(\omega'\leqq 2\pi, n_+)} B_\lambda(\omega')d\omega'; \qquad (1.22)$$

From a comparison of Eqs. 1.15-1.18, with account taken of Eqs. 1.21, 1.22, it follows that the sum of the coefficients is always unity:

$$\delta_s + \delta_f = 1. \qquad (1.23)$$

For spherical, Rayleigh's or other symmetrical form of the scattering indicatrix it is obvious that $\delta_s = \delta_f = 0.5$. In the case of directional irradiation with a parallel flux $\omega' \to d\omega'$ we have:

$$\delta_s(\theta',\phi')=\delta'_s=(1/4\pi) \int_{(2\pi, n_+)} \chi_\lambda(\gamma)d\omega; \qquad (1.24)$$

$$\delta_f(\theta',\phi')=\delta'_f=(1/4\pi) \int_{(2\pi, n_-)} \chi_\lambda(\gamma)d\omega. \qquad (1.25)$$

An analysis of a general case of irradiation of a volume element or a layer element from all directions ($\omega'=4\pi$) shows that we can obtain from Eq. 1.13 the following relationship [39] for m_λ:

$$m_\lambda = E_{o\lambda}/E_\lambda. \qquad (1.26)$$

Table 1.1 Terminology and the analytical relationships between the optical properties of a volume element or a layer element under different irradiation conditions.

Spectral optical properties	Symbol	Dimension	Directional (θ',ϕ') or spherical flux $(\omega'=4\pi)$	Diffuse flux $(\omega'<2\pi)$	Perfectly diffuse flux $(\omega'=2\pi)$
			Averaged basic characteristics		
Absorption coefficient	$\overline{k}_\lambda$	m^{-1}	k_λ	$m_\lambda k_\lambda$	$2k_\lambda$
Scattering coefficient	$\overline{\sigma}_\lambda$	m^{-1}	σ_λ	$m_\lambda \sigma_\lambda$	$2\sigma_\lambda$
Backward scattering coef.	s_λ	m^{-1}	$s'_\lambda=\delta'_s \sigma_\lambda$	$m_\lambda \delta_s \sigma_\lambda$	$2\delta_s \sigma_\lambda$
Forward scattering coef.	f_λ	m^{-1}	$f'_\lambda = \delta'_f \sigma_\lambda$	$m_\lambda \delta_f \sigma_\lambda$	$2\delta_f \sigma_\lambda$
			Averaged derivative characteristics		
Effective extinction coef.	$\varepsilon_{\lambda ef}$	m^{-1}	$\varepsilon'_{\lambda ef}=k_\lambda + s'_\lambda$	$k_\lambda + s_\lambda = m_\lambda(k_\lambda + \delta_s \sigma_\lambda)$	$2(k_\lambda + s'_\lambda)=2(k_\lambda + \delta_s \sigma_\lambda)$
Schuster number	$\Lambda_{\lambda ef}$	-	$\Lambda_{\lambda ef}=s'_\lambda/(k_\lambda + s'_\lambda)=s_\lambda/(\overline{k}_\lambda + s_\lambda)=\delta_s \sigma_\lambda/(k_\lambda + \delta_s \sigma_\lambda)$		
Effective specific absorption	$\beta_{\lambda ef}$	-	$\beta_{\lambda ef}=k_\lambda/s'_\lambda=\overline{k}_\lambda/s_\lambda=k_\lambda/\delta_s \sigma_\lambda=\delta_s^{-1}\beta_\lambda$		

Here, $E_{o\lambda}$ - spatial irradiance (watt/m^2) which has been defined [39] as follows:

$$E_{o\lambda} = \int_{(4\pi)} B_\lambda(l')d\omega = \pi^{-1} \int_{(4\pi)} E_\lambda d\omega. \qquad (1.27)$$

This means that $E_{o\lambda}$ is the total irradiance for a given surface from two directions, a positive and a negative direction of the normal with respect to the surface, i.e.:

$$E_{o\lambda} = E_+ + E_-. \qquad (1.28)$$

In the case of directional normal irradiation of the surface, $E_{o\lambda} = E_\lambda$; and in the case of perfectly diffuse radiation, $E_{o\lambda} = 2E_\lambda$.

With the aid of Eq. 1.26 we can rewrite the equations describing the fraction of a flux that is absorbed and scattered by a volume element in all directions in a simple form:

$$dF_{\lambda k} = k_\lambda E_o dV d\lambda,$$
$$dF_{\lambda \sigma} = \sigma_\lambda E_o dV d\lambda. \qquad (1.29)$$

Eq. 1.29 confirms that the coefficients k_λ and σ_λ are independent of the irradiation conditions. Therefore, these coefficients are characteristic only of the optical properties of a volume element or a layer element of a given substance. The spatial distribution of radiation scattered in all directions is determined by the scattering indicatrix $\chi_\lambda(\gamma)$.

As has been shown [33,64], the propagation and attenuation of radiation in foodstuffs can be described by the averaged absorption coefficient $\overline{k}_\lambda$ and the averaged backward scattering coefficient s_λ; both of these coefficients can be determined experimentally. Information regarding the third characteristic, $\chi_\lambda(\gamma)$, and the irradiation conditions are taken into account by the terms $\overline{k}_\lambda$

and s_λ with the aid of the coefficients δ_s and m_s.

The terminology used and the analytical relationships describing the main and averaged optical properties of radiation-absorbing and radiation-scattering substances are summarized in Table 1.1.

1.4 Thermal radiational characteristics of foodstuffs

By the thermal radiational characteristics of a substance we refer to its absorption, reflection, and transmission of incident radiation, as well as its emission of radiation. The first three characteristics depend both on the spectral composition of the radiation source and the degree of polarization and spatial properties of incident radiation, i.e., on the irradiation conditions, and on the state and the properties of the irradiated substance itself. As for the emission of radiation by a substance, it is determined solely by the state and the optical properties of the substance.

There are three basic types of incident spectral radiation, F_λ, or incident spectral integral radiation F: directional flux (Fig. 1.7) with a radiance $B_\lambda \theta$ or $B(\theta)$ within a solid angle $d\omega$; diffuse radiation with an arbitrary angular distribution of the radiance $B_\lambda(\omega)$ or $B(\omega)$ within a solid angle $\omega < 2\pi$; and hemispherical (uniformly diffuse) radiation with an angular distribution of the radiance $B_\lambda(2\pi)$ or $B(2\pi)$ according to the Lambert law within a solid angle $\omega = 2\pi$.

An analysis of the process of infrared irradiation of foodstuffs in industrial driers and thermal radiational equipment shows that in closed chambers irradiation takes place owing to diffuse and mixed (diffuse and directional) radiation [39]. Under arbitrary conditions of irradiation of the surface S of a substance with radiation F, the sum of the absorbed energy F_A, reflected energy F_R, and transmitted energy F_T per unit time, according to the law of conservation of energy, should be equal to the incident radiant power: $F_A + F_R + F_T = F$, or:

$$A + R + T = 1; \qquad\qquad\qquad (1.30)$$

where, $A = F_A/F$ - absorptance (absorption coefficient); $R = F_R/F$ - reflectance (reflection coefficient); $T = F_T/F$ - transmittance (transmission coefficient).

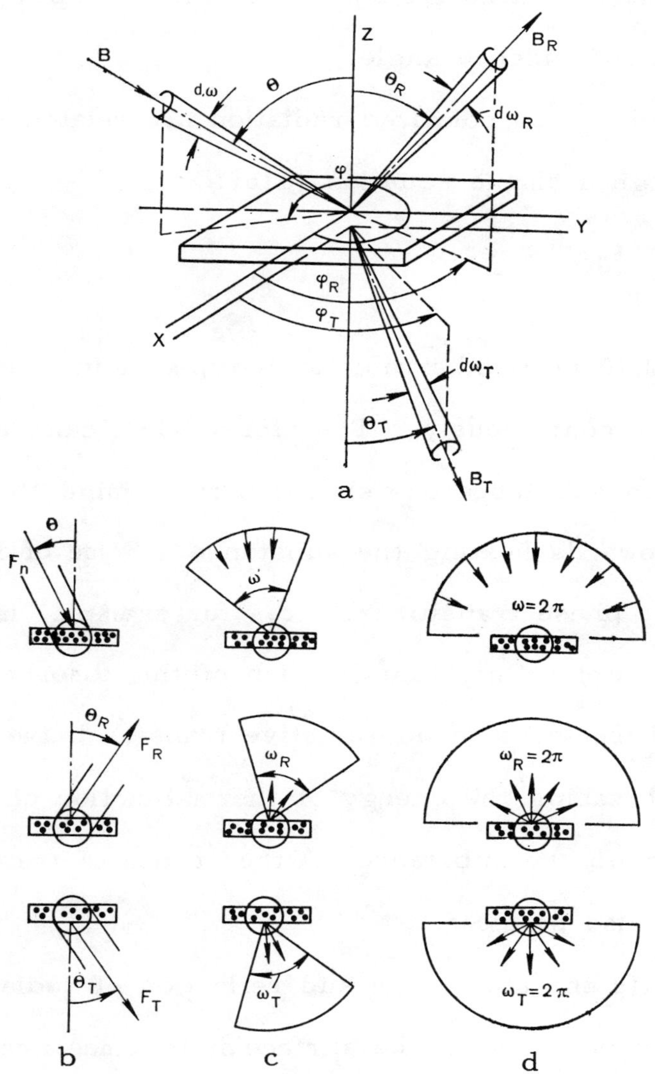

Figure 1.7 Spatial distribution of radiation incident upon the surface of a substance, reflected by the surface and transmitted through it under different irradiation conditions [39, 44]:
a,b - directional; c - diffuse; d - hemispherical.

The quantities A, R, and T depend on the state of polarization of the incident radiation. We have found that in the irradiation of bread, meat and other foodstuffs with plane-polarized radiation the magnitude of R_s for the perpendicular component $\vec{E}_s$ of light vector $\vec{E}$ is greater than the magnitude of R_p for the $\vec{E}_p$ component lying in the incidence plane. The difference between R_s and R_p is the greatest in the region corresponding to the absorption by the substance and depends on the incidence angle.

The components R_p and R_s for polarized radiation are related to R for nonpolarized radiation through a simple equation [116]:

$$R = 1/2(R_p + R_s). \qquad (1.31)$$

In foodstuffs subjected to infrared irradiation complex and irreversible biochemical processes take place continuously. Therefore, when calculating the heat transfer in an irradiated substance one should bear in mind that not all the absorbed energy goes towards heating the substance. Some of it is used up in endothermic reactions, phase transitions, etc. (for example, in dextrinization and denaturation processes). Only that fraction of the absorbed energy which is lost by the excited electrons in nonradiative transfer between the energy levels goes towards increasing the energy of thermal motion of microparticles, i.e., towards heating up the substance. Other types of transformation of the absorbed energy are also possible.

Since for most foodstuffs the absorption and reflection of radiation take place throughout a layer and not only on the surface it becomes necessary to define more precisely the concepts of diffuse and specular reflection.

From the point of view of wave optics the reflection of electromagnetic waves from a polydispersed medium means a partial or complete return of wave motion from a diffracting surface of the interface of two media to the half-space of the first medium. Specular reflection can be observed only when the radius of the curvature r of different inhomogeneities parallel to the interface changes by $\Delta r \ll \lambda$. The surface is considered optically rough when $\Delta r > \lambda$; in this case we have diffuse reflection [59,67]. In foodstuffs exposed to infrared radiation the incident electromagnetic waves cause the particles to oscillate and consequently to emit secondary waves which are coherent with the incident wave. What we have in fact is coherent Rayleigh scattering by atomic-molecular systems. The interference of incident and secondary waves results in the appearance of reflected and refracted waves.

We shall consider a layer of a substance which reflects (or transmits) radiation falling in the direction θ, ϕ (Fig. 1.7a) within a solid angle $d\omega$, and in the direction θ_R, ϕ_R (θ_T, ϕ_T) within a solid angle having the same dimensions, $d\omega_R = d\omega_T = d\omega$ (as determined by the laws of geometrical optics), to be a layer which specularly reflects radiation (or transmits it without scattering). And we shall consider a layer which reflects (or transmits) radiation falling in all directions into a half-space, with the energy in it being distributed in all directions according to Lambert's law, to be a layer which uniformly reflects (or transmits) diffuse radiation.

Let us suppose that the distribution of incident, reflected and transmitted radiant flux in space is defined by the functions $B_\omega(\theta,\phi)$ and $B_\omega(\theta,\phi; \theta_R, \phi_R)$, and $B_\omega(\theta,\phi; \theta_T,\phi_T)$, respectively. Then in a general case of diffuse radiation ($d\omega < \omega < 2\pi$) the radiant flux expressed through the function of radiant power is determined from the following equations [54,122]:

for radiant flux incident upon a surface S from a solid angle ω:

$$F(\omega) = S \int_\omega B_\omega(\theta,\phi)\cos\theta \, d\omega; \qquad (1.32)$$

for radiant flux reflected by a layer within a solid angle ω_R:

$$F_R(\omega,\omega_R) = S \int_\omega B_\omega(\theta,\phi; \theta_R,\phi_R)\cos\theta_R \, d\omega_R; \qquad (1.33)$$

for radiant flux transmitted through a layer within a solid angle ω_T:

$$F_T(\omega,\omega_T)=S\int_\omega B_\omega(\theta,\phi;\theta_T,\phi_T)\cos\theta_T d\omega_T. \tag{1.34}$$

By substituting the values of radiant power from Eqs. 1.32-1.34 into Eq. 1.30 we obtain a general relationship which describes the reflectance and transmittance of a layer:

$$R(\omega,\omega_R)=[\int_\omega B_\omega(\theta,\phi;\theta_R,\phi_R)\cos\theta_R d\omega_R]/[\int_\omega B_\omega(\theta,\phi)\cos\theta d\omega]. \tag{1.35}$$

In order to simplify Eq. 1.35 the term for incident power F is usually expressed through an isotropic flux with B_ω=const, which is independent of the direction (θ,ϕ) inside any solid angle ω' within the boundaries of the half-space $\omega'\leqq 2\pi$:

$$\tilde{B}_\omega=[\int_\omega B_\omega(\theta,\phi)\cos\theta d\omega]/[\int_{\omega'\leqq 2\pi}\cos\theta d\omega]. \tag{1.36}$$

The incident flux under different irradiation conditions will be as follows, according to Eq. 1.32 and with account taken of Eqs. 1.35 and 1.36 when calcating the integral with respect to the cosine within the limits $\omega=2\pi$:

$$F(\omega)=S\int B_\omega(\theta,\phi)\cos\theta d\omega=\pi\tilde{B}_\omega S. \tag{1.37}$$

By substituting Eq. 1.37 into Eq. 1.35 we can obtain   simpler relationships for R and T.

The considerations set forth above hold true for absolute measurements. However, very few methods have been developed for carrying out absolute measurements of R and T for light-scattering media.   Therefore, let us consider the theoretical aspects of the relative methods for measuring such quantities.

The relative methods are used for determining the relative intensity of radiation reflected by a sample or a standard in a given direction - (θ_R,ϕ_R) or (θ_T,ϕ_T).   For the radiation reflected by the standard we can introduce the equivalent intensity of reflection $B_{\omega R}$. It is analogous to the radiance

of the source B_ω (Eq. 1.36) and in a general case of arbitrary radiation can

be described through the following averaged function:

$$\widetilde{B}'_{\omega R}=[\int_{\omega_R} B'_\omega(\theta,\phi;\theta_R,\phi_R)\cos\theta_R d\omega_R]/[\int_{\omega_R}\cos\theta_R d\omega_R]. \tag{1.38}$$

For a standard which uniformly reflects diffuse radiation the integral with

respect to the cosine in the denominator should be taken within the limits

$\omega'_R=2\pi$; then, in place of Eq. 1.38 we obtain:

$$\widetilde{B}'_{\omega R}=\pi^{-1}\int_{\omega_R}B'_\omega(\theta,\phi;\theta_R,\phi_R)\cos\theta_R d\omega_R. \tag{1.39}$$

If the reflectance of the standard is unity ($R_s=1$), the reflected radiant

power $F'_R(\omega,\omega_R)$ will be equal numerically to that of the incident flux $F(\omega)$; and

we can write:

$$F'_R(\omega,\omega_R)=\pi\widetilde{B}'_{\omega R}S=F(\omega). \tag{1.40}$$

A comparison of Eq. 1.37 and 1.40 leads to an important conclusion on

which all methods of relative measurements of the reflectrance (or transmittance)

of foodstuffs and other light-scattering substances are based:

$$\widetilde{B}'_{\omega R}=\widetilde{B}_\omega(\omega'_R=\omega'). \tag{1.41}$$

In other words, the equivalent radiance of an ideal standard which fully re-

flects incident radiation (R=1) and the equivalent radiance of the correspond-

ing energy source are equal.

For relative measurements in a general case of arbitrary irradiation we

shall obtain a new expression for describing the reflectance of a sample by sub-

stituting F and F_R from Eqs. 1.33 and 1.40:

$$\tag{1.42}$$

Next we shall introduce an angularly averaged equivalent radiance $\tilde{B}_{\omega R}$ of the radiation reflected by the standard into the numerator under the integral sign. Then, in place of Eq. 1.42 we shall have the following relationship for a general case of arbitrary radiation:

$$R(\omega;\omega_R)=[\int_{\omega_R}\rho_\omega(\theta,\phi;\theta_R,\phi_R)\cos\theta_R d\omega_R]/[\int_{\omega_R}\cos\theta_R d\omega_R]; \qquad (1.43)$$

where, $\rho_\omega(\theta,\phi;\theta_R,\phi_R)=[B_\omega(\theta,\phi;\theta_R,\phi_R)]/\tilde{B}'_{\omega R}$ - radiance coefficient of a sample in the case of arbitrary irradiation. As can be seen, the quantity $\rho_\omega(\theta,\phi;\theta_R,\phi_R)$ depends on the irradiation conditions. For a given direction (θ_R,ϕ_R) under certain irradiation conditions the radiance coefficient is determined by the ratio of the reflected radiance of a sample to that of a standard (both the sample and the standard are exposed to the same radiation source, the reflectance of the standard being unity).

For directional infrared irradiation $F(\theta,\phi)$, diffuse infrared irradiation $F(\omega)$, and hemispherical infrared irradiation $F(2\pi)$ the corresponding radiance coefficients are introduced: $\rho(\theta,\phi;\theta_R,\phi_R)$, $\rho(\omega;\theta_R,\phi_R)$, and $\rho(2\pi;\theta_R,\phi_R)$, which are interrelated through Eqs. 1.52 and 1.53 (Table 1.2).

Relationships analogous to Eqs. 1.52 and 1.53 can be written for radiance coefficients in the case of radiation transmitted through the sample, $t_\omega(\theta,\phi;\theta_T,\phi_T)$. In practice it is impossible to measure the $t_\omega(\theta,\phi;\theta_T,\phi_T)$ of a sample relative to a hypothetically ideal transmitting standard having transmittance which equals unity $(T=1)$. For no such ideal standard, one which uniformly transmits diffuse radiation, exists. Therefore, measurements of $t_\omega(\theta,\phi;\theta_T,\phi_T)$ can be reduced to a measurement of the radiance transmitted by the sample and that reflected by the standard $(R=1)$. Then, t_ω has the following form:

$$t_\omega(\theta,\phi;\theta_T,\phi_T)=[B_\omega(\theta,\phi;\theta_T,\phi_T)]/\tilde{B}'_{\omega R}. \qquad (1.44)$$

The radiance coefficient for reflection $\rho_\omega(\theta_R,\phi_R)$ and that for transmission $t_\omega(\theta_T,\phi_T)$ characterize the spatial form of the indicatrix of reflectance and the indicatrix of transmission, respectively. The ratio of $R(\omega;\theta_R,\phi_R)$ to the total reflectance $R(\omega,\omega_R)$ for a given sample also describes the reflectance indicatrix $r(\omega;\theta_R,\phi_R)$ under the same irradiation conditions:

$$r(\omega;\theta_R,\phi_R)=[R(\omega;\theta_R,\phi_R)]/[R(\omega,\omega_R). \tag{1.45}$$

For a majority of light-scattering materials reflection and transmission are of a mixed character. This can be seen in Fig. 1.8 where the reflectance and transmittance indicatrixes of monochromatic radiation at different wavelengths for some foodstuffs are shown.

Since the R_λ and T_λ of foodstuffs depend not only on the latter's optical properties but also on the irradiation conditions, a double terminology is used to define these terms [39,44]. It characterizes the irradiation conditions and the optical properties of the substance which depend on the conditions under which the measurements are made (Table 1.2). Thus, the first part of a term defines the conditions of irradiation, and the second part - the conditions of measurements. For example, bihemispherical reflectance $R(2\pi;2\pi)$ means that the sample is irradiated from a hemisphere $\omega'=2\pi$, while measurements are made of the flux reflected into a hemisphere $\omega_R=2\pi$. The analytical equations describing the interrelationships between different types of reflectance and radiance coefficients, summarized in Table 1.2, have been derived from a general relationship (Eq. 1.43). The analytical relationship between the reflectance of a layer under different conditions of irradiation can be obtained with the aid of Eq. 1.35. This is done by substituting the term for the reflected radiance $B_\omega(\theta,\phi;\theta_R,\phi_R)$ with the product of incident radiance $B_\omega(\theta,\phi)$ multiplied by the reflectance $R(\theta,\phi;\omega_R)$ when irradiation is in the direction of angles θ and ϕ:

Table 1.2 Analytical equations for different types of reflectance and transmittance [54,163]

Types of reflectance and transmittance	Designation	Analytical equations* expressed through radiance coef.	
Bihemispherical	$R(2\pi;2\pi)$	$\pi^{-1}\int_{2\pi}\rho\,(2\pi;\theta_R,\phi_R)\cos\theta_R\,d\omega_R$	(1.46)
	$T(2\pi;2\pi)$		
Hemispherical-diffuse	$R(2\pi;\omega_R)$	$[\int_{\omega_R}\rho(2\pi;\theta_R,\phi_R)\cos\theta_R\,d\omega_R]/[\int_{\omega_R}\cos\theta_R\,d\omega_R]$	(1.47)
	$T(2\pi;\omega_T)$		
Hemispherical-directional	$R(2\pi;\theta_R,\phi_R)$	$\pi^{-1}\rho(2\pi;\theta_R,\phi_R)\cos\theta_R\,d\omega_R$	(1.48)
	$T(2\pi;\theta_T,\phi_T)$		
Diffuse-hemispherical	$R(\omega;2\pi)$	$\pi^{-1}\int_{2\pi}\rho(\omega;\theta_R,\phi_R)\cos\theta_R\,d\omega_R$	(1.49)
	$T(\omega;2\pi)$		
Bidiffuse	$R(\omega;\omega_R)$	$[\int_{\omega_R}\rho(\omega;\theta_R,\phi_R)\cos\theta_R,d\omega_R]/[\int_{\omega_R}\cos\theta_R\,d\omega_R]$	(1.50)
	$T(\omega;\omega_T)$		
Diffuse-directional	$R(\omega;\theta_R,\phi_R)$	$\pi^{-1}\rho(\omega;\theta_R,\phi_R)\cos\theta_R\,d\omega_R$	(1.51)
	$T(\omega;\theta_T,\phi_T)$		
Directional-hemispherical	$R(\theta,\phi;2\pi)$	$\pi^{-1}\int_{2\pi}\rho(\theta,\phi;\theta_R,\phi_R)\cos\theta_R\,d\omega_R=\rho(2\pi;\theta_R,\phi_R)$	(1.52)
	$T(\theta,\phi;2\pi)$		
Directional-diffuse	$R(\theta,\phi;\omega_R)$	$[\int_{\omega_R}\rho(\theta,\phi;\theta_R,)\cos\theta_R\,d\omega_R]/[\int_{\omega_R}\cos\theta_R\,d\omega_R]=$	(1.53)
	$T(\theta,\phi;\omega_T)$	$\rho(\omega;\theta_R,\phi_R)$	

Table 1.2 (cont)

Bidirectional	$R(\theta,\phi;\theta_R,\phi_R)$	$\pi^{-1}\rho(\theta,\phi;\theta_R,\phi_R)\cos\theta_R\,d\omega_R$	(1.54)
	$T(\theta,\phi;\theta_T,\phi_T)$		

---

*Equations for T are analogous in form in which ρ, ω_R, and ϕ_R are replaced by t, ω_T, θ_T, and ϕ_T.

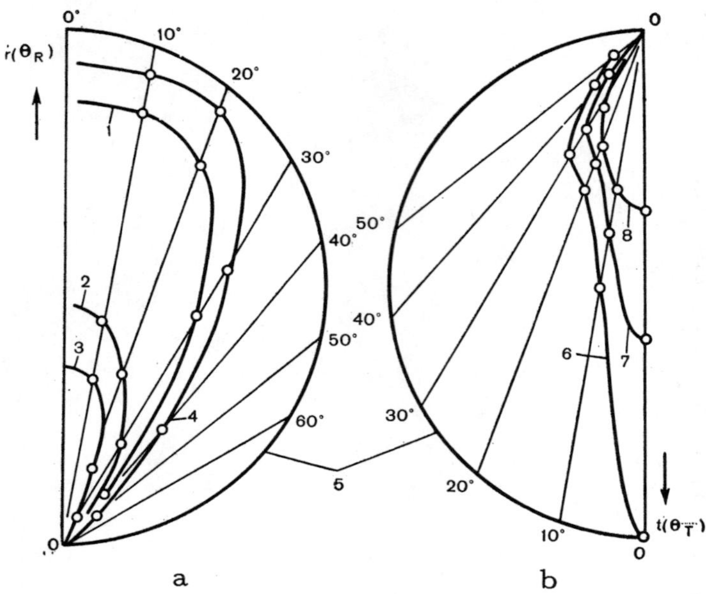

Figure 1.8   Reflectance (a) and transmittance (b) indicatrixes for some foodstuffs subjected to irradiation at different wavelengths:
    1,2 - air-dried starch (1 - λ=1.3 μm; 2 - λ=0.9-2.5 μm);
    3 - starch (moisture capacity (W) - 80%; λ=0.9-2.5 μm);
    4 - wheat flour (λ=1.3 μm)[28];
    5 - an ideal diffuse scatterer;
    6,7,8 - dough for vanilla-flavored zwieback (light source KG-1000 lamp;
         ℓ=0.15, 0.30, 0.40 mm, respectively) [88].

$$R(\omega;\omega_R)=[\int_\omega R(\theta;\phi;\omega_R)B_\omega(\theta,\phi)\cos\theta d\omega]/[\int_\omega B_\omega(\theta,\phi)\cos\theta d\omega] \qquad (1.55)$$

The use of the reversibility method simplifies considerably the analytical

relationships between the characteristics summarized in Table 1.2.   According

to the Helmholtz reversibility principle, if the same light is incident upon a re-

flecting object in the direction θ,ϕ (Fig. 1.7) and in the direction θ_R,ϕ_R and

if the reflected radiation is measured in the θ_R,ϕ_R and θ,ϕ directions, respec-

tively, then the measured reflected radiances will be equal, i.e.:

$$B(\theta,\phi;\theta_R)\cos\theta_R=B(\theta_R,\phi_R;\theta,\phi)\cos\theta , \text{ or}$$

$$\rho(\theta,\phi;\theta_R,\phi_R)=\rho(\theta_R,\phi_R;\theta,\phi). \qquad (1.56)$$

This property of the radiance coefficient makes it possible to obtain the following reciprocity relationships:

$$\rho_\omega(\theta,\phi;\theta_R,\phi_R)=R(\theta,\phi;\omega_R),$$

$$\rho_{2\pi}(\theta,\phi;\theta_R\phi_R)=R(\theta,\phi;2\pi),$$

$$R(\theta,\phi;\theta_R,\phi_R)=R(\theta_R,\phi_R;\theta,\phi),$$

$$R(2\pi;\omega_R)=R(\omega_R;2\pi),$$

$$R(\omega;2\pi)=R(2\pi;\omega_R),$$

$$R(\omega;\omega_R)=R(\omega_R;\omega).$$

$$(1.57)$$

Consequently, the hemispherical radiance coefficient $\rho(2\pi;\theta_R,\phi_R)$ of any sample is numerically equal to its directional-hemispherical reflectance $R(\theta,\phi;2\pi)$ when $\theta=\theta_R$ and $\phi=\phi_R$. Numerically $\rho_\omega(\theta,\phi;\theta_R,\phi_R)$ equals $R(\theta,\phi;\omega_R)$ when $\theta=\theta_R$, $\phi=\phi_R$, and $\omega_R=\omega$. A relationship also exists between the quantities $R(2\pi;\omega_R)$ and $R(\omega;2\pi)$. At the same time, when $\omega=\omega_R$ and approaches an infinitely small solid angle $d\omega$, the magnitude of these quantities approaches that of $R(2\pi;\theta_R,\phi_R)$ and $R(\theta,\phi;2\pi)$, respectively; between these latter characteristics there is no longer a reciprocity relationship, since $R(2\pi;\theta_R,\phi_R)$ is infinitely small and $R(\theta,\phi;2\pi)$ is the finite quantity (Table 1.2), i.e.:

$$R(2\pi;\theta_R,\phi_R)\neq R(\theta_R,\phi_R;2\pi),$$

$$R(\omega;\theta_R,\phi_R)\neq R(\theta_R,\phi_R;\omega).$$

$$(1.58)$$

$R(2\pi;\theta_R,\phi_R)$ quantity actually stands for the refraction of the radiation reflected in the direction θ_R,ϕ_R within a solid angle $d\omega_R$ under uniformly diffuse radiation conditions. $R(\omega;\theta_R,\phi_R)$ and $R(\theta,\phi;\omega_R)$ as well as the corresponding transmittance can be interpreted in a similar way.

The absorptance $A(\omega)$ of a plane layer under certain irradiation conditions, $d\omega<\omega<2\pi$, can be determined with the aid of Eq.1.30. In this equation the terms $R(\omega)$ and $T(\omega)$ always stand for hemispherical ($\omega_T=\omega_R\equiv2\pi$) reflectance and transmittance under any irradiation conditions. The types of absorptance are determined only by the irradiation conditions: directional-hemispherical $A(\theta,\phi; 2\pi)$, diffuse-hemispherical $A(\omega;2\pi)$, and bihemispherical $A(2\pi;2\pi)$. Thus, absorptance can be defined as the ratio of the radiant energy absorbed by a layer to that incident upon it.

To determine the integral relative emittance, the blackbody characteristic $\varepsilon(T)$ of a layer at a given temperature T, we can also use the data on the spectral absorptance A_λ:

$$\varepsilon(T)=[\textstyle\int_0^\infty R_\lambda(T)A_\lambda(T)d\lambda]/[\int_0^\infty R_\lambda(T)d\lambda]; \tag{1.59}$$

where, $A_\lambda(T)$ – bihemispherical absorptance of a layer $A_\lambda(2\pi;2\pi)$ at a given temperature T.

To determine the relative emittance of a layer $\varepsilon(T,\theta,\phi)$ in a certain direction confined by angles θ and ϕ it is necessary to substitute the term $A_\lambda(\theta,\phi;2\pi)$ in Eq. 1.59. The ratio of $\varepsilon(T,\theta,\phi)$ to total emittance $\varepsilon(T)$ for a given sample will be the emittance coefficient of radiation in the direction θ,ϕ.

Thus, the thermal radiating characteristics of a layer can be described by nine types of reflectance, nine types of transmittance, three types of absorptance and two types of relative emittance.

ENERGY TRANSFER IN FOODSTUFFS UNDER DIFFERENT CONDITIONS OF INFRARED IRRADIATION

When resorting to infrared irradiation in the production of high-quality foodstuffs the important thing is to achieve uniform heating of the surface of the articles being treated. Here much depends on the irradiation conditions maintained inside infrared installations. Thus, we need to know the principles governing energy transfer in foodstuffs and to determine their thermodynamic characteristics under different irradiation conditions.

2.1 Irradiation conditions in infrared installations for processing foodstuffs

Three main types of infrared irradiation may be identified (Fig. 1.7): directional irradiation at an angle θ with a low angular divergence; diffuse irradiation (imperfectly diffuse or mixed (directional-diffuse) radiation within a solid angle $\Delta\omega$; and hemispherical irradiation (perfectly diffuse irradiation) from a half-space $\Delta\omega=2\pi$. In industrial practice foodstuffs are usually exposed to mixed directional-diffuse radiation.

An example of directional irradiation is the drying of fruits, vegetables, cotton, tea leaves, tobacco leaves, peat, etc. under the sun. In closed thermal radiation chambers the distribution of the radiation flux on the surface of the foodstuffs is determined by the optico-geometrical parameters of the infrared generators, the reflectors, the foodstuffs being irradiated, and the enclosures of compartments, and by how all these are located in relation to one another. As a result of multiple reflection from different surfaces in such installations the irradiation conditions there correspond to those of mixed (directional-diffuse) or diffuse irradiation.

To increase the efficiency of infrared radiators various reflecting devices are used which help to provide uniform irradiation of foodstuffs. For directional irradiation different forms of reflectors are used - spherical, parabolic, hyperbolic, elliptical, and others (Figs. 2.1, 2.2). In practice, however, such reflectors are incapable of providing uniform directional radiation owing to the optical aberrations and the finite dimensions of the radiation source every surface element of which emits diffuse radiation.

When a parabolic-cylindrical reflector with a linear infrared source (Fig. 2.1,b) is used, we have mixed (directional-diffuse) radiation. In this case there is some divergence of the rays, which is due to the type of the infrared generator and reflector used. If an elliptical-cylindrical reflector (Fig. 2.1, a) is employed in combination with a point radiation source, we have diffuse radiation within a solid angle $\Delta\omega$.

To achieve maximum uniformity of radiant-flux density on the surface of the foodstuffs being irradiated there should be proper spacing(s) of the infrared generators and a proper distance (h) between the surface of the foodstuffs and the radiation source (Fig. 2.2,a). To this end blocks of two, four or more infrared generators are formed, the generators being located on a single plane parallel to a cylindrical or parabolic surface (Figs. 2.1, 2.2). The use of different types of reflectors (Fig. 2.1, 2) in combination with side reflecting surfaces (Fig. 2.1, 3) makes it possible to achieve directional-diffuse irradiation (Fig. 2.1, a, b; Fig. 2.2, d, e, f) or diffuse irradiation (Fig. 2.1, c, d, e; Fig. 2.2, a, b, c).

In locating infrared generators in thermal radiation installations attention

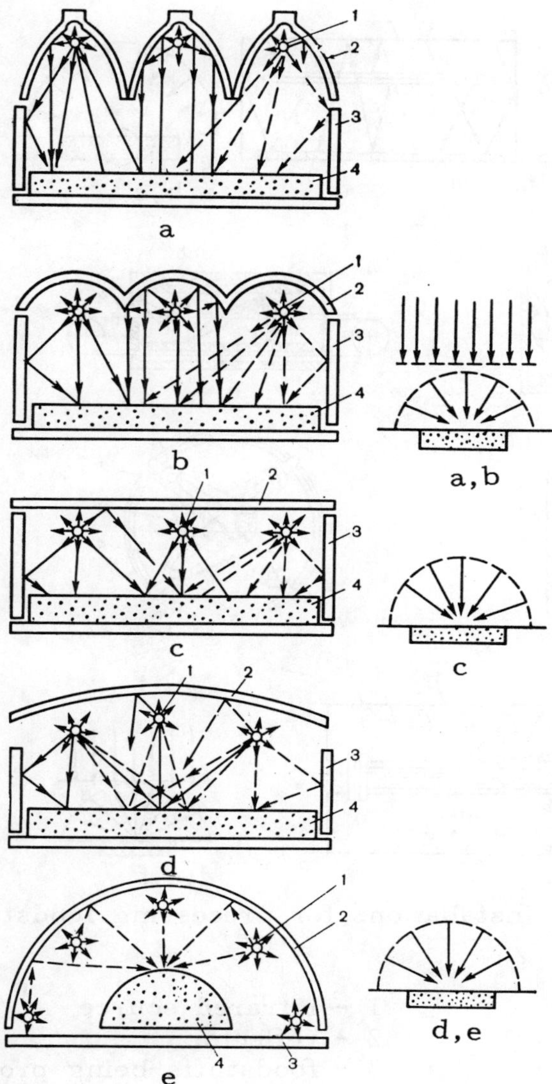

Figure 2.1 Diagrams showing the location of infrared generators and foodstuffs

in processing chambers with different types of reflectors [39]:
    a - elliptic-cylindrical   1 - infrared source
    b - parabolic-cylindrical   2 - reflector
    c - planar   3 - enclosure
    d,e - cylindrical   4 - foodstuffs being processed

should be paid to the form of the foodstuffs being treated and the specific

features of the technological process used. Diagrams of the possible location

of infrared generators and the foodstuffs being treated are shown in Fig. 2.2.

Thin layers of foodstuffs can be irradiated from one side (Fig. 2.2, a) or si-

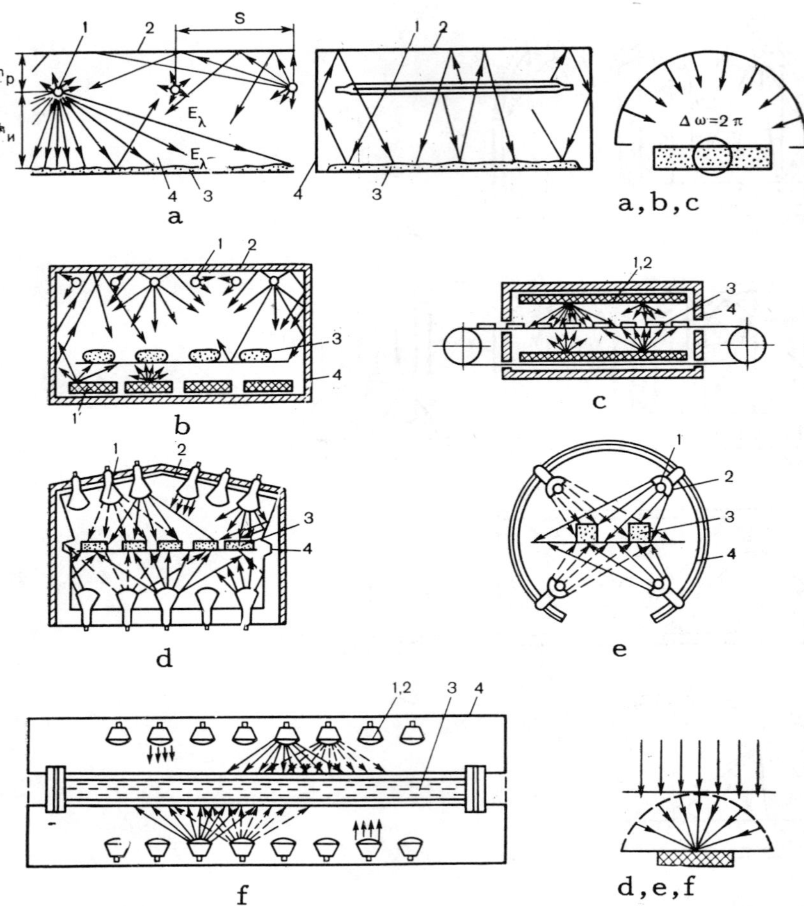

Figure 2.2 Industrial infrared installations for processing foodstuffs [39, 97, 98]:

a - drying
b,d - baking
e - thermal processing
c - blanching
f - pasteurization
a,b,c - diffuse irradiation
d,e,f - mixed directional-diffuse irradiation

1 - infrared source
2 - reflector
3 - foodstuffs being processed
4 - enclosure

multaneously from both sides (Fig. 2.2, b, c, d). Generally speaking, it is advisable that foodstuffs, whatever their form, should be irradiated from all sides (Fig. 2.2, e, f); this will help ensure their uniform heating.

The flux density of incident radiation is strongly affected by the enclosures inside the ovens. The enclosures are usually made of materials that are highly reflective in the infrared spectral region (e.g., polished aluminum and the like). Thus we can achieve high intensity of radiation and the necessary uniformity of irradiation, thereby improving the efficiency of the installations, by means of multiple reflection of radiation - from the surface of the enclosures and from the surface of the foodstuffs.

The designs of thermal radiation installations vary, depending on the technological processes used. But despite their differences they share certain common features. In Fig. 2.2 are shown the basic diagrams of ovens for baking bread and cakes, curing sausages, blanching fish for canning, and pasteurizing juices and milk by means of infrared irradiation.

Baking oven PIK-8 (Fig. 2.2, b) is equipped with a linear infrared generator KGT-220-1000 located above the object to be irradiated. According to the classification used in the USSR infrared generators are of three types: high-temperature lamps operating at $T \geq 1800$ K; tubular emitters with a spiral heating element operating at $1800 \geq T \geq 750$ K; and low-temperature panel emitters operating at $T \leq 750$ K [73]. We can assume with a fair degree of accuracy as regards engineering requirements that for a majority of emitters the distribution of the radiation flux conforms to Lambert's law [20, 88, 112]. Therefore, with the proper location of infrared generators, and thanks to the presence of side reflecting surfaces, conditions are created in PIK-8 ovens that approach those of diffuse irradiation [20, 80, 112]. Panel emitters placed at the bottom of the oven serve as a source of heat during baking.

An installation for continuous infrared blanching of fish for canning (Fig. 2.2, c) is equipped with panel infrared emitters operating at a maximum temperature of 850 K. The use of panel emitters along the entire length of a conveyer creates conditions of diffuse irradiation inside the installation.

Ovens for baking biscuits (Fig. 2.2, d) are equipped with reflector lamps so that radiation comes both from the top and the bottom [19, 97]. The biscuits move on a steel belt through the oven. The fraction of incident directional radiation coming straight from the infrared source is comparable to that of scattered diffuse radiation.

The thermal processing of sausages is carried out in cylindrical chambers (Fig. 2.2, e) equipped with infrared generators located at a 45° angle to the surface of the conveyer. Such a location of infrared generators emitting directional-diffuse radiation leads to multiple reflection of radiation – from the surface of the sausages, the reflectors, and the enclosures. The resulting diffuse radiation is comparable in its intensity to directional radiation.

A schematic diagram of a unit for the infrared pasteurization of foodstuffs in liquid form is shown in Fig. 2.2. The infrared reflector lamps are located radially around a pipe made of borosilicate glass through which flows the liquid substance to be processed. Such a location of emitters ensures directional-diffuse irradiation.

This brief discussion of the irradiation conditions in different infrared installations shows that the effect of infrared irradiation depends on many factors: the type of infrared emitters used; the design of the reflectors; the location of the infrared emitters and the irradiated objects; the form of the foodstuffs to be treated; and the presence of reflecting enclosures. Maximum uniformity of incident radiation flux is achieved when a plane layer of foodstuffs is exposed to perfectly diffuse radiation or to broad directional radiation (e.g., solar radiation). But such irradiation conditions are unobtainable in industrial infrared thermal radiation units. For infrared lamps emit radiation in all directions. Because of this, and owing to the use of special reflectors, the foodstuffs are irradiated at different angles. Furthermore, they are exposed to radiation repeatedly reflected from the side enclosures and from the bottom of the trays.

Thus, in closed thermal radiation chambers the foodstuffs are irradiated predominantly by diffuse radiation and by mixed diffuse-directional radiation.

2.2 Effect of multiple scattering of radiation in foodstuffs on radiant energy transfer

When foodstuffs are exposed to infrared radiation a process of energy transfer takes place. To study this process the foodstuffs, whether of plant or animal origin, can be regarded as absorbing-scattering polydispersed media consisting of relatively large particles whose indicatrix of scattering is strongly elongated in a forward direction. The number of particles encountered by a direct beam as it passes through a layer 0.1 mm thick can reach 100 or more. Since the distance between the particles is small there is multiple scattering of radiation in such an optically heterogeneous medium, with each particle scattering radiation and absorbing the radiation scattered by the other particles.

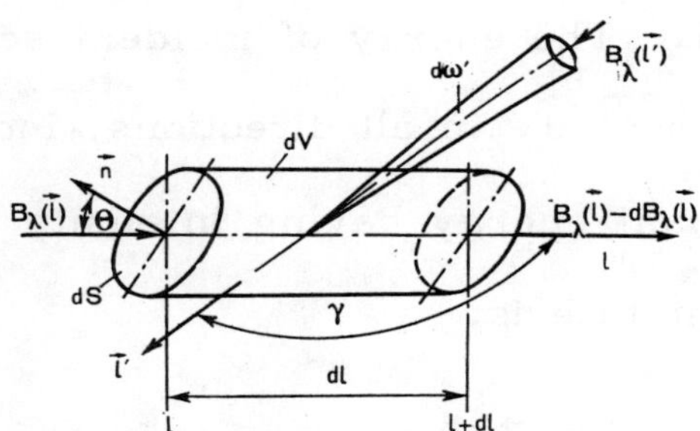

Figure 2.3 Diagram showing how multiple scattering of radiation affects energy transfer in foodstuffs [3, 44]

Let us consider the changes in the intensity of monochromatic radiation $B_\lambda(\vec{l})$ as it passes through a volume element dV in the direction $\vec{l}$ (Fig. 2.3). For the sake of simplicity we shall assume that the direction $\vec{l}$ is perpendicular ($\cos\theta=1$) to the base of a cylindrical volume element having dl length. Throughout this length the radiance will decrease owing to the absorption and scattering of radiation by the volume element dV, according to the Bouguer-Lambert law. The fraction of radiant energy corresponding to this decrease per unit time is defined as:

$$dF_{\lambda\varepsilon}=-(k_\lambda + \sigma_\lambda)B_\lambda(\vec{\ell})d\omega d\ell dSd\lambda. \tag{2.1}$$

A fraction of the radiant energy emitted spontaneously or as a result of induction per unit time by the volume element dV in the direction $\vec{\ell}$ into the solid angle $d\omega$ goes towards increasing $B_\lambda(\vec{\ell})$:

$$dF_{\lambda U}=k_\lambda n_\lambda^2 (c_\lambda U_\lambda/4\pi)[1 + (\lambda^5/2hc_\lambda^2)B_\lambda(\vec{\ell})]d\omega d\ell dSd\lambda. \tag{2.2}$$

In this equation the first term describes the spontaneous emission, and the second, the induced emission of radiant energy by the volume element dV; n_λ - refraction index of the medium.

The volume element dV, owing to multiple scattering of radiation within the absorbing-scattering medium, is irradiated from all sides, the intensity of the scattered radiation being $B_\lambda(\vec{\ell})$. The energy of incident scattered radiation is dispersed by the volume element dV in all directions, including the direction $\vec{\ell}$. The fraction of the radiant energy having intensity $B_\lambda(\vec{\ell})$ which is scattered in the direction $\vec{\ell}$ per unit time is:

$$dF_{\lambda\sigma}=(\sigma_\lambda/4\pi)d\ell d\omega dSd\lambda \int_{(4\pi)} B_\lambda(\vec{\ell'})\chi_\lambda(\vec{\ell},\vec{\ell'})d\omega'; \tag{2.3}$$

where, $\chi_\lambda(\vec{\ell},\vec{\ell'})$ - indicatrix of scattering for radiation at wavelength λ.

In a general case the radiant energy in the volume element dV changes also as a function of time owing to the above-mentioned processes which take place simultaneously. The total change in the radiant energy during period $d\tau$ is defined as follows [3, 39]:

$$\vec{\ell}\,\frac{dB_\lambda(\vec{\ell})}{dl}+\frac{1}{c_\lambda}\cdot\frac{dB_\lambda(\vec{\ell})}{d\tau} = n_\lambda^2 k_\lambda\,\frac{c_\lambda U_\lambda}{4\pi}\left[1+\frac{\lambda^5}{2hc_\lambda^2}B_\lambda(\vec{\ell})\right] -$$

$$-\,(k_\lambda+\sigma_\lambda)B_\lambda(\vec{\ell})+\frac{\sigma_\lambda}{4\pi}\int_{4\pi} B_\lambda(\vec{\ell'})\chi_\lambda(\vec{\ell},\vec{\ell'})d\omega'. $$

$$\tag{2.4}$$

This equation can serve as a basis for studying the transfer of radiant energy in absorbing-scattering media with account taken of the induced emission of radiation by the medium.

The left side of Eq. 2.4 describes the changes in the intensity of monochromatic radiation $B_\lambda(\vec{\ell})$ in the direction $\vec{\ell}$ as a function of time. The first term on the right side of the equation compensates for an increase in the radiance along the direction $\vec{\ell}$ due to the emission of radiation by the particles within the volume element. The second term on the right side of the equation describes a decrease in the radiance due to the absorption and scattering of radiation according to the Bouguer-Lambert law, with account taken of the induced emission of radiation by the medium.

Multiple scattering of radiation within a layer of the medium is taken into account by the scattering indicatrix, the third term on the right side of Eq. 2.4. This term defines the increase in the intensity $B_\lambda(\vec{\ell})$ in the direction $\vec{\ell}$ due to the scattering in the same direction by the volume element of the energy incident upon it from all other directions.

In a general case it is difficult to obtain a precise solution of the equation describing the transfer of radiant energy. Therefore, approximate differential and integral methods for solving this equation have been developed. These include differential-difference and differential-diffusion approximations, radiational thermal conductivity approximation, tensor and zonal methods, and others [3]. A detailed comparison of these methods has been made [134]. It shows that the tensor method is the most precise one, while the differential-difference method is the simplest but less precise [188, 189]. However, with regard to foodstuffs the basic equation describing the energy transfer cannot be solved by the tensor method, since no data on the optical properties and indicatrix of scattering for such media are available.

Also inapplicable here are the simple Bouguer and Bouguer-Lambert laws on the attenuation of radiation, which are still often used [19, 84, 94, 97] for determining the distribution of the absorbed energy of integral or monochromatic radiation throughout the thickness of a layer. However, the use of these laws in their present form $B_\lambda = B_{0\lambda} \exp(-\varepsilon_\lambda \ell)$, without account being taken of multiple scattering, involves a basic error. For the expression describing a decrease in the radiance B_λ in a scattering medium (Eq. 2.4) is fundamentally different from that given by the Bouguer-Lambert law:

$$dB_\lambda/dx = -(k_\lambda + \sigma_\lambda)B_\lambda(x) \tag{2.5}$$

In Eq. 2.4, on the other hand, the term on the right side takes into account secondary and multiple scattering that occurs during the propagation of radiation in foodstuffs:

$$(\sigma_\lambda/4\pi) \int_{(4\pi)} B_\lambda(\vec{\ell})\chi_\lambda(\gamma)d\omega.$$

The idea underlying the differential-difference method consists in the following. The net radiant-flux vector in a planer layer of the medium is represented as the difference between the densities of two counterfluxes: $q = E_+ - E_-$. The integral-differential relationship (Eq. 2.4) can be replaced by a system of two differential-linear equations of the first order (with account taken of the fact that $d\omega = \sin\theta \, d\theta \, d\phi$) through the integration of the radiance $B_\lambda(\vec{\ell})$ with respect to angle ϕ within the limits of 0 to 2π and of 2π to 4π, and with respect to angle θ within the limits of 0 to $\pi/2$ and of $\pi/2$ to π; i.e., through replacing the term $B_\lambda(\vec{\ell})$ by two hemispherical streams of radiant energy that flow in opposite directions. With this approach Eq. 2.4, if we are dealing with a unidimensional problem, is approximated by the following system of two averaged differential equations describing the transfer of radiant energy for E_+ and E_- [3, 44]:

$$dE_+/dx + (1/c_\lambda)(\partial E_+/\partial_\tau) = w_{c\lambda}/2 - m_+(k_\lambda + \delta_+\sigma_\lambda)E_+ + m_-\delta_-\sigma_\lambda E_-,$$

$$(2.6)$$

$$-dE_-/dx + (1/c_\lambda)(\partial E_-/\partial\tau) = w_{c\lambda}/2 - m_-(k_\lambda + \delta_-\sigma_\lambda)E_- + m_+\delta_+\sigma_\lambda E_+.$$

Here, the coefficients m_+ and m_-, which take into account the spatial distribution of the energy of E_+ and E_-, and the coefficients δ_+ and δ_-, which take into account the form of the scattering indicatrix of a layer element dx and which are numerically equal to the fractions of E_+ and E_- that are scattered backward, are determined with the aid of Eqs. 1.13 and 1.21, respectively. The considerable error ($\sim25\%$) in the Schuster-Schwarzschild method [188, 189] is due to the assumption that E_+ and E_- are isotropic in nature and exhibit ideal diffuse scattering. Such an assumption would be correct if the scattering indicatrix of the medium is spherical in form; i.e., if the following conditions are fulfilled:

$$m_+=m_-=2; \quad \delta_+=\delta_-=0.5. \tag{2.7}$$

Therefore, to make the Schuster-Schwarzschild approximation more precise one should disregard the assumption of the isotropic nature of the scattering, $\delta=0.5$, and the isotropic nature of the spatial distribution of the radiant fluxes inside the layer, $m=2$.

2.3 Transfer of energy of monochromatic radiation in foodstuffs exposed to diffuse radiation

When foodstuffs are exposed to diffuse radiation the following experimentally determined spectral coefficients can be taken as the optical properties. These coefficients describe the optical properties of a layer element dx having x coordinate (Fig. 2.49 [44, 188, 189]:

- averaged spectral coefficient of absorption $\bar{k}_\lambda$, which determines the fraction $\bar{k}_\lambda dx$ of the hemispherical radiation flux incident upon the layer dx absorbed by the medium as it passes through the layer;

- averaged spectral hemispherical coefficient of backward scattering s_λ, which determines the proportion $s_\lambda dx$ of the hemispherical radiation flux incident upon the layer dx that is scattered by the layer in a direction opposite to that of the incident flux.

The averaged coefficient k_λ takes into account the form of the unknown scattering indicatrix of the medium, and s_λ - the spatial distribution of the incident flux; they are related to the absorption coefficient k_λ and the scattering coefficient σ_λ, respectively (Table 1.1).

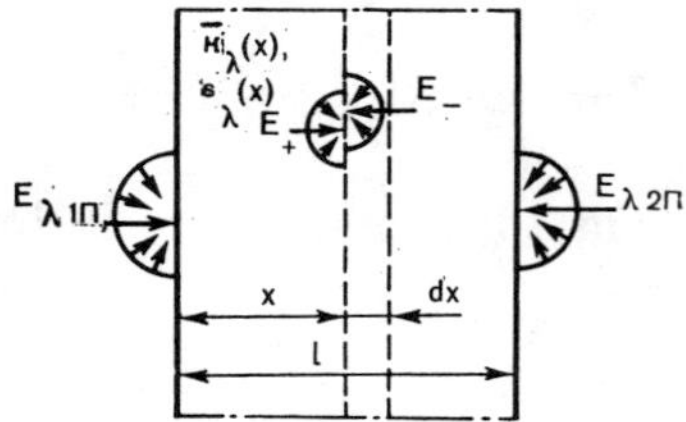

Figure 2.4   Radiation fluxes inside a plane layer of a selectively absorbing-scattering substance exposed to diffuse radiation [44, 188, 189]

In the infrared processing of foodstuffs the radiation emitted directly by the substance is much smaller than the incident radiant power, and the time τ is much greater than l/c_λ (l is the characteristic dimension of the irradiated particle). Therefore, the terms $1/c_\lambda(\partial E_\lambda/\partial\tau)$ and $w_{c\lambda}/2$ in Eq. 2.6, being smaller than the terms dE_λ/dx, $k_\lambda E_\lambda$ and $\sigma_\lambda E_\lambda$, can be discarded.

When a layer is exposed on both sides to monochromatic diffuse incident radiation at $E_{\lambda 1}$ and $E_{\lambda 2}$, the spatial energy distribution of E_+ can be considered to be the same as that of E_- since these two counter fluxes are interdependent. But the incident radiation is not ideally diffuse throughout the thickness of the layer, i.e.:

$$m_+ = m_- = m \leq 2, \quad 1 < m < \infty^{'} \quad \delta_+ = \delta_- = \delta \leq 1, \quad 0 < \delta \leq 1,$$

and therefore:

$$\overline{k}_{\lambda+} = \overline{k}_{\lambda-} = \overline{k}_\lambda; \quad s_{\lambda+} = s_{\lambda+} = s_{\lambda-} = s_\lambda. \tag{2.8}$$

In this case we obtain a simple two-constant system of linear differential equations describing two hemispherical counterfluxes E_+ and E_-:

$$\left.\begin{aligned}
\frac{dE_+}{dx} &= -(\overline{k}_\lambda + s_\lambda)E_+ + s_\lambda E_- \\[2mm]
-\frac{dE_-}{dx} &= -(\overline{k}_\lambda + s_\lambda)E_- + s_\lambda E_+
\end{aligned}\right\} 0 \leqslant x \leqslant l, \tag{2.9}$$

where, $\overline{k}_\lambda = m_\lambda k_\lambda$, $s_\lambda = m_\lambda \delta_s \sigma_\lambda$.

The boundary conditions in a general case of bilateral irradiation will be:

$$E_+(x=0) = E_{\lambda 1}, \quad E_-(x=\ell) = E_{\lambda 2}. \tag{2.10}$$

In earlier studies [3, 5, 15, 18, 22, 24, 26, 33, 44, 64, 85, 110, 114, 116, 150, 155, 165, 167, 168, 182, 188, 189, 197, 204, 205] it was assumed when solving Eq. 2.9 that the absorption constants $\overline{k}_\lambda = mk_\lambda = \mathrm{const}$ and backward scattering constants $s_\lambda = m\delta_s \sigma = \mathrm{const}$ averaged with respect to the hemisphere are independent of coordinate x.

Anisotropic scattering on large particles in object being investigated can take place when the scattering indicatrix of spatial distribution of radiance is strongly elongated forward (Fig. 1.6, d-i). In this case a change in the spatial distribution of the radiance $B_\lambda(\omega, x)$ with a change in coordinate x is possible even when the objects are exposed to diffuse radiation on both sides. Therefore, with a change in coordinate x there may be a change in the coefficients m and δ_s, which previously have been thought to be constant when solving Eq. 2.9.

According to our more precise differential-difference method the optical properties, depending on coordinate x and with account taken of the changes in the radiance $B_\lambda(\omega,x)$, are described by variable coefficients $\bar{k}_\lambda(x)$ and $s_\lambda(x)$. These coefficients, when the optical properties of the substance k_λ, σ_λ, and $\chi_\lambda(\gamma)$ are constant, contain the overall information on the spatial distribution of radiance $B_\lambda(\omega,x)$ as a function of the coordinate and on the form of the scattering indicatrix $\chi_\lambda(\gamma)$:

$$\bar{k}_\lambda(x)=m(x)k_\lambda;$$
$$s_\lambda(x)=m(x)\delta_s(x)\sigma_\lambda. \tag{2.11}$$

The parameters $m(x)$ and $\delta_s(x)$, which take into account the form of the scattering indicatrix and the change in the irradiation conditions of the layer element dx with a change in the coordinate, can be determined with the aid of Eqs. 1.13 and 1.21.

By replacing constants $\bar{k}_\lambda$ and s_λ in Eq. 2.9 with $\bar{k}_\lambda(x)$ and $s_\lambda(x)$ we obtain a relationship similar to Eq. 2.9 but with variable coefficients. To simplify this relationship we substitute the effective optical depth for variable x:

$$\tau=\int_0^x [\bar{k}_\lambda(x) + s_\lambda(x)]dx=\int_0^x \varepsilon(x)dx . \tag{2.12}$$

Furthermore, by substituting the effective optical thickness τ_ℓ, which is defined by Eq. 2.12 at $x=\ell$, for the layer's thickness ℓ, we have:

$$\tau_\ell=\int_0^\ell [\bar{k}_\lambda(x) + s_\lambda(x)]dx=\int_0^\ell \varepsilon(x)dx. \tag{2.13}$$

Next we introduce the Schuster number Λ_{ef} (Table 1.1):

$$\Lambda_{ef}(x)=[s_\lambda(x)]/[\bar{k}_\lambda(x) + s_\lambda(x)]=\delta_s(x)\Lambda/\{1 - [1 - \delta_s(x)]\Lambda\}. \tag{2.14}$$

By replacing the variable x with τ and dx with $d\tau=[\bar{k}_\lambda(x) + s_\lambda(x)]dx$

in Eq. 2.10, and by taking into account Eqs. 2.11-2.14, we obtain a new system of differential equations of the first order relative to the dimensionless quantity of the counterfluxes E_+^* and E_-^*. This system of equations contains one variable coefficient $\Lambda_{ef}(\tau)$:

$$dE_+^*/d\tau = -E_+^* + \Lambda_{ef}(\tau)E_-^*$$
$$-dE_-^*/d\tau = -E_-^* + \Lambda_{ef}(\tau)E_+^* \qquad 0 \leq \tau \leq \tau_\ell; \qquad\qquad (2.15)$$

where, E_i^* - dimensionless quantity of the radiation flux.

The boundary conditions in a general case of bilateral radiation will be:

$$E_+^*(\tau=0)=E_{\lambda 1}^*, \quad E_-^*(\tau=\tau_\ell)=E_{\lambda 2}^*; \qquad\qquad (2.16)$$

where, $E_{\lambda i}^*$ - dimensionless quantity of the incident radiation flux referred to the radiation flux emitted by the blackbody $E_{\lambda a}$ at wavelength λ and absolute temperature of infrared radiation T_e: $E_{\lambda i}^*=E_{\lambda i}/E_{\lambda a}$.

When determining the incident fluxes in Eq. 2.16 account should be taken of the multiple reflection and the absorption of infrared radiation in the vapor-air medium in the oven according to a method which will be described in Chapter 7.

A solution of Eq. 2.15 under the boundary conditions (Eq. 2.16) which we have developed for a general case involving the variable $\Lambda_{ef}(\tau)$ will be given in Chapter 4. At the same time there is a simpler method for solving Eq. 2.15. According to this method the changes in the conditions of irradiation of the volume element having dx thickness with a change in coordinate x are taken into account by using $\overline{k}_\lambda(x)$ and $s_\lambda(x)$, which are determined from Eq. 2.11.

It can be seen from the determination of the Schuster number and of its analytical relationship (Eq. 2.14) with the survival probability of the photon and with the backward scattering coefficient, that Λ_{ef} is independent of m; that

is, Λ_{ef} is independent of the conditions of irradiation of the layer element dx.
The dependence of Λ_{ef} on x is due solely to $\delta_s(x)$, which is a function of the
form of the backward scattering indicatrix. According to Eq. 1.21 the quantity
$\delta_s(x)$ determines the fraction of radiation flux that is scattered backward by the
layer element dx. This quantity is determined by averaging the scattering indicatrix with respect to the hemisphere $(2\pi, n_+)$ of incident radiance B_λ.

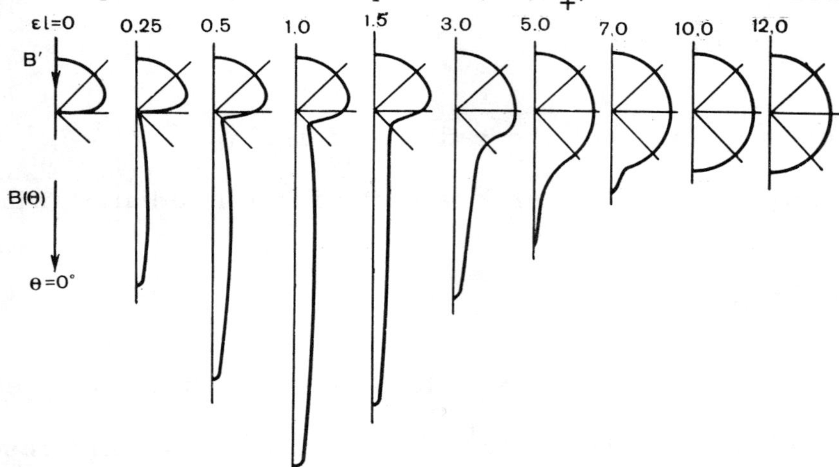

Figure 2.5 Polar diagrams illustrating the angular distribution of radiance at
different optical depths in a semi-infinite uniformly scattering layer consisting
of large particles at $\rho=20$ (radiation which is scattered is not shown) [100].

Experimental data [92, 93, 100] show that the backward scattering indicatrix is almost diffuse in form even in those cases where the indicatrix of scattering of radiation on large particles is strongly elongated. The angular distribution of the intensity of scattered radiation (Fig. 2.5) at different optical
depths in a semi-infinite layer of a scattering medium consisting of large particles ($\rho=20$) indicates the diffuse form of the backward scattering indicatrix.
Therefore, practically in all cases of scattering on small and large particles
(Figs. 1.6, 2.5) there should be an insignificant deviation from the value δ_s
when there is a change in the irradiation conditions, namely, from directional
to diffuse spatial distribution of the incident radiation flux. The deviation can
be 10-20% of the absolute value of δ_s. Furthermore, for a whole number of

substances belonging to group III and IV according to the classification based on their optical properties (cf. Chapter 6), the quantity Λ_{ef} in Eq. 2.15 can be considered to be constant when a high-temperature infrared source is used in the irradiation process. This is due to the fact that the quantity Λ_{ef} (Eq. 2.14) is slightly affected by a change in δ_s when $\Lambda > 0.9$, $\sigma_\lambda > k_\lambda$ ($s_\lambda > \bar{k}_\lambda$).

For starch, cereals, bread and flour products, flour, fruit candy, and so on the magnitude of $\Lambda > 0.9$ and $s_\lambda \gg \bar{k}_\lambda$ in the spectral range 0.7-2.5 µm. With a change in the magnitude of δ_s up to 20% the magnitude of Λ_{ef} according to Eq. 2.14 deviates from its initial level by not more than 1.0%. Under conditions of diffuse or mixed irradiation the change in the magnitude of δ_s will obviously be considerably smaller. This makes it possible to substitute in Eq. 2.15 the constant quantity Λ_{ef} averaged with respect to the layer for the variable quantity $\Lambda_{ef}(\tau)$. In this case a change in the conditions of irradiation of the layer dx as a function of the depth x, and the anisotropic nature of the scattering indicatrix are taken into account with the aid of τ, which depends on $m(x)$ and $\delta_s(x)$ according to Eqs. 2.12 and 2.13, respectively.

By taking into account the changes in $\bar{k}_\lambda(x)$ and $s_\lambda(x)$ with a change in the coordinate, which are caused by changes in the spatial distribution of the radiance $B(\omega,x)$ under the limiting conditions defined by Eq. 2.16, we obtain a general solution of Eq. 2.15 [39]:

$$E_+^* = E_{\lambda 1}^* / (1-\psi_\lambda^2) [\exp(-L_\lambda^*\tau) - \psi_\lambda^2 \exp(L_\lambda^*\tau)] +$$
$$+ E_{\lambda 2}^* \psi_\lambda / (1-\psi_\lambda^2) [\exp(L_\lambda^*\tau) - \exp(-L_\lambda^*\tau)]; \tag{2.17}$$

$$E_-^* = E_{\lambda 2}^* / (1-\psi_\lambda^2) \{\exp[-L_\lambda^*(\tau_\ell - \tau] - \psi_\lambda^2 \exp[L_\lambda^*(\tau_\ell - \tau')]\} +$$
$$+ E_{\lambda 1}^* \psi_\lambda / (1-\psi_\lambda^2) \{\exp[L_\lambda^*(\tau_\ell - \tau)] - \exp[-L_\lambda^*(\tau_\ell - \tau)]\}. \tag{2.18}$$

In Eqs. 2.17 and 2.18 the following designations are used:

$$\psi_\lambda = R_{\lambda\infty} \exp(-L_\lambda^*\tau_\ell); \tag{2.19}$$

$$R_{\lambda\infty}=\Lambda_{ef}^{-1}[1-(1-\Lambda_{ef}^{2})^{\frac{1}{2}}]; \qquad (2.20)$$

$$L_{\lambda}^{*}=(1-\Lambda_{ef}^{2})^{\frac{1}{2}}=L_{\lambda}/\varepsilon_{ef\lambda}. \qquad (2.21)$$

The obtained solutions (Eqs. 2.17-2.21) of Eq. 2.15 enable us to determine the monochromatic radiation flux at depth x in a layer having thickness ℓ. They are more precise since they take into account the changes in the optical properties $\bar{k}_{\lambda}(x)$ and $s_{\lambda}(x)$. Furthermore, they are derived without introducing special assumptions as regards the form of the scattering indicatrix. The magnitudes of m and δ_{s}, differing from those in the Schuster-Schwarzschild approximation (where m=2, δ=0.5), can vary within the possible limits shown in Eq. 2.8.

The dimensionless quantity of the net radiant flux is determined from the following equation:

$$q_{\lambda}^{*}=E_{+}^{*}-E_{-}^{*}=E_{\lambda 1}^{*}(1-R_{\lambda\infty})/(1-\psi_{\lambda}^{2})[\exp(-L_{\lambda}^{*}\tau)+\psi_{\lambda}^{2}/R_{\lambda\infty}\exp(L_{\lambda}^{*}\tau)]-$$
$$-E_{\lambda 2}^{*}(1-R_{\lambda\infty})/(1-\psi_{\lambda}^{2})\{\exp[-L_{\lambda}^{*}(\tau_{\ell}-\tau)]\}. \qquad (2.22)$$

The amount of energy absorbed per unit time at depth x by a volume element is determined from the energy conservation equation with account taken of Eq. 2.15:

$$w^{*}(\tau)=-dq_{\lambda}^{*}/d\tau=-[(dE_{+}^{*}/d\tau)-(dE_{-}^{*}/d\tau)]=(1-\Lambda_{ef})(E_{+}^{*}+E_{-}^{*}). \qquad (2.23)$$

By using Eqs. 2.17 and 2.18 for the dimensionless quantity of the energy absorbed per unit time we obtain:

$$w_{\lambda}^{*}(\tau)=(1-\Lambda_{ef})E_{\lambda 1}^{*}(1+R_{\lambda\infty})/(1-\psi_{\lambda}^{2})\{[\exp(-L_{\lambda}^{*}\tau)-(\psi_{\lambda}^{2}/R_{\lambda\infty})\cdot$$
$$\cdot\exp(L_{\lambda}^{*}\tau)]\}+(1-\Lambda_{ef})E_{\lambda 2}^{*}(1+R_{\lambda\infty})/(1-\psi_{\lambda}^{2})\{\exp[-L_{\lambda}^{*}(\tau_{\ell}-\tau)]-$$
$$-(\psi_{\lambda}^{2}/R_{\lambda\infty})\exp[L_{\lambda}^{*}(\tau_{\ell}-\tau)]\}. \qquad (2.24)$$

The expressions next to the terms $E_{\lambda 1}^{*}$ and $E_{\lambda 2}^{*}$ stand for the function of the thickness of the layer and the wavelength only. These expressions determine the dimensionless quantity of the spatial irradiance $E_{\lambda 0}^*$, which is independent of the incident fluxes $E_{\lambda 1}^*$ and $E_{\lambda 2}^*$. Therefore, we can write:

$$w_\lambda^*(x)=(1-\Lambda_{ef})E_{\lambda 1}^* E_{\lambda 01}^*(x)+(1-\Lambda_{ef})E_{\lambda 2}^* E_{\lambda 02}^*(x). \qquad (2.25)$$

The functions $E_{\lambda 01}^*(x)$ and $E_{\lambda 02}^*(x)$ are spectral functions describing the distribution of absorbed radiant energy throughout the thickness of the layer. They can be determined beforehand if we know the optical properties of the substance $\bar{k}_\lambda$ and s_λ. These properties can be determined with the aid of the experimentally obtained thermal radiational characteristics of a layer having a finite thickness and by using the functional relationship between $\bar{k}_\lambda$, s_λ and R_λ, T_λ.

Next we shall consider four particular cases.

1. When the incident fluxes are equal, $E_{\lambda 1}^*=E_{\lambda 2}^*$, Eq. 2.24 becomes considerably simplified:

$$w_\lambda^*(\tau)=(1-\Lambda_{ef})E_\lambda^*(1+R_{\lambda\infty})/(1+\psi_\lambda^2)\{\exp(-L_\lambda^*\tau)+\exp[-L_\lambda^*(\tau_\ell-\tau)]\}. \qquad (2.26)$$

2. When a layer without a supporting base and having a finite thickness is irradiated from one side, $E_{\lambda 2}^*$, the amount of radiant energy absorbed per unit time at depth τ is:

$$w_\lambda^*(\tau)=(1-\Lambda_{ef})E_\lambda^*(1+R_{\lambda\infty})/(1-\psi_\lambda^2)[\exp(-L_\lambda^*\tau)-(\psi_\lambda^2/R_{\lambda\infty})\exp(L_\lambda^*\tau)]. \qquad (2.27)$$

3. When a plane layer supported by a "cold" base is irradiated from one side (i.e., with no account being taken of the radiation coming from the supporting base), the boundary conditions for Eq. 2.15 are as follows:

$$E_+^*(\tau=0)=E_\lambda^*; \quad E_-^*(\tau=\tau_\ell)=R_\lambda E_+^*(\tau=\tau_\ell); \qquad (2.28)$$

where, R_λ – reflectance of the supporting base when irradiation is carried out with a radiation flux $E_+(\tau_\ell)$ at depth τ_ℓ. The solution of Eq. 2.15 under the limiting   conditions set by Eq. 2.28 yields the following relationship describing the radiant energy absorbed per unit time at depth τ:

$$w_\lambda^*(\tau)=(1-\Lambda_{ef})E_\lambda^*(1+R_{\lambda\infty})/(1-R_{\lambda ef}\psi_\lambda^2)[\exp(-L_\lambda^*\tau)-(R_{\lambda ef}\psi_\lambda^2/R_{\lambda\infty})\exp(L_\lambda^*\tau)];\tag{2.29}$$

where, $R_{\lambda ef}=(R_{\lambda\infty}-R_\lambda)/R_{\lambda\infty}(1-R_\lambda R_{\lambda\infty}).$ $\hspace{2cm}$ (2.30)

From Eq. 2.29 it follows that the reflection from the supporting base can be disregarded at $L_\lambda^*\tau_\ell\geqq4.$

4.   In an optically infinitely thick layer $(L_\lambda^*\tau_\ell\to\infty)$ the amount of radiant energy absorbed per unit time at depth τ is:

$$w_\lambda^*(\tau)=(1-\Lambda_{ef})E_\lambda^*(1+R_{\lambda\infty})\exp(-L_\lambda^*\tau)=(1-\Lambda_{ef})E_\lambda^*E_{\lambda0}^*(x).\tag{2.31}$$

The scalar quantity of the net radiant flux vector in this case, as determined by Eq. 2.22, is as follows:

$$q_\lambda^*=E_\lambda^*(1-R_{\lambda\infty})\exp(-L_\lambda^*\tau).\tag{2.32}$$

Since $E_\lambda^*(1-R_{\lambda\infty})$ is equal to the dimensionless quantity of the net radiant flux near the surface at $\tau=0$, Eq. 2.32 for the coordinate τ can be rewritten as:

$$q_\lambda^*(\tau_1)=q_\lambda^*(\tau=0)\exp(-L_\lambda^*\tau_1).\tag{2.33}$$

From this it follows:

$$L_\lambda^*=(1/\tau_1)\ln\{[q_\lambda^*(\tau=0)]/q_\lambda^*(\tau_1)\}.\tag{2.34}$$

Thus, the dimensionless coefficient L_λ^* describes the attenuation of the spectral radiant flux as it propagates through an optically infinitely thick layer. If at an optical depth τ_1 the ratio $[q_\lambda^*(\tau=0)]/[q_\lambda^*(\tau_1)]$ equals e, then Eq. 2.34 yields $L_\lambda^*=\tau_1^{-1}.$ Therefore. the coefficient L_λ^*, which we refer to as the dimensionless coefficient of effective attenuation of a radiant flux, is numerically equal to the reciprocal quantity of the optical depth of a layer. During

its passage through this layer the net radiant flux q_λ^* decreases e times.

In the case where the constants δ_s and m are independent of the coordi-

nate x, $\overline{k}_\lambda$ and s_λ will also be constant. And the analytical expressions ob-

tained above are in full agreement with the known equations [44, 45]. At $\overline{k}_\lambda=$

const and $s_\lambda=$const Eq. 2.13 yields the following relationship between the opti-

cal thickness τ_ℓ and the geometrical thickness ℓ of the layer:

$$\tau_\ell=(\overline{k}_\lambda + s_\lambda)\ell=\varepsilon_{ef}\ell, \tag{2.35}$$

$$\tau=\varepsilon_{ef}x. \tag{2.36}$$

With account taken of Eqs. 2.21, 2.35, and 2.36, the relationship between

the dimensionless quantity $L_\lambda^*\tau_\ell$ and the product of the effective attenuation co-

efficient L_λ (Table 1.1) and the thickness of layer ℓ is:

$$L_\lambda^*\tau_\ell=\varepsilon_{ef}(1-\Lambda_{ef}^2)^{\frac{1}{2}}\ell=L_\lambda\ell, \tag{2.37}$$

$$L_\lambda^*\tau=L_\lambda x.$$

The substitution of $L_\lambda\ell$ for $L_\lambda^*\tau_\ell$ in Eqs. 2.17-2.33, according to Eq. 2.37,

yields relationships which are in complete agreement with the previously de-

rived ones [45]. In this case, at constant $\overline{k}_\lambda$ and s_λ, the amount of radiant

energy absorbed per unit time at depth x, as determined on the basis of Eq.

2.24, with account taken of Eq. 2.35, fully accords with published results [44]:

$$w_\lambda(x)=\overline{k}_\lambda E_{\lambda 1}[(1+R_{\lambda\infty})/(1-\psi_\lambda^2)][\exp(-L_\lambda x)-(\psi_\lambda^2/R_{\lambda\infty})\exp(L_\lambda x)]+$$
$$\overline{k}_\lambda E_{\lambda 2}[(1+R_{\lambda\infty})/(1-\psi_\lambda^2)]\{\exp[-L_\lambda(\ell-x)]-(\psi_\lambda^2/R_{\lambda\infty})\exp[L_\lambda(\ell-x)]\}. \tag{2.38}$$

For an optically infinitely thick layer Eq. 2.31, with account taken of Eq. 2.37,

yields:

$$w_\lambda(x)=\overline{k}_\lambda E_\lambda(1+R_{\lambda\infty})\exp(-L_\lambda x)=\overline{k}_\lambda E_\lambda E_{\lambda 0}^*. \tag{2.39}$$

The spatial irradiance $E_{\lambda 0}$ of a volume element having dx thickness at depth x of a semi-infinite medium is:

$$E_{\lambda 0} = \dot{E}_\lambda (1 + R_{\lambda \infty}) \exp(-L_\lambda x). \tag{2.40}$$

An equation of this type was developed for the first time [18] to describe the total irradiance $E_{\lambda 0}$ in weakly absorbing but strongly scattering media (e.g., opal glass). Afterwards analogous equations were obtained [24, 26, 33, 155]; they are based on the assumption that the optical factors $\bar{k}_\lambda$ and s_λ are constant at any values of x.

The derived analytical expressions, Eqs. 2.17-2.33, can also be used directly for calculating q_λ and $w_\lambda(x)$ in the case where $\bar{k}_\lambda$ and s_λ are constant. To do this it is only necessary to substitute in the index of exponent the dimensionless complexes $L_\lambda x$ and $L_\lambda \ell$ for $L_\lambda^* \tau$ and $L_\lambda^* \tau_\ell$, respectively. The validity of such a substitution is confirmed by results shown in Fig. 2.6. It can be seen that the dependence of $L_\lambda^* \tau_\ell$ on the layer's thickness ℓ can be satisfactorily approximated by a linear dependence. Therefore, the quantities L_λ, $\bar{k}_\lambda$, and s_λ can almost be considered to be constant inside the layer and independent of the depth x for the given substances.

To calculate $w_\lambda(x)$ by using Eqs. 2.38 and 2.39 we need to know three constants of the irradiated substance: $\bar{k}_\lambda$, L_λ, and $R_{\lambda \infty}$. By substituting $L_\lambda(1 - R_{\lambda \infty})$ for $\bar{k}_\lambda(1 + R_{\lambda \infty})$ in accordance with the relationship between these quantities (cf. Chapter 3) we can calculate $w_\lambda(x)$ with the aid of two constants, L_λ and $R_{\lambda \infty}$. Thus, Eq. 2.39 acquires the following form:

$$w_\lambda(x) = L_\lambda E_\lambda (1 - R_{\lambda \infty}) \exp(-L_\lambda x) = L_\lambda q_\lambda(x). \tag{2.41}$$

This form is more convenient for calculations, although Eq. 2.39 reflects more fully the physical meaning of the absorption of radiation at depth x.

To evaluate how precise our differential-difference method is for investigating radiant energy transfer in capillary-porous and colloidal substances it is necessary to determine experimentally the radiant power at the interfaces for different substances. In the case of unilateral irradiation the magnitude of the dimensionless radiation flux E_+^* at the interface $\tau=\tau_\ell$ is equal to the transmittance of a layer of substance, while the magnitude of the radiation flux E_-^* at the interface $\tau=0$ is equal to the reflectance of the layer. An assessment of the effectiveness of this method is given in Chapter 3; there, in Figs. 3.2 and 3.3

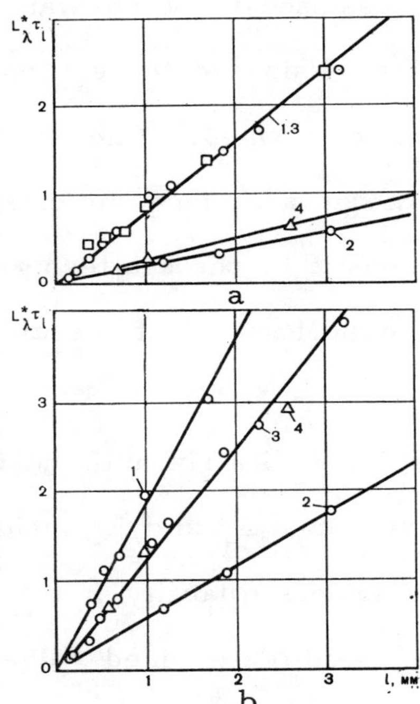

Figure 2.6 Dependence of the dimensionless quantity $L_\lambda^* \tau_\ell$ on the thickness of a layer at different wavelengths, $\lambda=0.8$ μm (a) and $\lambda=1.2$ μm (b), for some substances [39, 44]:
    1 - potato starch, W=11.8%;   3 - pine, W=6.2%;
    2 - potato, W=80.5%;   4 - dough (first-grade flour), W=48.0%

are shown the relationships between the spectral reflectance $R_\lambda(2\pi;2\pi)$, the transmittance $T_\lambda(2\pi;2\pi)$, and the absorptance $A_\lambda(2\pi;2\pi)$ for starch and other substances. These relationships are calculated with the aid of derived equations and are also determined experimentally.

It can be seen from Figs. 3.2 and 3.3 that values for the flux density of radiation at the interfaces of a layer that are determined experimentally differ from those that are calculated by ±10% on the average. This indicates that the proposed method is sufficiently precise. The higher level of precision of this method as compared to the usual differential-difference method of Schuster-Schwarzschild is explained by the following circumstances. In our method we exclude from the differential equations describing the counter fluxes E_+ and E_- the coefficients $\bar{k}_\lambda$ and s_λ, which are assumed to be constant, and the unknown indicatrix $\chi_\lambda(\gamma)$, whose form is usually considered to be symmetrical, $\delta_s = 0.5$ (this corresponds to a case of uniform scattering). And the multiple scattering of radiation in a medium with an unknown scattering indicatrix is taken into account through the averaged characteristic of the scattering properties of the medium Λ_{ef}, which can be determined experimentally for each particular substance.

The unknown form of the scattering indicatrix and the possible changes in the spatial distribution of the radiation fluxes E_+ and E_- inside a layer are taken into account with the aid of the dimensionless quantity $L_\lambda^* \tau$. This quantity is calculated on the basis of the experimentally determined values for R_λ, T_λ, and $R_{\lambda\infty}$. It includes the unknown coefficients $m(x)$ and $\delta_s(x)$.

Thus, the theoretical expressions, Eqs. 2.17-2.33, are more precise in comparison with the known relationships [3, 18, 26, 33, 44] for characterizing the distribution of monochromatic radiation fluxes in foodstuffs and other absorbing-scattering substances. The possible changes in the conditions of irradiation of the volume element inside a layer with a change in the coordinate, and the form of the scattering indicatrix are taken into account in these equations with the aid of the variable coefficients $m(x)$ and $\delta_s(x)$.

2.4 Attenuation of infrared radiation in foodstuffs and the angle of incidence

The thermal processing of foodstuffs is usually carried out under conditions of directional or mixed (directional-diffuse) radiation (Figs. 2.1, 2.2). Therefore, we need to know the laws governing the attenuation of radiation directed at a certain angle. In the case of solar irradiation the angle of incidence changes all the time. Thus, it is also necessary to know how the attenuation of infrared radiation takes place in foodstuffs in relation to the angle of incidence [120, 129, 173].

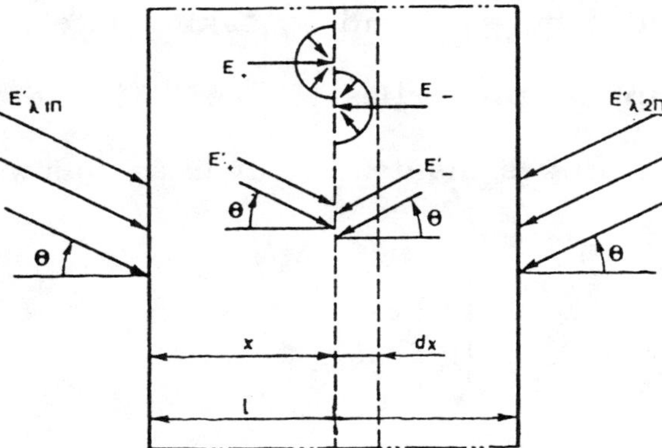

Figure 2.7 Distribution of directional radiation at a certain angle in a plane layer of an absorbing-scattering substance [39, 42, 44]

Experimental investigations [44] have shown that in most foodstuffs (such as dough, flour, starch, etc) directional radiation, as it propagates throughout the layer adjacent to the surface of the irradiated substance, quickly changes into diffuse (or scattered) radiation. Therefore, within a certain zone adjacent to the irradiated surface at depth x there are present a directional radiation flux E'_λ and a newly formed diffuse radiation flux E_λ, which are comparable in intensity.

Let us consider the transfer of energy of monochromatic radiation in a plane layer of an isotropic selective attenuating medium having ℓ thickness. The layer is exposed on both sides to incident directional radiation fluxes $E'_{\lambda 1}$ and $E'_{\lambda 2}$ at an angle $\theta = arc\ cos\mu$ (fig. 2.7). We shall consider a layer element having dx thickness at depth x. Its optical properties will be defined by spectral coefficients that can be determined experimentally. For this it is not necessary to know the exact form of the scattering indicatrix or of the spatial distribution of radiation fluxes inside the layer. For directional radiation we use the spectral absorption coefficient k_λ and the scattering coefficient σ_λ, and for diffuse radiation - the absorption coefficient $\bar{k}_\lambda$, the forward scattering coefficient f_λ and the backward scattering coefficient s_λ, which are averaged with respect to the incidence angle.

By assuming $m_+=m_-=m$ and $\delta_+=\delta_-=\delta_s$, and by taking into consideration the interrelationship between the optical properties (Table 1.1), we can describe the directional and diffuse radiation fluxes inside the layer as follows:

$$\left.\begin{aligned}
\frac{dE'_+}{dx} &= -\frac{1}{\mu}(k_\lambda + \sigma_\lambda)E'_+ = -\frac{\varepsilon_\lambda}{\mu}E'_+ \\
-\frac{dE'_-}{dx} &= -\frac{1}{\mu}(k_\lambda + \sigma_\lambda)E'_- = -\frac{\varepsilon_\lambda}{\mu}E'_-
\end{aligned}\right\} \qquad (2.42)$$

$$\left.\begin{aligned}
\frac{dE_+}{dx} &= -(\bar{k}_\lambda + s_\lambda)E_+ + s_\lambda E_- + \frac{f'_\lambda}{\mu}E'_+ + \frac{s'_\lambda}{\mu}E'_- \\
-\frac{dE_-}{dx} &= -(\bar{k}_\lambda + s_\lambda)E_- + s_\lambda E_+ + \frac{s'_\lambda}{\mu}E'_+ + \frac{f'_\lambda}{\mu}E'_-
\end{aligned}\right\}, \qquad (2.43)$$

where, $\mu=\cos\theta$.

Multiple scattering in a plane layer in the case of directional irradiation is taken into account by a system of two relationships, Eq. 2.43. The attenuation of directional radiant flux is described by Eq. 2.42, which complements the Bouguer-Lambert law. The solution of Eqs. 2.42 and 2.43 under the boundary conditions:

$$x=0, \quad E'_+(0)=E'_{\lambda 1}, \quad E_+(0)=0 \qquad (2.44)$$

$$x=\ell, \quad E'_-(\ell)=E'_{\lambda 2}, \quad E_-(\ell)=0 \qquad (2.45)$$

obtained by the authors for a general case [44] is a very complex one. Therefore, we shall write only the expression for $w'_\lambda(x)$, which is of considerable

practical importance. In the heat transfer equation $w'_\lambda(x)$ is a function of the

energy sources due to the absorption of infrared radiation in the irradiated sub-

stance.

Now let us consider several particular cases which are of practical interest

in calculating heat transfer in substances subjected to directional infrared irra-

diation.

1. When the incident radiation fluxes are equal, $E'_{\lambda 1}=E'_{\lambda 2}=E'_\lambda$, the equation

for $w'_\lambda(x)$ is considerably simplified:

$$w'_\lambda(x)=\bar{k}_\lambda(1+R_{\lambda\infty})/(1+\psi_\lambda)\{\exp(-L_\lambda x)+\exp[-L_\lambda(\ell-x)]\}[C_2+C_1\exp(-\varepsilon_\lambda\ell/\mu)]-$$
$$-\bar{k}_\lambda E'_\lambda(C_1+C_2)\{\exp(-\varepsilon_\lambda x/\mu)+\exp[-(\varepsilon_\lambda/\mu)(\ell-x)]\}+k_\lambda E'_\lambda\{\exp(-\varepsilon_\lambda x/\mu)+$$
$$+\exp[-(\varepsilon_\lambda/\mu)(\ell-x)]\}. \tag{2.46}$$

2. When irradiation is unilateral $E'_{\lambda 2}=0$, the quantity $w'_\lambda(x)$ is determined

by the following equation:

$$w'_\lambda(x)=[\bar{k}_\lambda E'_{\lambda 1}(1+R_{\lambda\infty})/(1-\psi_\lambda^2)]\{C_2[\exp(-L_\lambda x)-(\psi_\lambda^2/R_{\lambda\infty})\exp(L_\lambda x)]+$$
$$+C_1\exp(-\varepsilon_\lambda\ell/\mu)[\exp\{-L_\lambda(\ell-x)\}-(\psi_\lambda^2/R_{\lambda\infty})\exp\{L_\lambda(\ell 0x)\}]\}-$$
$$-\bar{k}_\lambda E'_{\lambda 1}(C_1+C_2)\exp(-\varepsilon_\lambda x/\mu)+k_\lambda E'_{\lambda 1}\exp(-\varepsilon_\lambda x/\mu). \tag{2.47}$$

In Eqs. 2.46 and 2.47 C_1 and C_2 are the parameters which determine the optical

properties of the medium and the spatial distribution of the energy of the

fluxes E_+ and E_- at depth x. With account taken of the conditions of irradia-

ting the layer these parameters are:

$$C_1=(\mu s'_\lambda \varepsilon_{\lambda ef}-s'_\lambda \varepsilon_\lambda+\mu f'_\lambda s_\lambda)/(\varepsilon_\lambda^2-\mu^2 L_\lambda^2); \tag{2.48}$$
$$C_2=(\mu f'_\lambda \varepsilon_{\lambda ef}+f'_\lambda \varepsilon_\lambda+\mu s'_\lambda s_\lambda)/(\varepsilon_\lambda^1-\mu^2 L_\lambda^2). \tag{2.49}$$

3. In an optically infinitely thick layer ($\varepsilon_\lambda\ell\to\infty, L_\lambda\ell\to\infty$) the net radiant

flux vector and the vector of the absorbed energy are:

$$q_\lambda'(x)=E_\lambda' \, [(1-R_{\lambda\infty})C_2\exp(-L_\lambda x)+(1+C_1-C_2)\exp(-\varepsilon_\lambda x/\mu)]; \tag{2.50}$$

$$w_\lambda'(x)=\bar{k}_\lambda E_\lambda'[(1+R_{\lambda\infty})C_2\exp(-L_\lambda x)-(C_1+C_2)\exp(-\varepsilon_\lambda x/\mu)]+$$

$$+k_\lambda E_\lambda'\exp(-\varepsilon_\lambda x/\mu). \tag{2.51}$$

Equations 2.46-2.51 determine the values of q_λ' and w_λ' at any point on a plane layer exposed to a broad directional radiant flux at an angle θ' normal to the surface.

The known relationships for E_+, E_-, q_λ, and w_λ [39, 44] make it possible to analyze the radiation field inside a plane layer of substance exposed to a broad radiant flux. Such an analysis can be carried out with the aid of data in Fig. 2.8 for a case of a symmetrical scattering indicatrix ($\delta_f=\delta_s=0.5$). It shows the dependence of the dimensionless quantities $E_+^*=E_+/E_\lambda'$ and $E_-^*=E_-/E_\lambda'$ for diffuse counterfluxes and of the spatial irradiance $E_{\lambda 0}^{*'}=E_{\lambda 0}'/E_\lambda' + E_{\lambda 0s}/E_\lambda'$ on the scattering properties of the medium $\Lambda=\sigma_\lambda/\varepsilon_\lambda$ and the optical depth $\varepsilon_\lambda x$ in a semi-infinite medium under irradiation conditions ($\mu=1$). Every curve is complex in nature, differing considerably in the case of small optical thickness from the exponential type and nearly exponential type of curves in the case of large values of $\varepsilon_\lambda x$.

Eq. 2.51, derived for the dimensionless quantity of the energy absorbed per unit volume and per unit time at depth x, makes it possible to analyze the distribution of the absorbed energy in a layer in the case of directional irradiation:

$$w_\lambda^{*'}(x)=A_1\exp(-L_\lambda x)-A_2\exp(-\varepsilon_\lambda\mu). \tag{2.52}$$

In this equation we use the following definitions of the constants A_1 and A_2 which depend on the optical properties of the irradiated subctance and the irradiation conditions:

$$A_1=(1-\Lambda_{ef})(1+R_{\lambda\infty})C_2; \tag{2.53}$$

$$A_2 = (1-\Lambda_{ef})(1-\Lambda+C_1+C_2). \qquad\qquad (2.54)$$

An example of how the obtained relationships can be used in calculating the radiation fields in foodstuffs subjected to directional irradiation is shown in Fig. 2.9. As can be seen, with an increase in x the magnitude of $w_\lambda^{*\prime}$ increases, reaching a maximum at x_m=0.16 mm, and then decreases according to a complex law. At x>0.4 mm the dependence of $w_\lambda^{*\prime}$ on x in the case of directional irradiation and the dependence of w_λ^{*} on x in the case of diffuse irradiation (dashed line) coincide.

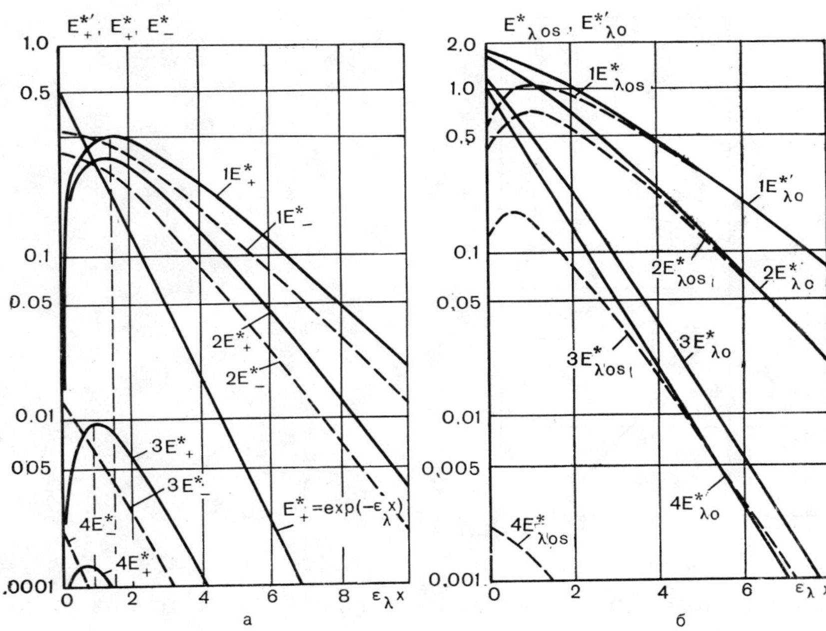

Figure 2.8   Dependences of the flux densities of dimensionless radiation $E_+^{*\prime}$ and scattered radiation E_+^*, E_-^* (a), the spatial irradiance of scattered radiation $E_{\lambda 0}^*$, and the integral spatial irradiance $E_{\lambda 0}^{*\prime}$ on the optical thickness $\varepsilon_\lambda x$ (b) at different values of Λ [39, 44]:

    1 - 0.95;   3 - 0.5;
    2 - 0.9;   4 - 0.1.

At the same time at x<0.4 mm these dependences differ from one another. In the case of diffuse irradiation the magnitude of w_λ^* reaches a maximum at x=0, and then decreases according to the exponential law.

The position of the maximum of the absorbed energy $w_\lambda^{*\prime}$ in the case of directional irradiation is determined from Eq. 2.51 by assuming that the first derivative $dw_\lambda^{*\prime}/dx$ is equal to zero:

$$\varepsilon_\lambda x_m(w_{\lambda max}^{*\prime})=[\mu/(1-\mu K)]\ln[(1/\mu K)(1-\Lambda+C_1+C_2)/(1+R_{\lambda\infty})C_2]. \qquad (2.55)$$

The presence of the maximum of $w_\lambda^{*\prime}$ in this case is due to the fact that the value of the absorption coefficient k_λ for directional radiation differs from that of the absorption coefficient $\bar{k}_\lambda=mk_\lambda$ for scattered radiation. Since the

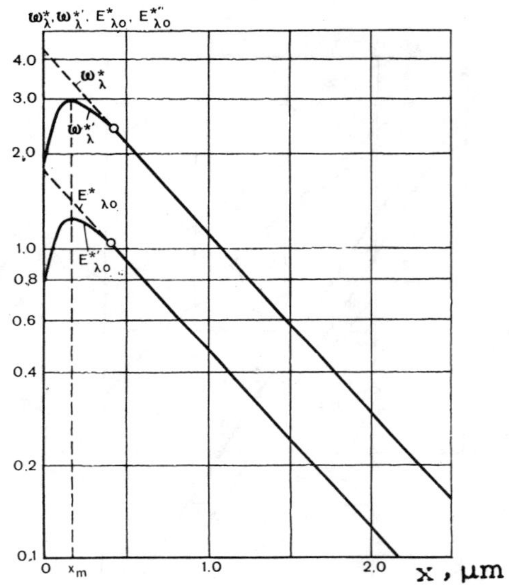

Figure 2.9   Spatial irradiance and the absorbed energy as a function of x in the case of diffuse radiation, $E_{\lambda0}^*, w_\lambda^*$, and of directional radiation, $E_{\lambda0}^{*\prime}, w_\lambda^{*\prime}$, for wheat-flour dough (W=48.0%) irradiated at $\lambda=1.1$ μm [39].

spatial irradiance of the volume element inside the layer is due mainly to the scattered radiation $E_{\lambda0s}^*$ (Fig. 2.8, b), and since $\bar{k}_\lambda>k_\lambda$, the dependence of w_λ^* on x is similar to that of $E_{\lambda0s}^*$ on the depth, and $w_\lambda^{*\prime}$ has a maximum. The integral spatial irradiance $E_{\lambda0}^*$ both for directional and diffuse radiation exhibit no extremum. Fig. 2.8, a shows that with an increase in $\varepsilon_\lambda x$ at any values of Λ

the magnitude of the scattered radiation E_+^* in a boundary zone increases, reaching a maximum at a certain value of $\varepsilon_\lambda x_m$, and then decreases according to a complex law. By assuming that the first derivative dE_+/dx is equal to zero we obtain [39]:

$$e_\lambda x_m(E_+^*)=[\mu/(1-\mu K)]\ln(\mu K); \tag{2.56}$$

where, $K=L_\lambda/\varepsilon_\lambda$.

The spatial irradiance $E_{\lambda 0s}^*=E_+^*+E_-^*$ of scattered radiation (Fig. 2.9, b) rapidly increases in a boundary zone of the layer, reaching a maximum at depth x_m:

$$\varepsilon_\lambda x_m(E_{\lambda 0s}^*)=[\mu/(\mu K-1)]\ln\{\mu K[(1+R_{\lambda\infty})C_2)/(C_2+C_1]\}. \tag{2.57}$$

Here, the magnitude of $E_{\lambda 0s}$ can exceed that of E_+' twofold or more. In the case of strongly scattering media ($\Lambda>0.9$, Fig. 2.8, curves 1 and 2) the spatial irradiance $E_{\lambda 0}^*$ and the radiation flux E_+^* at depth $\varepsilon_\lambda x>\varepsilon_\lambda x_m$ are determined mainly by the scattered radiation. And in this case the magnitude of the directional radiation $E_+^{*'}$ is small as compared with that of E_+^*.

To evaluate the precision of the method considered above for investigating radiant energy transfer in scattering-absorbing substances subjected to directional irradiation it is necessary to determine the values of the radiant fluxes at the layer's interfaces and to compare them with the results obtained by other methods. In the case of unilateral irradiation the dimensionless quantity of the sum of the radiation fluxes $(E_+^*+E_+^{*'})$ at the layer's boundary $x=\ell$ is equal to the transmittance $T_\lambda(\mu,2\pi)$ of the layer, while the magnitude of E_-^* at the layer's boundary $x=0$ is equal to the reflectance $R_\lambda(\mu,2\pi)$ of the layer.

In Fig. 3.4 and Tables 3.2–3.4 are summarized the magnitudes of $R_\lambda(\mu; 2\pi)$, $T_\lambda(\mu;2\pi)$ and $R_{\lambda\infty}(\mu;2\pi)$ as calculated by different methods. As can be seen, the magnitude of the reflectance and transmittance calculated by our method for discrete fluxes differ by 1-2% from those obtained by numerical methods [144, 158, 192, 199] at different incidence angles θ and with different scattering properties Λ of the medium and different optical thicknesses $k_\lambda \ell$ of the layer. This indicates that our method is sufficiently precise.

Thus, with the aid of the averaged characteristics, which can be obtained without a knowledge of the form of the scattering indicatrix $\chi_\lambda(\gamma)$, our method can be used for determining the radiation field inside capillary-porous and colloidal media subjected to directional irradiation at angle θ.

2.5 Multiple reflections from the interfaces of a layer of an irradiated substance

In the theory of energy transfer in opaque media it is assumed that there is no reflection from a layer's interface [33, 36, 64, 11, 115, 126, 142, 143, 148]. In some instances this assumption is justified on the ground that the magnitude of the reflection coefficient of a substance is very small; in other instances it is thought that the layer consists of scattering-absorbing particles suspended in air [24, 44, 125, 127, 155, 167]. The applicability of the second consideration to foodstuffs has been shown experimentally. Indeed, at low moisture content the pores and capillaries inside a layer contain air and the surface of the substance is moisture-free. Therefore, it can be generally considered that such a layer consists of particles suspended in air, no account being taken of the boundary reflection.

In the case of substances rich in moisture or of frozen foods the situation is different. Here, the outer surface is the surface of a liquid or ice.

The magnitude of the reflection coefficient at the air-medium interface, $\rho'_{\lambda 0}$, for an external directional radiant flux is not the same as that at the medium-air interface, ρ'_λ, for a directional radiant flux emerging from the layer; both depend on the refraction index n_λ, the absorption coefficient k_λ, and the in-

incident angle θ, and are determined by Fresnel's formulas (Fig. 2.10). For a

directional radiant flux normal to the surface ($\theta=0$) of an absorbing medium the

reflection coefficient $\rho'_{\lambda 0}$ is determined according to the formula:

$$\rho'_{\lambda 0}(\theta=0)=[(n_\lambda-1)^2+k_\lambda^2]/[(n_\lambda+1)^2+k_\lambda^2]. \tag{2.56}$$

Depending on the incidence angle θ the coefficients $\rho'_{\lambda 0}$ and ρ'_λ change in

a complex way at different values of the refractive index n_λ and of the absorp-

tion coefficient k_λ. It can be seen from Fig. 2.10 that the surface has a low

reflectance $\rho'_{\lambda 0}$ for the incoming incident radiant flux both for an absorbing me-

dium ($k_\lambda>0$) and an non-absorbing ($k_\lambda=0$) medium. However, for a radiant flux

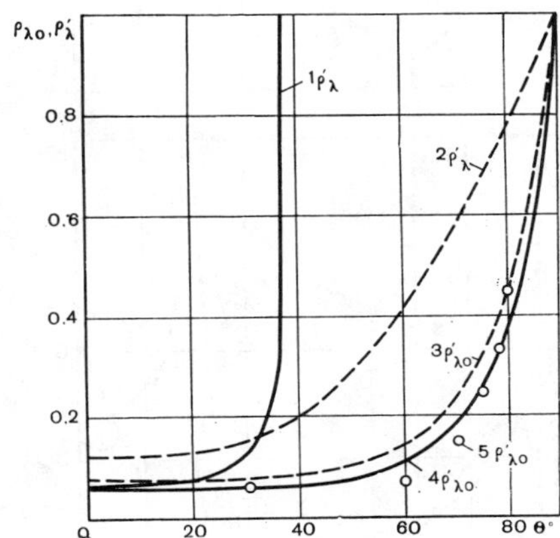

Figure 2.10 Reflection coefficients at the interface of two media as a function of

the incidence angle θ at different values of the refractive index n_λ and of the ab-

sorption coefficient k_λ [60, 130. 138]:

    1,2 – $n_\lambda=0.6$;   1,4 – $k_\lambda=0$;

    3,4 – $n_\lambda=1.6$;   2,3 – $k_\lambda=0.4$;

             5 – $\rho'_{\lambda 0}$ for ice at $\lambda=3.05$ μm

emerging from a layer through the interface the reflectance is much greater.

And with a decrease in k_λ the magnitude of ρ'_λ increases at larger incidence

angles θ. The beams incident upon the interface at an angle θ that is greater

than arc $\sin 1/n_\lambda$ are completely reflected inside the layer ($\rho'_\lambda=1$). In this connection it is necessary to know the optical constants for water and ice. In Fig. 2.11 are shown the spectral dependences of k_λ and n_λ for water at 25°C and of k_λ n_λ, and $\rho'_{\lambda 0}$ for ice at -7°C. It can be seen that at wavelengths 0.4-2.0 μm the refractive index of water and ice changes very little, within 1.300-1.344.

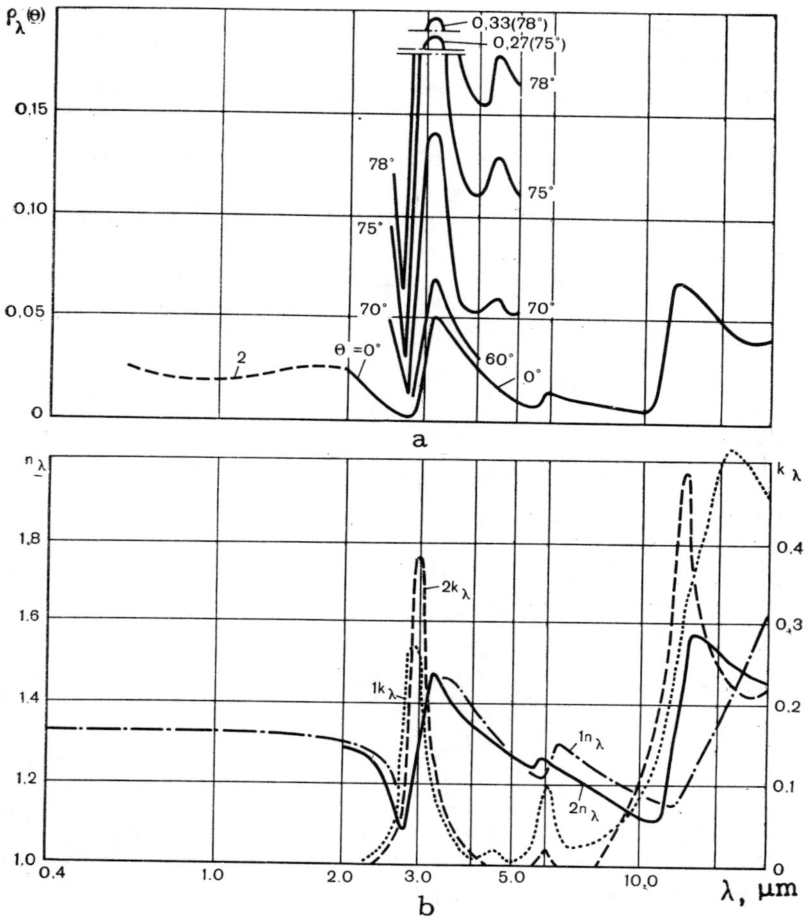

Figure 2.11   Reflective coefficient of ice $\rho'_\lambda(\theta)$ as a function of wavelength λ at different incidence angles θ (a); absorption coefficient k_λ and refractive index n_λ of ice and water as a function of wavelength λ (b) at different temperatures [60, 130, 138, 179]:
      1 - water at 24°C;
      2 - ice at -7°C.

In this spectral range the absorption coefficient $k_\lambda \ll 1$ and does not exceed $1.1 \cdot 10^{-3}$. The characteristic absorption bands of water and ice are observed near

3.0, 6.0, 12.0, and 15.0 μm. Therefore, the reflectance averaged with respect to two polarization components, depending on the incidence angle θ, can be determined (when $k_\lambda \ll n_\lambda$) with the aid of the following equation:

$$\rho'_{\lambda 0} = \frac{1}{2}\left[\left(\frac{\sqrt{n_\lambda^2 - \sin^2\theta} - \cos\theta}{\sqrt{n_\lambda^2 - \sin^2\theta} + \cos\theta}\right)^2 + \left(\frac{n_\lambda^2\cos\theta - \sqrt{n_\lambda^2 - \sin^2\theta}}{n_\lambda^2\cos\theta + \sqrt{n_\lambda^2 - \sin^2\theta}}\right)^2\right],$$

(2.59)

where, n_λ - relative refractive index of the irradiated substance in respect to the surrounding medium.

For a radiant flux emerging from a substance the boundary reflection coefficient ρ'_λ can be determined with the aid of Eq. 2.59, with n_λ in the equation being replaced by $1/n_\lambda$ for all θ angles that are smaller than θ_{lim}=arc sin$1/n_\lambda$. For θ angles that are larger than θ_{lim} the radiation undergoes complete internal reflection and ρ'_λ=1.

In the case of diffuse radiation flux the reflection coefficient $\rho_{\lambda 0}$ averaged with respect to the hemisphere will be greater than $\rho'_{\lambda 0}$ at normal incident flux (θ=0°). For a radiant flux emerging from a substance through the interface the reflection coefficient ρ_λ will also be greater than ρ'_λ(θ=0°). It should be noted that the reflection coefficients $\rho_{\lambda 0}$ and ρ_λ are not equal in magnitude and are interlinked through the following relationship [18, 33]:

$$(1-\rho_{\lambda 0})=n_\lambda^2(1-\rho_\lambda).$$

(2.60)

This equation makes it possible to calculate ρ_λ from known values of $\rho_{\lambda 0}$ and n_λ. From Eq. 2.60 it follows that at n_λ>1 the magnitude of ρ_λ for the outgoing radiation considerably exceeds that of $\rho_{\lambda 0}$.

For water and ice the magnitude of k_λ reaches 0.5 and that of n_λ,

1.1-1.6 in the range of wavelengths 2.0-20.0 μm (Fig. 2.11). Thus, the magnitude of $\rho_{\lambda 0}$ should be less than 2-10%; consequently, it is advisable to disregard the external reflection in many calculations. However, for radiation emerging from a layer, according to Eq. 2.60, the magnitude of ρ_λ can be as large as 65%, which obviously should be taken into account. It is to be noted that in a strict analysis the internal reflection should also be taken into account in the case of substances with a low moisture content.

At the air-surface layer interface the magnitude of ρ_λ can be quite large. This means that the presence of a medium-air interface decreases the probability of the radiation escaping to the layer. And the absorptance A_λ, calculated with no account being taken of the internal reflection, will be underestimated. The boundary reflection strongly affects the magnitude of R_λ and T_λ, as well as the distribution of the absorbed radiant energy $w_\lambda(x)$.

The considerable effect of internal reflection on the energy distribution inside the irradiated substance and on the outgoing radiation is illustrated in Fig. 2.12. When a fractured grain of rice is irradiated no radiation pentrates the half that is not irradiated.[*] The fracture forms a dividing line between the irradiated and the nonirradiated part of the grain. This is explained by the fact that the fracture is filled with air, and the radiation scattered on the endosperm undergoes complete internal reflection at the endosperm-air interface at incidence angles $\theta > \mathrm{arc}\ \sin 1/n_\lambda$; where, n_λ - relative refractive index of the endosperm. Many investigations of the reflection at the boundary interface have been carried out [15, 44, 47, 87, 105, 125, 132, 136, 160, 161, 165, 171, 200, 201].

---

[*]This phenomenon has been utilized to modify the diaphanoscope for the purpose of determining fracturing in grains [58].

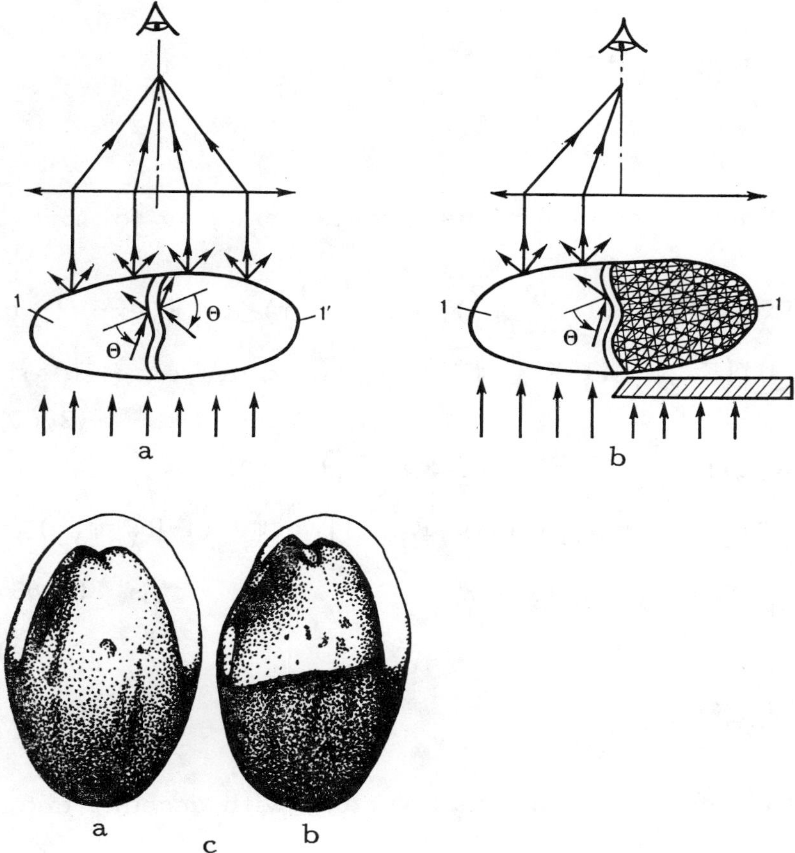

Figure 2.12 Effect of internal reflection at the endosperm-air interface in

an irradiated fractured grain of rice on the energy distribution in the

grain [58]:
    a - when both halves of the grain are irradiated;
    b - when one half of the grain is irradiated;
    c - the appearance of the grain on a modified diaphanoscope under ir-
        radiation conditions a and b.

The effect of external and internal boundary reflection can also be

taken into account when solving differential equations for energy transfer,

Eqs. 2.15, 2.42, and 2.43, by setting the corresponding boundary condi-

tions.

Next we shall consider a case of bilateral diffuse irradiation by dimensionless radiation fluxes $E_{\lambda 1}^{*}$ and $E_{\lambda 2}^{*}$ of a plane scattering layer having boundary reflection coefficients $\rho_{\lambda 0}$ and ρ_{λ}. We shall define the limiting conditions in Eq. 2.15 as follows:

$$\text{when } x=0, \ E_{+}^{*}(0)=(1-\rho_{\lambda 0})E_{\lambda 1}^{*} + \rho_{\lambda}E_{-}^{*}(0)$$
$$\text{when } x=\ell, \ E_{-}^{*}(\ell)=(1-\rho_{\lambda 0})E_{\lambda 2}^{*} + \rho_{\lambda}E_{+}^{*}(\ell). \tag{2.61}$$

A solution of Eq. 2.15 under these conditions has the following form:

$$E_{+}^{*}=[(1-\rho_{\lambda 0})/(1-\rho_{\lambda}R_{\lambda\infty})][E_{\lambda 1}^{*}/(1-R_{\lambda\rho}^{2}\psi_{\lambda}^{2})][\exp(-L_{\lambda}^{*}\tau)-R_{\lambda\rho}\psi_{\lambda}^{2}\exp(L_{\lambda}^{*}\tau)+$$
$$+[(1-\rho_{\lambda 0})/(1-\rho_{\lambda}R_{\lambda\infty})][E_{\lambda 2}\psi_{\lambda}/(1-R_{\lambda\rho}^{2}\ \psi_{\lambda}^{2})][\exp(L_{\lambda}^{*}\tau)-R_{\lambda\rho}\exp(-L_{\lambda}^{*}\tau)]; \tag{2.62}$$

$$E_{-}^{*}=[(1-\rho_{\lambda 0})/(1-\rho_{\lambda}R_{\lambda\infty})][E_{\lambda 2}^{*}/(1-R_{\lambda\rho}^{2})]\{\exp[-L_{\lambda}^{*}(\tau_{\ell}-\tau)-$$
$$-R_{\lambda\rho}\psi_{\lambda}^{2}\exp[L_{\lambda}^{*}(\tau_{\ell}-\tau)]\}+[(1-\rho_{\lambda 0})/(1-\rho_{\lambda}R_{\lambda\infty})][E_{\lambda 1}^{*}\psi_{\lambda}/(1-R_{\lambda\rho}^{2}\psi_{\lambda}^{2})$$
$$\{\exp[L_{\lambda}^{*}(\tau_{\ell}-\tau)]-R_{\lambda\rho}\exp[-L_{\lambda}^{*}(\tau_{\ell}-\tau)]\}; \tag{2.63}$$

where, $\psi_{\lambda}=R_{\lambda\infty}\exp(-L_{\lambda}^{*}\tau_{\ell});$ $\hspace{2cm}$ (2.64)

$$R_{\lambda\rho}=[(r_{\lambda\infty}-\rho_{\lambda})/R_{\lambda\infty}(1-\rho_{\lambda}R_{\lambda\infty}). \tag{2.65}$$

The dimensionless quantity of the radiant flux vector, with account taken of the boundary reflection, is determined as follows:

$$q_{\lambda}^{*}(\tau)=E_{+}^{*} - E_{-}^{*}. \tag{2.66}$$

And the dimensionless radiant energy absorbed by the substance at depth τ per unit time is:

$$w_{\lambda}^{*}(\tau)=(1-\Lambda_{ef})(E_{+}^{*} + E_{-}^{*}). \tag{2.67}$$

Now let us consider some particular cases.

1. When the incident fluxes are equal, $E_{\lambda 1}^{*}=E_{\lambda 2}^{*}=E_{\lambda}^{*}$, Eqs. 2.66 and 2.67 are considerably simplified:

$$q_\lambda^*(\tau)=E_\lambda^*[(1-\rho_{\lambda 0})/(1-\rho_\lambda R_{\lambda\infty})][(1-R_{\lambda\infty})/(1+R_{\lambda\rho}\psi_\lambda)]\{\exp(-L_\lambda^*\tau)-$$
$$-\exp[-L_\lambda^*(\tau_\ell-\tau)]\};\qquad(2.68)$$

$$w_\lambda^*(\tau)=(1-\Lambda)E_\lambda^*[(1-\rho_{\lambda 0})/(1-\rho_\lambda R_{\lambda\infty})][(1+R_{\lambda\infty})/(1+R_{\lambda\infty}\psi_\lambda)]\cdot$$
$$\cdot\{\exp(-L_\lambda^*\tau)+\exp[-L_\lambda^*(\tau_\ell-\tau)]\}.\qquad(2.69)$$

2.   In the case of unilateral irradiation, $E^*_{\lambda 2}=0$, of a layer without a supporting base and having finite thickness Eqs. 2.62 and 2.63 are in complete accord with the known equations [18] for E^*_+ and E^*_-.

3.   In the case of unilateral irradiation of a layer supported by a "cold" base (i.e., with no account being taken of the radiation from the supporting base) the limiting conditions for Eq. 2.15 are:

$$\text{when } x=0, \ \tau=0 \quad E_+^*(0)=(1-\rho_{\lambda 0})E_\lambda^*+\rho_\lambda E_-^*(0),$$
$$\text{when } x=\ell, \ \tau=\tau_\ell \quad E_-^*(\tau_\ell)=R_\lambda E_+^*(\tau_\ell);\qquad(2.70)$$

where, R_λ — bihemispherical reflectance of the supporting base at the medium-base interface.

By solving Eq. 2.15 under these conditions we obtain the following equations for the dimensionless net radiant flux vector q_λ^* and for the energy w_λ^* absorbed per unit time at depth τ:

$$q_\lambda^*=E_\lambda^*[(1-\rho_{\lambda 0})(1-R_{\lambda\infty})/(1-\rho_\lambda R_{\lambda\infty})(1-R_{\lambda ef}R_{\lambda\rho}\psi_\lambda^2)][\exp(-L_\lambda^*\tau)+$$
$$+(R_{\lambda ef}/R_{\lambda\infty})\psi_\lambda^2\exp(L_\lambda^*\tau)];\qquad(2.71)$$

$$w_\lambda^*=(1-\Lambda_{ef})E_\lambda^*[(1-\rho_{\lambda 0})(1+R_{\lambda\infty})/(1-\rho_\lambda R_{\lambda\infty})(1-R_{\lambda ef}R_{\lambda\rho}\psi_\lambda^2)][\exp(-L_\lambda^*\tau)-$$
$$-R_{\lambda ef}(\psi_\lambda^2/R_{\lambda\infty})\exp(L_\lambda^*\tau)];\qquad(2.72)$$

where, $R_{\lambda ef}=(R_{\lambda\infty}-R_\lambda)/R_{\lambda\infty}(1-R_\lambda R_{\lambda\infty}).\qquad(2.73)$

For an optically infinitely thick layer ($L_\lambda^*\tau_\ell\to\infty$), according to Eqs. 2.66 and 2.67. with account taken of Eqs. 2.62 and 2.63, the dimensionless quantities of the radiant flux and the energy absorbed per unit time are described by the following relationships, respectively:

$$q^*_\lambda = E^*_\lambda (1-\rho_{\lambda 0})/(1-\rho_\lambda R_{\lambda \infty})[(1-R_{\lambda \infty})\exp(-L^*_\lambda \tau)]; \qquad (2.74)$$

$$w^*_\lambda = [(1-\Lambda_{ef})E^*_\lambda(1-\rho_{\lambda o})/(1-\rho_\lambda R_{\lambda \infty})](1+R_{\lambda \infty})\exp(-L^*_\lambda \tau). \qquad (2.75)$$

Eqs. 2.62-2.75 are more precise compared with the known relationships [44]. The former take into account the possible changes with a change in the coordinate in the conditions of irradiation of a volume element inside the layer and in the form of the scattering indicatrix with the aid of the dimensionless term $L^*_\lambda \tau$ which includes the variables $m_\lambda(x)$ and $\delta_s(x)$.

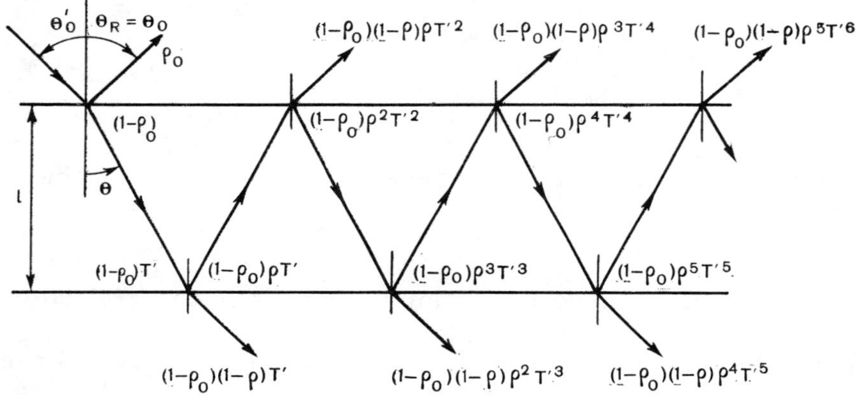

Figure 2.13 Attenuation of the un-scattered component of a radiant flux in a plane layer of substance exposed to directional irradiation at angle θ [33, 39, 44].

In the case of broad directional irradiation it is necessary to take into account the multiple reflection inside the un-scattered component of the incoming radiation and the multiple reflection inside the radiation component that is scattered in a layer. The path taken by the un-scattered component of the radiation entering the layer at an angle θ when there is multiple scattering inside the layer is shown in Fig. 2.13. The direction θ of the radiation entering the layer, depending on the direction of the incident radiation, is determined by the Snellius law on refraction:

$$(1-\mu^2)^{\frac{1}{2}}=(1/n_\lambda)(1-\mu'^2)^{\frac{1}{2}}, \text{ or } \sin\theta'/\sin\theta=n_\lambda ; \tag{2.76}$$

where, $\mu\cos\theta$, $\mu'=\cos\theta'$.

Summing up the events of multiple reflections of the un-scattered and scattered components of the radiation flux at $x=0$ and $x=\ell$, we obtain the following boundary conditions for Eqs. 2.42 and 2.43:

$$\text{when } x=0, \ E_+(0)=\rho_\lambda E_-(0); \tag{2.77}$$

$$E_+(0)=\{(1-\rho_{\lambda 0})[E_{\lambda 1}+\rho_\lambda E_{\lambda 2}\exp(-\varepsilon_\lambda\ell/\mu)]\}/[1-\rho_\lambda^2\exp(-2\varepsilon_\lambda\ell/\mu)]=$$

$$=E_{\lambda 1ef}; \tag{2.78}$$

$$\text{when } x=\ell \ E_-(\ell)=\rho_\lambda E_+(\ell); \tag{2.79}$$

$$E_-(\ell)=\{(1-\rho_{\lambda 0})[E_{\lambda 2}+\rho_\lambda E_{\lambda 1}\exp(-\varepsilon_\lambda\ell/\mu)]\}/[1-\rho_\lambda^2\exp(-2\varepsilon_\lambda\ell/\mu)]=$$

$$=E_{\lambda 2ef}. \tag{2.80}$$

The solution of Eqs. 2.42 and 2.43 under these conditions [44] makes it possible to calculate q_λ and w_λ for each concrete case.

In an optically infinitely thick layer ($\varepsilon_\lambda\ell\to\infty$, $L_\lambda\ell\to\infty$) the energy absorbed at depth x per unit time is:

$$w_\lambda(x)=\bar{k}(1-\rho_{\lambda 0})E_\lambda[(1+R_{\lambda\infty})C_{2ef}\exp(-L_\lambda x)-(C_1+C_2)\exp(-\varepsilon_\lambda x/\mu)]+$$

$$+k_\lambda(1-\rho_{\lambda 0})E_\lambda\exp(-\varepsilon_\lambda x/\mu); \tag{2.81}$$

$$\text{where, } C_{2ef}=(C_2-\rho_\lambda C_1)/(1-\rho_\lambda R_{\lambda\infty}). \tag{2.82}$$

The effect of reflection at the external interface $\rho_{\lambda 0}$ and of the total internal reflection ρ_λ on radiant energy transfer inside the substance can be estimated by using as an example the function of the distribution of the energy $w_\lambda^*(x)$ absorbed per unit time at depth x.

In the case of diffuse irradiation of a non-reflecting surface ($\rho_{\lambda 0} \equiv \rho_{\lambda}=0$) Eq. 2.75 changes into Eq. 2.31. Then the ratio of $w_{\lambda}^{*}(\rho_{\lambda 0},\rho_{\lambda})$ to w_{λ}^{*} is equal to the magnitude of the correction of the function w_{λ}^{*}, which takes into account the reflection at the interface in the case of diffuse irradiation:

$$[w_{\lambda}^{*}(\rho_{\lambda 0},\rho_{\lambda})]/w_{\lambda}^{*}=(1-\rho_{\lambda 0})/(1-\rho_{\lambda}R_{\lambda\infty}). \qquad (2.83)$$

To determine the magnitude of the correction in the case of directional irradiation it is necessary to compare Eqs. 2.81 and 2.51. For strongly scattering substances ($\Lambda>0.9$) at sufficiently large depths ($k_{\lambda}\ell>4$, Fig. 2.9), where the directional radiation flux $E_{\lambda 0}$ is negligibly small in comparison with the scattered radiation flux $E_{\lambda os}$, the correction for w_{λ} can be determined from the relationship:

$$w_{\lambda}(\lambda 0,\rho_{\lambda})/w_{\lambda}=(1-\rho_{\lambda 0})(C_{2}-\rho_{\lambda}C_{1})/(1-\rho_{\lambda}R_{\lambda\infty}). \qquad (2.84)$$

The ratios in Eqs. 2.83 and 2.84 depend on the refractive index n_{λ} and the absorption coefficient k_{λ} which determine $\rho_{\lambda 0}'$, $\rho_{\lambda o}$, and ρ_{λ}, and on the scattering characteristics of the medium, $\chi_{\lambda}(\gamma)$ and Λ which determine C_{1}, C_{2}, and $R_{\lambda\infty}$.

The ratios in Eqs. 2.83 and 2.84 increase with an increase in n_{λ} and Λ. Thus, at $n_{\lambda}=1.5$ (for example, for water in the vicinity of $\lambda=3.0$ and 18.0 μm, and for ice - $\lambda=3.5$ and 14.0 μm, see Fig. 2.11), with a spherical scattering indicatrix $\delta_{f}=0.5$, the ratio of $w_{\lambda}^{*}(\rho_{\lambda 0},\rho)$ to w_{λ}^{*} (in Eq. 2.83) for weakly scattering media ($\Lambda<0.5$, $R_{\lambda\infty}<0.2$) lies within 0.9-1.2; for middle-range scattering media ($0.5<\Lambda<0.9$; $0.2<R_{\lambda\infty}<0.5$) - 1.2-1.4; and for strongly scattering media ($\Lambda>0.9$, $R_{\lambda\infty}>0.5$) - 1.4-2.5. The ratio of $w_{\lambda}\cdot(\rho_{\lambda 0},\rho)$ to w_{λ} (in Eq. 2.84) at $\Lambda=0.9$ reaches 3.0, and at $\Lambda=0.95$ exceeds

3.0. Since the refractive indices for water and ice in the 1-15 μm range lie within 1.1-1.6, and the reflectance of foodstuffs $R_{\lambda\infty}$ can reach 0.9 or greater, the error that arises in determining the functions w_λ and w_λ', with no account being taken of the reflection at the interface, can be considerable.

Thus, the reflection at the layer's interface should be taken into account when investigating radiant energy transfer in substances with a high moisture content and in frozen foods.

2.6 Scattering and spreading of a directional radiant flux having a narrow cross section

The above-derived equations describing the transfer of energy in irradiated foodstuffs make it possible to calculate the functions of the distribution of radiation fluxes E_λ, q_λ and of the absorbed energy $w_\lambda(x)$ under different irradiation conditions - those of directional and diffuse irradiation. At the same time, when measuring the thermal radiational characteristics of foodstuffs by spectral methods (Chapter 5), use is made of directional radiant fluxes having a narrow cross section and an angular divergence θ= 5-10°. It thus becomes necessary to know not only the laws governing the attenuation of radiation but also the regularities of the spatial and zonal distribution of the energy which is reflected and transmitted by the sample across a given surface.

Now we shall consider the changes in the spatial distribution and in the cross section of a directional radiant flux in a plane layer of an absorbing-scattering substance. Let us assume that a directional radiant flux, F, with a cross section having d_0 diameter falls upon a layer having ℓ thickness at a low angular divergence ($\theta\to0$) (Fig. 2.14,a). As has been shown above, a characteristic feature of the propagation of a directional radiant

flux having a narrow cross section in foodstuffs is the rapid and almost

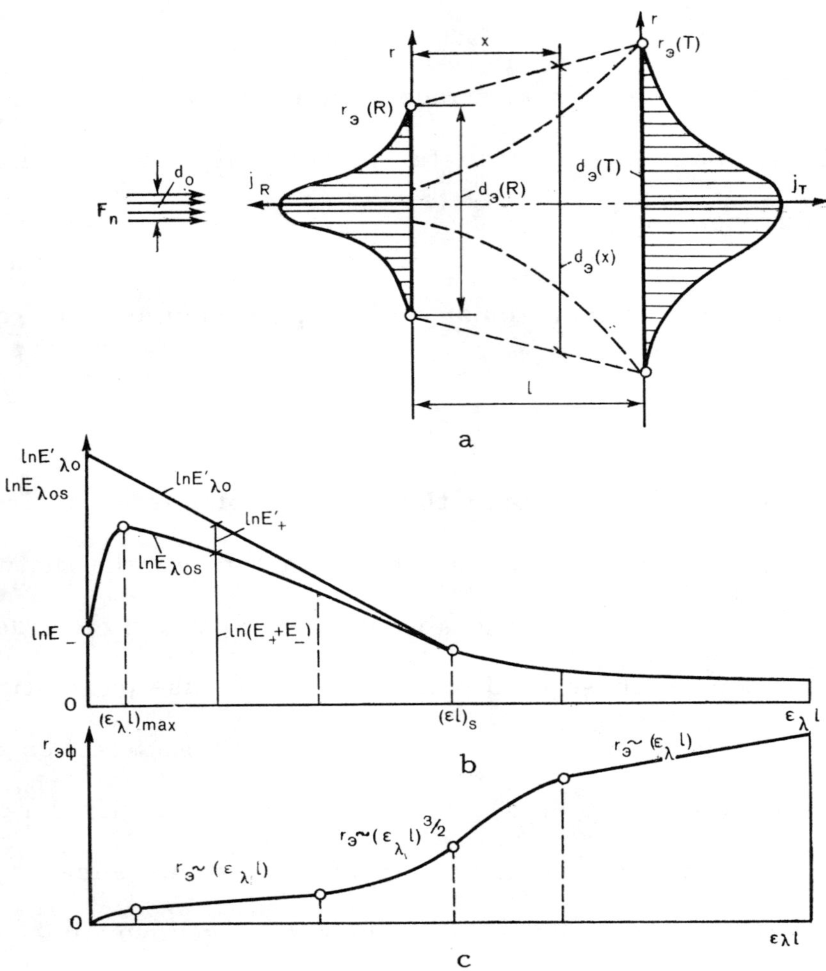

Figure 2.14　Change in the basic characteristics of a directional radiant

flux in foodstuffs and in other scattering-absorbing substances [27]:
  a - the spreading of the effective cross section of a radiant flux, and
    the distribution in the effective cross section of the flux densities
    of the reflected and transmitted radiation;
  b - distribution of scattered and un-scattered radiation throughout the
    thickness of a layer;
  c - dependence of the radius of the effective cross section r_{ef} on the
    depth $\varepsilon_\lambda \ell$ in a semi-infinite layer ($\varepsilon_\lambda \ell \rightarrow \infty$).

complete transformation of the radiant flux into diffuse radiation.　Such a transformation is due to the multiple scattering of radiation on the optical inhomogeneities in a given layer of substance (Fig. 2.8). And as the radiant flux propagates in the layer there is extensive spreading of the narrow cross section of the radiant flux, while the cross section of the flux emitted from the sample increases considerably (Fig. 2.14, a, c).

In the propagation of an initially narrow directional flux in a semi-infinite opaque medium with a strongly forward elongated scattering indicatrix the r_{ef} of the effective cross section of the flux changes in proportion to $r_{ef} \sim x^{3/2}$ at small values of the depth (kx<1.0). At large values of x, where the effect of the depth is felt (kx>5; Figs. 2.8, 2.14, b) and where the directional flux is absent, r_{ef} is proportional to the depth x ($r_{ef} \sim x$). However, these conclusions hold true only when the energy transfer equations are solved for a semi-infinite opaque medium under approximation of very small angles: the scattering indicatrix differs from zero within the limits $0 \leq \gamma \leq \gamma_0 < 1$. Experiments have been carried out [125] for determining the radial dependence of a the radiation flux passing through a model medium (small spheres made of polymethylmethacrylate). The obtained results are in good agreement with results calculated on the basis of an equation describing energy transfer according to diffuse approximation. The flux density of radiation emitted from the opposite side of a given layer having l thickness when its outer surface is irradiated by a point-diffuse source is:

$$j\,(r,\,l) = 2Dl \sum_{i=0}^{\infty} \exp\left[-\varkappa \sqrt{r^2 + (2i+1)^2\,l^2}\right] \frac{1 + \varkappa \sqrt{r^2 + (2i+1)^2\,l^2}}{[r^2 + (2i+1)^2\,l^2]^{3/2}}, \tag{2.85}$$

where, χ - absorption index; D - coefficient of photon diffusion; r - distance from the symmetry axis of the point source to the point of the surface of the layer under consideration.

There are no published works dealing with the radial dependence of a reflected radiant flux emerging from a layer of substance.

The first experimental investigations into the spreading of the narrow cross section of a directional radiant flux passing through a given layer of foodstuffs were undertaken by the present authors in 1969 [39, 50, 118]. A method has been developed for measuring the effective cross section of an outgoing radiant flux with the aid of an iris diaphragm and an integrating sphere. The need and the possibility of carrying out experimental research into the spreading of the cross section of a radiant flux have been pointed out. Strong multiple scattering of radiation inside a given layer of substance should lead to an enlargement of the cross section of the reflected radiant flux. The methods for investigating this problem are described in Chapter 5.

The radial dependence of the flux density of reflected radiation and that of transmitted radiation j_T for different foodstuffs irradiated at $\lambda=0.63$ μm is shown in Fig. 2.15.

In the case of bread (the soft part) subjected to directional irradiation the effective radius of the reflected radiation and that of the transmitted radiation are 11 mm and more than 14 mm, respectively. This means that the cross section of the reflected and the transmitted radiation versus that of incident radiation increases approximately 120 and 165 times, respectively. With an increase in the thickness of the sample the effective radius of the cross section for the transmitted radiation increases, while that for the reflected radiation decreases to a certain limiting value corresponding to the optically thick sample. This behavior was observed for all investigated samples (Fig. 2.15, b). The decrease in the effective radius of the cross section of the spreading in the case of the reflected flux is due to a decrease in the fraction of the flux undergoing multiple reflection from the lower surface of the sample with an increase in its thickness.

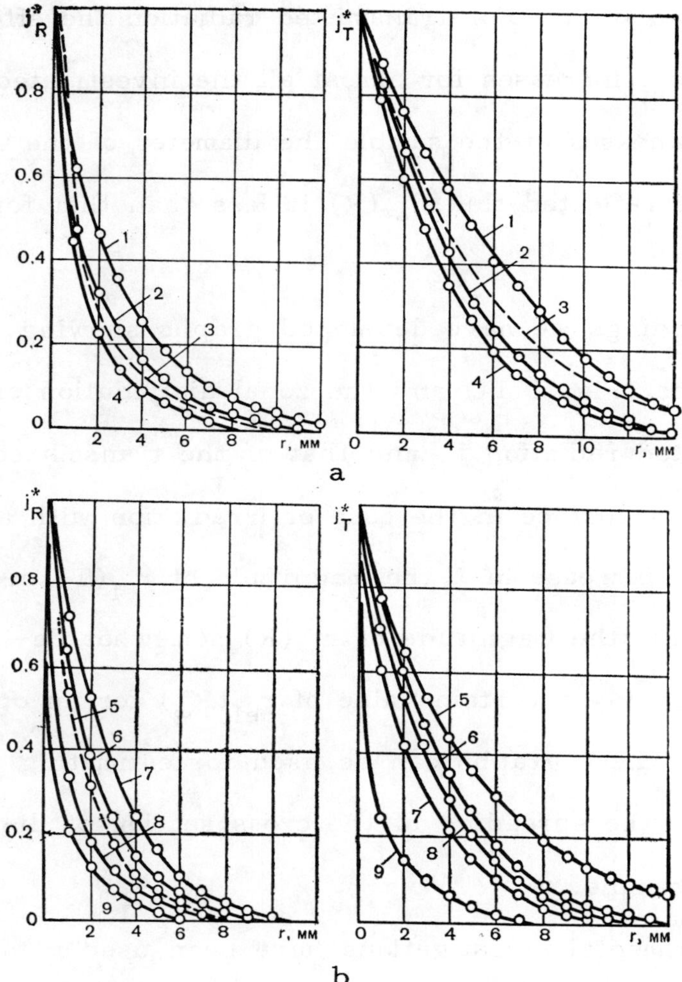

Figure 2.15 Dependences of the dimensionless flux density of the reflected

radiation j_R^* and that of transmitted radiation j_T^* at $\lambda=.0.63$ μm in the case

of a layer having finite thickness for some foodstuffs [39, 118]:
    a - fruits and vegetables, r=1.0 mm:
        1 - apple, W=85.2%;   3 - potato, W=75.5%;
        2 - carrot, W=85.4%;   4 - beet, W=84.5%;
    b - bakery products (inner soft part of a loaf of bread made of
       grade I wheat flour, W^C=12.0%):
       5 - 10.0 mm thick; 8 - bread crumbs, W^C==11.8%, 3.0 mm in di-
       6 - 5.5 mm thick;   ameter;
       7 - 3.5 mm thick;   9 - bread crust, W^C=12.0%, 3.0 mm in diameter

An analysis of the spreading in the cross section of a narrow parallel

radiant flux in different foodstuffs shows that the degree of spreading depends on the thickness of the sample and is different for different substances. Thus, for the inner soft part of a loaf of bread an increase in the thickness of the sample increases only slightly the effective diameter d_{ef} in the case of reflected radiation. For transmitted radiation the effect of the sample's thickness on d_{ef} increases for almost all the investigated foodstuffs. At different thicknesses of the sample the diameter of the effective cross section for the reflected flux $d_{ef}(R)$ is less than that for the transmitted flux $d_{ef}(T)$.

On the basis of these findings we have developed graphs showing the spreading of the cross sections (Fig. 2.14) and the zonal distribution of the flux density of the reflected radiation j_R and that of the transmitted radiation j_T across the sample's surface in the case of irradiation with a narrow point source. With an increase in ℓ the magnitude of $r_{ef}(T)$ increases (Fig. 2.15, a, b), while the magnitude of $r_{ef}(R)$ somewhat decreases asymptotically, approaching a certain value of $r_{ef}(R_\infty)$ for an optically infinitely thick layer. These graphs can be used for estimating errors in measurements due to the spreading of the cross section of directional radiant flux in the sample.

The results of our experimental investigations have been used in developing a method for measuring the thermal radiational characteristics of foodstuffs. We have found that when various foodstuffs (bread (the inner soft part), apples, potatoes, etc) are irradiated with a narrow directional source the diameter of the effective cross section for the outgoing transmitted radiation flux is more than 29 mm. This information should be taken into account when selecting the minimal dimensions of a sample in order to prevent energy losses due to the spreading of the cross section of the transmitted radiant flux.

2.7 Deviation from the basic laws on the absorption of light

There have been numerous experimental investigations into the transmittance of a layer having a finite thickness in different foodstuffs [19, 75, 84, 94, 97, 106, 107]. On the basis of these investigations researchers have come to the conclusion that Bouguer's law, the Bouguer-Lambert law and the Beer-Lambert-Bouguer law are not valid here. At the same time they compare the dependence of the transmittance T_λ of a layer having a finite thickness ℓ with the laws on light attenuation which are expressed in the following form:

$$\text{Bouguer's law: } B_\lambda(\ell)=B_\lambda(0)\exp(-k_\lambda\ell); \tag{2.86}$$

$$\text{the Bouguer-Lambert law: } B_\lambda(\ell)=B_\lambda(0)\exp(-\varepsilon_\lambda\ell); \tag{2.87}$$

$$\text{the Beer-Lambert-Bouguer law: } B_\lambda(\ell)=B_\lambda(0)\exp(\varepsilon_\lambda c\ell); \tag{2.88}$$

where, $B_\lambda(\ell)$ and $B_\lambda(0)$ - radiance of a flux passing through a layer having ℓ thickness and incident upon the layer, respectively; c - concentration of the substance causing attenuation of the absorbed and scattered radiation. The transmittance T_λ is taken to be:

$$T_\lambda=B_\lambda(\ell)/B_\lambda(0)=\exp(-\varepsilon_\lambda\ell). \tag{2.89}$$

The natural logarithm of this equation is in linear dependence on the thickness of the layer:

$$\ln[B_\lambda(\ell)/B_\lambda(0)]=-\varepsilon_\lambda\ell=-(k_\lambda+\sigma_\lambda)\ell. \tag{2.90}$$

It is further assumed that k_λ and ε_λ are independent of the concentration and of the intensity $B_\lambda(0)$ of the incident radiation.

Numerous investigations [19, 84, 97, 106, 107] of the transmittance in foodstuffs (dough, bread, fruit-paste candy, potato, tomato paste, etc) subjected to directional integral irradiation and directional spectral irradiation have shown that the dependence of the transmittance on the thickness of a layer does not follow the exponential law.

It should be pointed out that for infrared irradiation of foodstuffs and other absorbing-scattering substances Bouguer's law holds true. Moreover, the theory of the transfer of energy in opaque nedia is based precisely on this law [13].

In the basic equation (Eq. 2.4) describing radiant transfer the decrease in the intensity of the radiant flux due to its absorption and scattering by the volume element dV is determined according to Bouguer's law (Eq. 2.1). The first of Bouguer's laws (1729) dealing with the absorption of light says that the flux density dB_λ of directional radiation in an infinitely thin layer dx decreases in proportion to the thickness of the layer:

for an absorbing medium:

$$dB_\lambda = -k_\lambda B_\lambda dx; \tag{2.91}$$

for an absorbing-scattering medium:

$$dB_\lambda = -(k_\lambda + \sigma_\lambda) B_\lambda dx = -\varepsilon_\lambda B_\lambda dx. \tag{2.92}$$

Eqs. 2.89 and 2.90 are obtained by integrating Eqs. 2.91 and 2.92 within the limits $0 \leq x \leq \ell$ at the boundary conditions:

$$B_\lambda(x=0) = B_\lambda(0), \quad B_\lambda(x=\ell) = 0. \tag{2.93}$$

A comparison of Eq. 2.92 with the basic equation (Eq. 2.4) describing radiant energy transfer and of Eq. 2.89 with the relationships (Eqs. 3.24 and 3.2) describing transmittance (see Chapter 3) shows the fundamental difference between them. Eq. 2.4 differs from Eq. 2.92 in that the former contains a term on the right side of the equation which accounts for secondary and multiple scattering of radiation on the optical inhomogeneities in the irradiated substances.

Thus, when applying Bouguer's law to strongly scattering media, account should be taken of multiple scattering; and when choosing the boundary conditions (Eq. 2.93), account should be taken of boundary reflection and reflection inside the sample. It would be more correct in this case, therefore, to speak not of violation of Bouguer's law, but of the deviation of the dependence of $T_\lambda(\ell)$ on ℓ from the exponential law in the form given in Eq. 2.89.

The main reason for the deviation from the Bouguer-Lambert law in the form expressed in Eq. 2.89, as regards the dependence of the transmittance of a strongly scattering plane layer on its thickness ℓ, is multiple scattering inside the layer.

The attenuation of a directional radiant flux as it passes through a plane layer in a light-scattering medium proceeds according to a complex relationship, Eq. 3.24 (see Chapter 3). The attenuation is characterized not only by the usual extinction coefficient $\varepsilon_\lambda = k_\lambda + \sigma_\lambda$, but also by the effective attenuation coefficient L_λ whose physical meaning has been elucidated above (see Eq. 2.34). The coefficient L_λ, which is related to ε_λ through Eq. 3.19, depending on the scattering properties of the medium, i.e., on Λ, can be either greater or smaller than ε_λ. This is due to the fact that ε_λ takes into account only single scattering, while L_λ takes into account multiple scattering of radiation. The latter provides for additional pumping of the energy of the transmitted radiant flux, thereby causing deviation from the exponential law of attenuation. The multiple scattering of radiation in a plane layer in the case of directional irradiation can be taken into account with the aid of Eq. 2.43. This equation suppliments the Bouguer-Lambert law (Eq. 2.42), which defines the attenuation of a direct radiant flux in a layer element having dx thickness.

If all the scattering particles inside a layer having ℓ thickness are located in a row, we have single scattering, and in the case of directional irradiation the Bouguer–Lambert law holds true according to Eq. 2.89.

Most foodstuffs are opaque media with anisotropic scattering. In Figs. 2.5, 2.8, and 2.14,b are shown polar diagrams of the angular distribution of radiance in the case of scattering on large particles, and the basic attenuation characteristics of a directional radiant flux in an opaque medium. As can be seen from Fig. 2.8, in the layer adjacent to the boundary $(\varkappa_\lambda \ell < 1.0)$ the main contribution to the radiance comes from low-level scattering, with the direct luminous flux being predominant. At large depths $(k_\lambda \ell > 1.0)$ the intensity of scattered radiation predominates (Fig. 2.9), and there is spreading of the cross section of radiant flux (Fig. 2.14, b). Multiple scattering and the spreading of the cross section of the radiant flux in irradiated substances lead to the deviation of the regularities of radiation energy transfer from the simple Bouguer relationships (Eqs. 2.86–2.88).

Violation of the Beer–Lambert–Bouguer law has also been observed in gaseous and liquid media [39, 44]. However, experimentally observed deviations from the basic laws of radiation absorption are not always real deviations and can be related to the properties of the given substance and the experimental methods used. All deviations can be attributed to two sets of factors: physico-chemical factors (causing real deviations) and instrumental factors (causing apparent deviations) [14, 18, 24, 27, 33, 35, 39, 64, 96, 100, 105, 111, 114, 115, 119, 124, 127, 155, 159].

In deriving the Bouguer–Lambert law it is assumed that k_λ and ε_λ are independent, first, of the concentration of the substance irand, second, of the intensity of incident radiation. The second assumption has been subjected to rigorous experimental verification [14]. And it has been found to be valid for liquid solutions of dyes when $B_\lambda(0)$ is increased almost by 20 orders of magnitude (from the range of visible radiation to the energy of direct solar radiation). In this case the electron transitions correspond to the life-span τ of atoms in an excited state. The dependence of k_λ on B_λ should be expected in the case of powerful laser radiation. This dependence should be different for different spectral lines.

Thus, when using standard infrared generators (lamps; tubular, panel and other types of generators) the absorption coefficient k_λ for most radiation-scattering substances can be considered to be independent of the incident radiance $B_\lambda(0)$.

To sum up, Bouguer's law (Eq. 2.92) is valid and serves as a basis for the fundamental equation of energy transfer (Eq. 2.4). The Bouguer-Lambert law in the form expressed in Eqs. 2.86-2.88 can be considered to be sufficiently precise for technical purposes only in the case of weakly scattering opaque media consisting of particles in a vacuum, and when the optical thickness of a given layer is small ($k_\lambda \ell < 3.0$, $\Lambda < 0.1$). In general, the law on the attenuation of energy in a scattering medium is of a non-exponential nature, which can be established by complex relationships (Eqs. 2.29 and 2.47). For a hemisphere Eq. 2.29 changes into an exponential relationship (Eq. 2.31); in this relationship, instead of the extinction coefficient ε_λ the effective attenuation coefficient L_λ is used which takes into account multiple scattering in a layer of substance.

OPTICAL AND THERMAL RADIATIONAL CHARACTERISTICS OF FOODSTUFFS

In the last two chapters we have set forth theoretical relationships describing the distribution of monochromatic radiation inside capillary-porous colloidal systems. And we have put forward a hypothesis according to which the spectral absorption coefficient k_λ and the backward scattering coefficient s_λ are independent of the coordinate x. Both the theoretical equations and the hypothesis require experimental verification. To do this it is necessary to establish the relationship between the optical properties and the thermal radiational characteristics of the given substances.

3.1 Spectral optical and thermal radiational characteristics of a plane layer exposed to diffuse radiation

From Eq. 2.18, which describes a dimensionless oncoming radiant flux scattered backward under conditions of unilateral diffuse irradiation, we derive an equation for determining the reflectance:

$$R_\lambda = \{E_-^* \left|{}^{E_{\lambda 2}=0}_{\tau=0}\right|\}/E_{\lambda 1} = R_{\lambda\infty}[1-\exp(-2L_\lambda^* \tau_\ell)]/[1-R_{\lambda\infty}^2 \exp(-2L_\lambda^* \tau_\ell)]. \qquad (3.1)$$

In the case of unilateral diffuse irradiation the transmittance is derived from Eq. 2.17:

$$T_\lambda = \{E_+^* \left|{}^{E_{\lambda 2}=0}_{\tau=\tau_\ell}\right|\}/E_{\lambda 1} = [(1-R_{\lambda\infty}^2)\exp(-L_\lambda^* \tau_\ell)]/[1-R_{\lambda\infty}^2 \exp(-2L_\lambda^* \tau_\ell)]. \qquad (3.2)$$

Eqs. 3.1 and 3.2 are more general in comparison with those reported in literature [3, 18, 26, 33, 44, 204] for constants $\bar{k}_\lambda$ and s_λ, which are independent of x. The exponents in these equations include the quantities L_λ^* and τ_ℓ. These quantities are determined from Eqs. 2.13 and 2.21, which take into account the possible dependence of the coefficients $\bar{k}_\lambda(x)$ and $s_\lambda(x)$ on x, according to Eqs. 2.11 and 2.13 and with the aid of the variable parameters $m(x)$ and $\delta_s(x)$. For a particular case involving the constants $\bar{k}_\lambda$ and s_λ, with account taken of Eqs. 2.13 and 2.21, we obtain $L_\lambda^* \tau_\ell \equiv L_\lambda \ell$. And we find that Eqs. 3.1 and 3.2 are in complete accord with the known relationships for R_λ and T_λ.

For an absorbing medium ($s_\lambda=0$), by assuming m=cost and δ_s=const, we have $L_\lambda = k_\lambda$, $R_{\lambda\infty}=0$; and Eqs. 3.1 and 3.2 are transformed into the known relationships:

$$R_\lambda(s_\lambda=0)=0; \tag{3.3}$$

$$T_\lambda(s_\lambda=0)=\exp(-\bar{k}_\lambda \ell). \tag{3.4}$$

For a scattering medium ($\bar{k}_\lambda=0$) we have:

$$R_\lambda(\bar{k}_\lambda=0)=s_\lambda \ell/(1+s_\lambda \ell); \tag{3.5}$$

$$T_\lambda(\bar{k}_\lambda=0)=1/(1+s_\lambda \ell). \tag{3.6}$$

Eqs. 3.1 and 3.2 for a particular case involving the constants $\bar{k}_\lambda$ and s_λ when $L_\lambda^* \tau_\ell \equiv L_\lambda \ell$ were first derived by Stokes for a series of weakly absorbing non-scattering plates and were later used in the case of a layer of scattering, weakly absorbing particles ($\bar{k}_\lambda \ell \ll 1$) in a vacuum [6, 7, 16, 17, 23, 39, 44]. These equations can be applied to a majority of foodstuffs and other absorbing-scattering substances [23, 39, 44].

The fraction of radiation absorbed by a layer having a finite thickness can be determined from the law on the conservation of energy:

$$A_\lambda=1-(R_\lambda+T_\lambda)=1-[R_{\lambda\infty}+\exp(-L_\lambda^* \tau_\ell)]/[1+R_{\lambda\infty}\exp(-L_\lambda^* \tau_\ell)]. \tag{3.7}$$

For limiting cases we obtain the known relationships:

$$A_\lambda(\bar{k}_\lambda=0)=0; \quad A_\lambda(s_\lambda=0)=1-\exp(-\bar{k}_\lambda\ell). \tag{3.8}$$

For a layer having a finite thickness Eqs. 3.1 and 3.2, according to the Schuster-Schwarzschild approximation ($m=2$, $\delta_f=0.5$), yield values of R_λ that are too high and of T_λ that are too low. These values, obtained by different methods, are summarized in Table 3.1.

Table 3.1 Values of reflectance and transmittance obtained by different methods

Methods ($\bar{k}_\lambda\ell=1.0$, $m=2$, $\delta_f=0.5$)	$\Lambda=0.1$		$\Lambda=0.5$		$\Lambda=0.9$		Ref.
	R_λ	T_λ	R_λ	T_λ	R_λ	T_λ	
Eqs. 3.1 and 3.2	0.025	0.150	0.162	0.236	0.403	o.422	-
Diffusion method	-	-	-	-	0.398	0.431	154;199
Method of moments	0.024	0.233	0.136	0.308	0.352	0.471	153
Eddington approximation	-	-	0.158	0.284	-	-	90;151
Six-flux approximation	-	-	-	-	0.413	0.375	144

The simultaneous solution of Eqs. 3.1 and 3.2 for $\overset{*}{L}_\lambda\tau_\ell$ yields the relationship between the optical properties and the thermal radiational characteristics of a layer:

$$\overset{*}{L}_\lambda\tau_\ell = \ln[(1-R_\lambda R_{\lambda\infty})/T_\lambda]. \tag{3.9}$$

Since $\overset{*}{L}_\lambda\tau_\ell$ is in linear dependence on ℓ (Fig. 2.6) we can assume that L_λ is independent of x. In this case we find on the basis of Eq. 2.13 that $\tau_\ell = =\varepsilon_{ef}\ell$ and $\overset{*}{L}\tau_\ell \equiv L_\lambda\ell$. And by taking into account Eq. 3.9 we obtain the relationship between the optical properties of the medium and the thermal radiational characteristics of a layer:

$$L_\lambda=(1/\ell)\ln[(1-R_\lambda R_{\lambda\infty})/T_\lambda]. \tag{3.10}$$

The magnitudes of $\bar{k}_\lambda$ and s_λ are determined from Eq. 3.15, with account taken of Eq. 3.13:

$$s_\lambda=2R_{\lambda\infty}L_\lambda/(1-R_{\lambda\infty}^2); \tag{3.11}$$

$$\bar{k}_\lambda=[(1-R_{\lambda\infty})L_\lambda]/(1+R_{\lambda\infty}). \tag{3.12}$$

Eqs. 3.10-3.12 can be used for determining the spectral optical properties of a substance on the basis of the experimentally determined values of R_λ and T_λ for a layer having finite thickness ℓ and of $R_{\lambda\infty}$ for an optically infinitely thick layer.

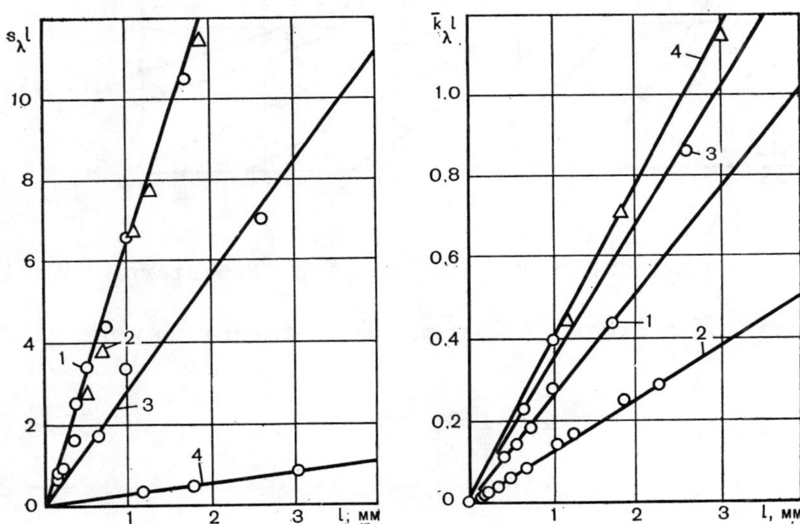

Figure 3.1 Backward scattering coefficient $s_\lambda\ell$ and absorption coefficient $\bar{k}_\lambda\ell$ as a function of the thickness of a layer for some substances at $\lambda=1.2$ μm [39, 44]:

      1 - potato starch, W=11.8%;   3 - wheat flour dough, W=48.0%;
      2 - pinewood, W=6.2%;   4 - potato, W=80.5%.

The results in Fig. 3.1 are calculated on the basis of experimentally obtained data and by using Eqs. 3.11 and 3.12. It is important to note that in the range of wavelengths used the dependence of $\bar{k}_\lambda\ell$ and $s_\lambda\ell$ on the layer's thickness is linear in nature for all the substances investigated. This indicates that we can accept the hypothesis according to which the coefficients $\bar{k}_\lambda$ and s_λ are independent of the coordinate. Thus, the equations for radiant energy transfer in absorbing-scattering media can be simplified by replacing $L_\lambda^* \tau_\ell$ with $L_\lambda \ell$.

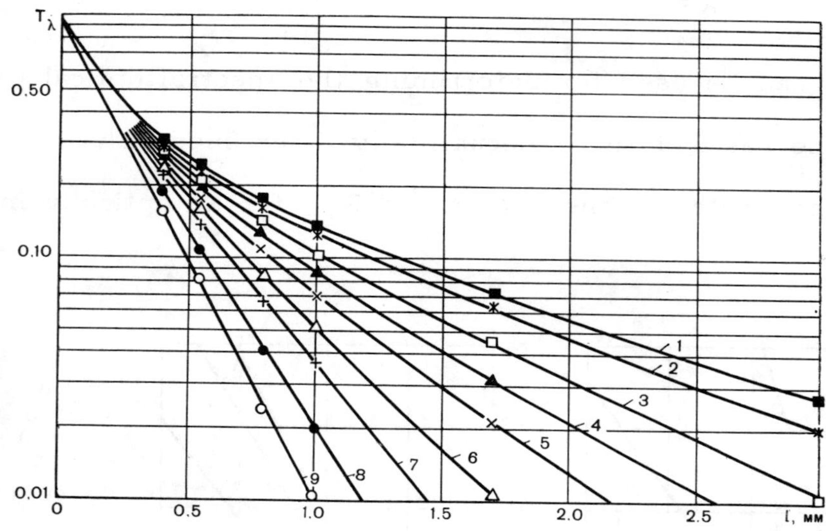

Figure 3.2   Spectral hemispherical transmittance of potato starch (W=11.8%) as a function of the thickness of a layer (ℓ in mm) at different wavelengths (λ in μm) [39]:

1 - 0.85;	4 - 1.3;	7 - 2.2;
2 - 1.1;	5 - 1.2;	8 - 2.9;
3 - 0.7;	6 - 0.4;	9 - 2.98.

In Figs. 3.2 and 3.3 are shown the spectral dependences of T_λ, R_λ, and A_λ on the thickness of a layer for different substances. The results are both calculated with the aid of Eqs. 3.1, 3.2, and 3.7 and obtained experimentally. It can be seen that the experimentally determined values of the radiant flux at the boundary of a layer differ from the calculated values by $\pm 1.0\%$ on an average. This shows that the proposed method is sufficiently precise for calculating R_λ, T_λ, and A_λ for different foodstuffs.

3.2 Calculating the thermal radiational characteristics of an optically infinitely thick layer exposed to diffuse radiation

For purposes of technical calculations a layer of an irradiated substance having finite thickness ℓ can be considered to be optically infinitely thick when its transmittance $T_\lambda < 0.01$ or when $L_\lambda^* \tau_\ell > 5.0$ and $L_\lambda > 5.0$. In this case $\psi_\lambda^2 \ll 1$, and Eqs. 2.17, 2.18, and 2.24 acquire the form of Eqs. 2.31 and 2.32. These two equations correspond to the condition of an optically infinitely thick layer $L_\lambda^* \tau_\ell \to \infty$ or $L_\lambda \ell \to \infty$. From Figs. 3.2 and 3.3 we can determine the thickness $\ell = \ell_\infty$ at $\lambda = 1.1$ μm from the intersection of the $T_\lambda(\ell)$ curve when $T_\lambda = 0.01$. At this thickness various substances can be considered to be optically infinitely thick. Thus, for flour $\ell_\infty = 1.4$ mm, for starch - 5 mm, for fruit candy - 9.2 mm, etc. The depth $\ell = \ell_\infty$, which corresponds to $T_\lambda = 0.01$, is conditionally referred to as the infrared penetration depth of foodstuffs. This definition is conditional since the penetration of infrared radiation into matter is unlimited according to theoretical conclusions (see Figs. 3.2, 3.3, and Eq. 3.2). Here only the fraction of the penetrating radiation decreases - to 0.01, 0.001, 0.0001, etc.

Eqs. 3.1, 3.2, 3.7, and 3.8 can be used to determine the spectral thermal radiational characteristics of an optically infinitely thick layer, $L_\lambda^* \tau_\ell \to \infty$:

$$R_\lambda\,(\tau_l \to \infty) = \frac{\overline{k}_\lambda + s_\lambda - L_\lambda}{s_\lambda} = 1 + \frac{\overline{k}_\lambda}{s_\lambda} - \sqrt{\frac{\overline{k}_\lambda^2}{s_\lambda^2} + 2\,\frac{\overline{k}_\lambda}{s_\lambda}} = R_{\lambda\infty}; \tag{3.13}$$

$$T_\lambda\,(\tau_l \to \infty) = 0; \quad A_\lambda\,(l \to \infty) = \sqrt{\frac{\overline{k}_\lambda^2}{s_\lambda^2} + 2\,\frac{\overline{k}_\lambda}{s_\lambda}} - \frac{\overline{k}_\lambda}{s_\lambda} = A_{\lambda\infty}. \tag{3.14}$$

From Eq. 3.13 a relationship has been obtained which is widely used in the spectroscopic studies of light-scattering media:

$$\overline{k}_\lambda / s_\lambda = mk_\lambda / m\delta_s \sigma_\lambda = k_\lambda / s_\lambda' = (1 - R_{\lambda\infty})^2 / 2R_{\lambda\infty}. \tag{3.15}$$

The ratio of the backward scattering coefficient s_λ to the effective extinction coefficient $\varepsilon_{\lambda ef}$ is known as the Schuster number $\Lambda_{ef} = s_\lambda / \varepsilon_{\lambda ef}$. The effective extinction coefficient differs from the extinction coefficient $\varepsilon_\lambda = k_\lambda + \sigma_\lambda$ as follows:

$$\varepsilon_{\lambda ef} = \bar{k}_\lambda + s_\lambda = m(k_\lambda + \delta_s \sigma_\lambda) = m\varepsilon_\lambda(1 - \delta_f \Lambda). \tag{3.16}$$

From this equation it follows that the coefficient $\varepsilon_{\lambda ef}$ in a case of scattering in a backward direction only ($\delta_f = 0$, $\delta_s = 1$, $s_\lambda \sigma_\lambda$) is equal to the hemispherical extinction coefficient $\varepsilon_{\lambda ef} = m\varepsilon_\lambda = \varepsilon_\lambda$.

The relationship between $R_{\lambda\infty}$ and L_λ, on the one hand, snd the effective hemispherical extinction coefficient and the Schuster number, on the other, is then determined from the following equations:

$$L_\lambda = \varepsilon_{\lambda ef}(1 - s_\lambda^2/\varepsilon_{\lambda ef}^2)^{\frac{1}{2}} = \varepsilon_{\lambda ef}(1 - \Lambda_{ef}^2)^{\frac{1}{2}} = [\bar{k}_\lambda(\bar{k}_\lambda + 2s_\lambda)]^{\frac{1}{2}}; \tag{3.17}$$

$$R_{\lambda\infty} = (\varepsilon_{\lambda ef}/s_\lambda)[1 - (1 - s_\lambda^2/\varepsilon_{\lambda ef}^2)^{\frac{1}{2}}] = (1/\Lambda_{ef})[1 - (1 - \Lambda_{ef}^2)^{\frac{1}{2}}]. \tag{3.18}$$

Here the characteristics L_λ and $R_{\lambda\infty}$ depend on the form of the scattering indicatrix $\chi_\lambda(\gamma)$ and the parameter Λ. The relationship between L_λ and R_λ, on the one hand, and the extinction coefficient $\varepsilon_\lambda = k_\lambda + \sigma_\lambda$ and the survival probability of the quantum $\Lambda = \sigma_\lambda/(k_\lambda + \sigma_\lambda)$, on the other, is determined from Eqs. 3.17 and 3.18, with account taken of the data in Table 3.1:

$$L_\lambda = m\varepsilon_\lambda\{(1 - \Lambda)[1 + \Lambda(1 - 2\delta_f)]\}^{\frac{1}{2}} = K\varepsilon_\lambda; \tag{3.19}$$

$$R_{\lambda\infty} = \{1 - \delta_f \Lambda - [(1 - \Lambda)(1 + \Lambda(1 - 2\delta_f)]^{\frac{1}{2}}\}/(1 - \delta_f)\Lambda; \tag{3.20}$$

where, K – dimensionless parameter which is characteristic of the optical properties of a substance and of the spatial distribution of the diffuse radiation fluxes E_+ and E_- inside the layer, and which is equal to the ratio of the effective attenuation coefficient to the extinction coefficient:

$$K = L_\lambda/\varepsilon_\lambda = m\{(-\Lambda)[1 + \Lambda(1 - 2\delta_f)]\}^{\frac{1}{2}}. \tag{3.21}$$

In the case of a maximum-elongated backward scattering indicatrix ($\delta_f = 0$)

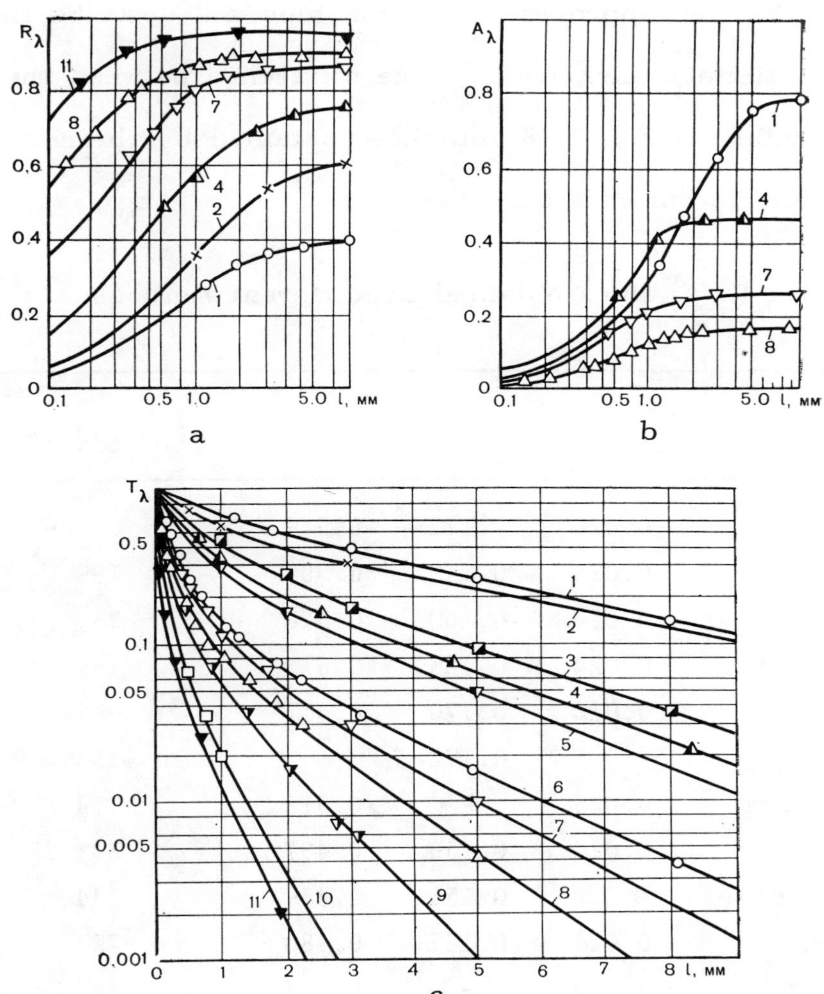

Figure 3.3 Spectral hemispherical reflectance (λ=1.1 µm (a)), absorptance (λ=

1.2 µm (b)), and transmittance (λ=1.1 µm (c)) as a function of the thickness

ℓ of a layer of different substances [39, 44]:

1 – potato, W=80.5%;	6 – pine, W=6.2%;
2 – beet, W=85.5%;	7 – starch, W=11.8%;
3 – cornmeal, W=14.3%;	8 – beech wood, W=7.1%;
4 – wheat flour dough, W=50.0%;	9 – birch wood (pressed), W= 4.5%;
5 – fruit candy, W=30.0%;	10 – wheat flour, grade I, W=8.2%;

11 – paper, 75g/m^2, W=5.2%.

the parameter $K=m(1-\Lambda^2)^{\frac{1}{2}}$, and Eq. 3.20 becomes:

$$R_{\lambda\infty}=(1/\Lambda)-[(1/\Lambda^2)-1]^{\frac{1}{2}}. \tag{3.22}$$

A comparison of the values of $R_{\lambda\infty}$ calculated according to Eq. 3.18 and reported data is given in Table 3.2. The data, obtained for a spherical indicatrix (δ_f=0.5), indicate that the approximate relationships will describe the reflectance of an optically infinitely thick layer. The difference between the values of $R_{\lambda\infty}$ obtained according to Eq. 3.18 and those obtained by the most precise numerical methods at different Λ is only ±1-2%.

Table 3.2 A comparison of $R_{\lambda\infty}$ values obtained by different methods

Methods	Λ			References
	0.1	0.5	0.9	
According to Eq. 3.18	0.026	0.172	0.519	-
Numerical	0.024	0.149	0.508	199
Numerical	0.024	0.160	0.510	33, 111
Diffusion	0.024	0.165	0.513	154
Diffusion	0.026	0.170	0.517	199
Eddington approximation	-	0.172	-	154, 199
Spherical harmonics, 3rd approx.	0.025	-	0.512	174
Method of moments	0.024	0.156	0.487	33, 37, 38
Method of moments, 2nd approx.	0.024	0.156	0.489	144
Method of moments	0.024	0.147	0.480	72
Six-flux approximation	0.020	-	0.484	144

3.3 Calculating the thermal radiational characteristics of foodstuffs under different irradiation conditions

It has been shown in Chapter 1 that the thermal radiational characteristics of irradiated substances depend not only on their optical properties but also on the irradiation conditions: the direction and the spatial distribution of the incident radiation. In the case of directional irradiation at a certain angle the thermal radiational characteristics depend on the incidence angle θ=arc cosμ.

We have obtained relationships [44] for determining the hemispherical characteristics R_λ', T_λ', and A_λ' of a plane layer of a medium with an arbitrary scattering indicatrix in the case of directional irradiation by a parallel flux at $\theta=$ arc $\cos\mu$.

We shall designate a fraction of a direct radiation flux E_+', defined according to Bouguer's law, as $T_{\lambda B}=\exp(-\varepsilon_\lambda \ell/\mu)$. By using Eqs. 3.1 and 3.2 we can represent the relationships between the thermal radiational characteristics for directional (R_λ', T_λ', and A_λ') and for diffuse radiation ($R\lambda$, T_λ, A_λ) as follows:

$$R_\lambda'=C_2 R_\lambda - C_1(1-T_{\lambda B}'T_\lambda'); \tag{3.23}$$

$$T_\lambda'=C_2(T_\lambda-T_{\lambda B})+T_{\lambda B}(1+C_1 R_\lambda); \tag{3.24}$$

$$A_\lambda'=1-\{C_2(R_\lambda+T_\lambda-T_{\lambda B})-C_1[1-T_{\lambda B}'(R_\lambda+T_\lambda)]+T_{\lambda B}'\}. \tag{3.25}$$

The spectral thermal radiational characteristics of an optically infinitely thick layer ($\varepsilon_\lambda \ell\to\infty$) can be determined from the relationships:

$$R_{\lambda\infty}'=C_2 R_{\lambda\infty}-C_1; \tag{3.26}$$

$$T_\lambda'(\varepsilon_\lambda \ell\to\infty)=0; \quad A_{\lambda\infty}'=1-(C_2 R_{\lambda\infty}-C_1). \tag{3.27}$$

Eqs. 2.48 and 2.49 can be represented in a more convenient form for their analysis with the aid of Eqs. 3.16 and 3.19:

$$C_1=[(1-\delta_f)(m\mu-1)\Lambda]/(1-\mu^2 K^2); \tag{3.28}$$

$$C_2=[1+m\mu)\delta_f\Lambda+m\mu(1-2\delta_f)\Lambda^2]/(1-\mu^2 K^2); \tag{3.29}$$

where, K - parameter determined according to Eq. 3.21.

From Eqs. 3.28 and 3.29, with account taken of the condition expressed in Eq. 3.21, we find that with $\mu=0.5$ (with irradiation of a layer at an angle $\theta=60°$ to the normal) for an arbitrary scattering indicatrix (i.e., at any δ_f) and with $m=2$, the parameters C_1 and C_2 are as follows:

$$C_1(\mu=0.5)=0; \quad C_2(\mu=0.5)=1. \tag{3.30}$$

These values correspond to the values for diffuse radiation. That is, with irradiation at $\theta = 60°$ the magnitudes of the diffuse counterfluxes E_+ and E_- and thus also of R'_λ, T'_λ, and A'_λ of a layer having ℓ thickness are equal to those of diffuse irradiation.

In the case of uniform diffusion of E_+ and E_- ($m=2$) and a spherical or symmetrical scattering indicatrix (e.g., of the Rayleigh type) the equations $\delta'_f = 0.5$, $\chi_\lambda(\gamma) = 1$, and the expressions for C_1, C_2, and $R_{\lambda\infty}$ become simplified:

$$C_1 = [(2\mu-1)\Lambda]/2(1-\mu^2 K_0^2); \tag{3.31}$$

$$C_2 = (1+2\mu)\Lambda/2(1-\mu^2 K_0^2); \tag{3.32}$$

$$R'_{\lambda\infty} = [1-(1-\Lambda)^{\frac{1}{2}}]/[1+2\mu(1-\Lambda)^{\frac{1}{2}}] \qquad (\text{see } [105]) \tag{3.33}$$

In the case of strongly scattering media ($\Lambda > 0.9$, Fig. 2.8, curves 1 and 2) the spatial irradiance $E^*_{\lambda 0}$ and the radiation flux E^*_+ at depth $k_\lambda x > k_\lambda x_m$ are determined mainly by the scattered radiant flux. The magnitude of the direct flux $E^{*'}_+$ is small in comparison with that of E^*_+; i.e. $T'_B \ll T'_\lambda$. In that case Eqs. 3.23 and 3.24 become simplified:

$$R'_\lambda = C_2 R_\lambda - C_1; \tag{3.34}$$

$$T'_\lambda = C_2 T_\lambda. \tag{3.35}$$

These equations, which are in a more general form than those obtained for a case where $\mu = 1$ [150, 169], make it possible to establish the relationship between the thermal radiational characteristics of a layer R'_λ and T'_λ and the optical properties of the medium k_λ, $\bar{k}_\lambda$, and s_λ, and to determine the parameters C_1 and C_2 from the experimentally obtained values of T'_λ and T_λ, R'_λ and R_λ for directional and diffuse irradiation.

The simultaneous solution of Eqs. 3.23 and 3.24 with respect to $L_\lambda \ell$ for a case where $\exp(-\varepsilon_\lambda \ell/\mu) \to 0$ yields the following equations. These equations establish the relationship between the optical properties of the medium, on the one hand, and on the other, the thermal radiational characteristics of a layer, its reflectance and transmittance, and the reflectance of an optically infinitely thick layer in the case of directional irradiation:

$$L_\lambda=(1/\ell)\ln[(R'_{\lambda\infty}-R_\lambda)(R'_{\lambda\infty}-C_1)/C_2 T'_\lambda];\tag{3.36}$$

$$\bar{k}_\lambda=\{[C_2-(R'_{\lambda\infty}-C_1)]/[C_2+(R'_{\lambda\infty}-C_1]\}L_\lambda;\tag{3.37}$$

$$s_\lambda=\{2(R'_{\lambda\infty}-C_1)/[C_2-(R'_{\lambda\infty}-C_1)^2\}L_\lambda.\tag{3.38}$$

The extinction coefficient ε_λ can be determined from Eqs. 3.23 and 3.24 only in the case of small optical thicknesses when $T'_B \gg 0$:

$$\varepsilon_\lambda=(1/\ell)\ln[C_1 T_\lambda/(R'_\lambda + C_1-C_2 R_\lambda)],\tag{3.39}$$

or

$$\varepsilon_\lambda=(1/\ell)\ln[(1+C_1 R_\lambda-C_2)/(T'_\lambda - C_2 T_\lambda)].\tag{3.40}$$

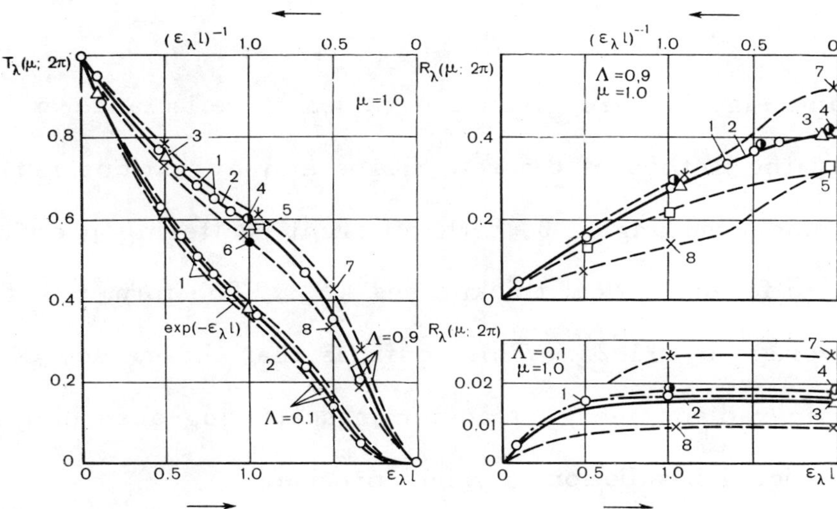

Figure 3.4 Hemispherical reflectance and transmittance as a function of the thickness of a layer at fixed values of Λ obtained by different methods (the numbers marking the points on the curves correspond to those in Table 3.3) [39, 44, 45].

Table 3.3 Values of the reflectance $R_\lambda(\mu;2\pi)$ and transmittance $T_\lambda(\mu;2\pi)$ of an irradiated layer obtained by different methods ($k_\lambda \ell=1.0$; $\Lambda=0.9$; $\chi_\lambda(\gamma)=1$)

Methods	$\mu=\cos\theta$						Refs
	0		0.5		1.0		
	R_λ'	T_λ'	R_λ'	T_λ'	R_λ'	T_λ'	
Eqs. 3.23 and 3.24	0.632	0.189	0.403	0.422	0.274	0.601	–
Numerical	0.635	0.182	0.393	0.415	0.267	0.593	199
Numerical	0.636	0.173	0.400	0.418	0.282	0.585	158
Diffusion	0.600	0.220	0.398	0.431	0.273	0.602	199
Spherical harmonics (1st approximation)	0.630	0.170	0.390	0.400	0.275	0.540	36
Method of moments (2nd approximation)	0.630	0.180	0.390	0.415	0.270	0.590	144
Six-flux approx.	0.533	0.181	0.375	0.413	0.137	0.562	199
Two-flux approx.	0.519	0.000	0.421	0.420	0.284	0.621	199

In Tables 3.3–3.5 and Fig. 3.4 are given the values of reflectance and transmittance obtained by the method of discrete fluxes and by various numerical methods at different incidences angles θ, with different scattering properties of the medium Λ, and at different optical thicknesses $k_\lambda \ell$. The former differ from the latter on the average by ±1-2%. This confirms that the method discussed above for investigating radiant energy transfer in scattering-absorbing substances subjected to directional irradiation is fairly precise.

As can be seen from these data, in the case of an infinitely thick layer the values for the dependence of $R_{\lambda\infty}(\mu;2\pi)$ on Λ and μ are also close (±1-2%) to the precise values.

With the aid of Eqs. 3.26 and 3.27 we can estimate the error involved in the use of $A_{\lambda\infty}'$ instead of $A_{\lambda\infty}$ when calculating radiant heat exchange. A comparison

Table 3.4 Reflectance $(R_{\lambda\infty}(\mu;2\pi)$; $\delta_f=0.5$; $\chi_\lambda(\gamma)=1)$ as determined by different methods

Methods	$\mu=0$	$\Lambda=0.1$ $\mu=0.5$	$\mu=1.0$	$\mu=0$	$\Lambda=0.9$ $\mu=0.5$	$\mu=1.0$	Refs
Eq. 3.33	0.051	0.026	0.017	0.683	0.518	0.435	-
Numerical	0.051	0.024	0.016	0.684	0.508	0.415	199
Numerical	-	-	-	0.680	0.510	0.430	36, 111
Diffusion	0.048	0.026	0.018	0.659	0.517	0.426	199
Diffusion	0.051	0.024	0.016	0.697	0.513	0.418	154
Spherical harmonics (3rd approximation)	0.049	0.025	0.017	0.673	0.512	0.418	174
Method of moments (2nd approximation)	0.051	0.024	0.016	0.671	0.489	0.403	144
Six-flux approx.	0.026	0.020	0.009	0.569	0.484	0.333	199
Two-flux approx.	0.026	0.026	0.026	0.519	0.519	0.519	199

of the magnitudes of $A'_{\lambda\infty}$ for directional radiation and those of $A_{\lambda\infty}$ for diffuse radiation (Table 1.8) in the case of isotropic and anisotropic scattering shows that when $A_{\lambda\infty}$ is replaced with $A'_{\lambda\infty}$ the relative error ε can exceed $\pm20\%$. Therefore, the calculated values of the absorbed energy per unit time determined from the product $A'E$ are 24.3-24.7% too high in comparison with the real value of AE in the case of diffuse irradiation. The data in Table 3.6 show that for an elongated indicatrix the error is bigger. As is known (see Chapter 1), in foodstuffs scattering takes place on large particles for which the scattering indicatrix is strongly elongated forward. From this it follows that for foodstuffs the errors can be greater than those shown in Table 3.6. Thus it is necessary to determine experimentally the thermal radiational characteristics of substances under different irradiation conditions.

Table 3.5 Reflectance $R_{\lambda\infty}(\mu;2\pi)$ ($\mu=1$, $\chi_\lambda(\gamma)=1+\cos\gamma$ obtained by different methods

Methods	0.4	0.5	Λ 0.6	0.7	0.8	0.9	Refs
Eq. 3.26	0.052	0.073	0.107	0.145	0.212	0.338	-
Numerical	0.04	0.06	0.09	0.13	0.20	0.33	33, 111
Differential-difference	0.05	0.08	0.11	0.15	0.22	0.32	164
Method of moments	0.053	0.076	0.109	0.151	0.223	0.349	33, 37

Table 3.6 Relative error ε at different values of the survival probability of the photon Λ in the case of isotropic scattering $\chi_\lambda(\gamma)=1$ and anisotropic scattering $\chi_\lambda(\gamma)=1+\cos\gamma$ [39]

Survival probability	Isotropic scattering			Anisotropic scattering		
	$A'_{\lambda\infty}(\theta;2\pi)$	$A_{\lambda\infty}(2\pi;2\pi)$	$\varepsilon(\%)$	$A'_{\lambda\infty}(\theta;2\pi)$	$A_{\lambda\infty}(2\pi;2\pi)$	$\varepsilon(\%)$
0	1	1	0	1	1	0
0.2	0.965	0.954	1.2	0.985	0.970	1.5
0.5	0.885	0.853	3.8	0.938	0.896	4.7
0.8	0.715	0.658	8.7	0.799	0.726	10.1
0.9	0.585	0.522	12.1	0.673	0.592	13.7
0.95	0.464	0.403	15.2	0.544	0.467	16.5
0.99	0.247	0.205	20.5	0.297	0.245	21.2
0.995	0.183	0.150	22.0	0.221	0.181	22.1
0.999	0.087	0.070	24.3	0.106	0.085	24.7

3.4 Taking into account the reflection of radiation from the boundaries of a layer of substance when determining the optical and thermal radiational characteristics of foodstuffs

When determining the optical properties of foodstuffs with a high moisture content it is necessary to consider the reflection of radiation at the air-substance interface (external reflection) and at the substance-air interface (internal reflection). This can be done in two ways:

- by calculating the true magnitude of R_λ, T_λ, and $R_{\lambda\infty}$ or R'_λ, T'_λ, and $R_{\lambda\infty}$ with the aid of special relationships which take into account the boundary effects and by calculating the sought-for optical properties with the aid of the method described above;

- by direct estimation of the reflection of radiation from the boundaries of a layer with the aid of special equations describing optical properties.

To determine the true values of R_λ and T_λ it is desirable to employ a known method [165] which uses the experimentally determined quantities $R_{\lambda ef}$ $(\rho_{0\lambda},\rho_\lambda)$ and $T_{\lambda ef}(\rho_{0\lambda},\rho_\lambda)$ for a sample without a supporting base and those of $R_{\lambda B}(\rho_{\lambda 0},\rho_\lambda)$ for a sample with such a base. With the aid of these quantities R_λ, T_λ, and ρ_λ are determined as follows:

$$R_\lambda=\{A_1[(1-\rho_{\lambda 0})-(1-R_{\lambda B})]\}/\{(1-\rho_{\lambda 0})[T^2_{\lambda ef}-(R_{\lambda ef}-R_{\lambda B})^2]\}; \qquad (3.41)$$

$$T_\lambda=\{A_1[A_1-(1-\rho_\lambda)(R_{\lambda ef}-R_{\lambda B})]\}/\{(1-\rho_{\lambda 0})T_{\lambda ef}[T^2_{\lambda ef}-(R_{\lambda ef}-R_{\lambda B})^2]\}; \qquad (3.42)$$

$$\rho_\lambda=[(1-\rho_{\lambda 0})(R_{\lambda ef}-R_{\lambda B})]/A_1; \qquad (3.43)$$

where, $A_1=(1-R_{\lambda ef})(R_{\lambda ef}-R_{\lambda B})+T^2_{\lambda ef}.$

The quantity $R_{\lambda\infty}$ can be calculated from Eq. 3.57 by using ρ_λ and the measured quantity $R_{\lambda\infty ef}(\rho_{\lambda 0},\rho_\lambda).$

The obtained values of R_λ, T_λ, and $R_{\lambda\infty}$ are used in calculating L_λ or in determining L_λ with the aid of nomograms. The quantity L_λ can also be determined with the aid of $R_{\lambda ef}$, $T_{\lambda ef}$, and $R_{\lambda\infty ef}$, which have been established experimentally.

By solving simultaneously Eqs. 3.47 and 3.48 with regard to $L_\lambda\ell$, and by taking into consideration Eq. 3.57, we obtain the following expression interrelating the optical properties and the thermal radiational characteristics, with account taken of the reflection at the interface:

$$L_\lambda=(1/\ell)\ln\{\,(1-\rho_{\lambda0})[1-\rho_{\lambda0}-\rho_\lambda(1-R_{\lambda\infty ef})]/(1-\rho_{\lambda0})T_{\lambda ef}\}-$$
$$(R_{\lambda ef}-\rho_{\lambda0})[R_{\lambda\infty ef}(1+\rho_\lambda)-\rho_{\lambda0}-\rho_\lambda]/(1-\rho_{\lambda0})T_{\lambda ef}\}. \tag{3.44}$$

This equation includes $R_{\lambda ef}$, $R_{\lambda\infty ef}$, and $T_{\lambda ef}$, which are measured directly under real conditions. The coefficients $\bar{k}_\lambda$ and s_λ are determined from Eqs. 3.37 and 3.38 by substituting into them the quantity L_λ which is determined from Eq. 3.44.

Thus, when determining the optical properties of substances, the reflection from the interface can be taken into account by using the methods considered above.

The thermal radiational characteristics of a layer of substance subjected to unilateral directional irradiation ($E'_{\lambda2}=0$), with account taken of the reflection from its boundaries, are determined from equations for E_+, E'_+, E_-, and E'_- [44]:

$$R'_\lambda=\rho'_{\lambda0}+(1-\rho_\lambda)[E_-(x=0)/E'_\lambda]+(1-\rho_\lambda)[E'_-(x=0)/E'_\lambda]; \tag{3.45}$$
$$T'_\lambda=(1-\rho_\lambda)E_+(x=0)/E'_\lambda+(1-\rho_\lambda)E'_+(x=\ell)/E'_\lambda. \tag{3.46}$$

In the case of a strongly scattering medium ($\Lambda>0.9$) and a thick layer $[\exp(-\varepsilon_\lambda\ell/\mu)\to0]$ Eqs. 3.45 and 3.46 become simplified:

$$R'_\lambda=\rho'_{\lambda0}+\frac{(1-\rho'_{\lambda0})(1-\rho_\lambda)}{1-\rho_\lambda R_{\lambda\infty}}\left[R_{\lambda\infty}\frac{1-R_{\lambda\rho}\exp(-2L_\lambda l)}{1-R_{\lambda\rho}^2\psi_\lambda^2}(C_2-\rho_\lambda C_1)-C_1\right]; \tag{3.47}$$

$$T'_\lambda=\frac{(1-\rho'_{\lambda0})(1-\rho_\lambda)}{1-\rho_\lambda R_{\lambda\infty}}\cdot\frac{(1-R_{\lambda\rho}R_{\lambda\infty}^2)\exp(-L_\lambda l)}{1-R_{\lambda\rho}^2\psi_\lambda^2}(C_2-\rho_\lambda C_1). \tag{3.48}$$

In this case the absorptance is expressed as follows:

$$A'_\lambda = 1 - \left(R'_\lambda + T'_\lambda\right) = 1 - \rho'_{\lambda 0} - \frac{\left(1 - \rho'_{\lambda 0}\right)\left(1 - \rho_\lambda\right)}{1 - \rho_\lambda\, R_{\lambda \infty}} \times$$

$$\times \left[\frac{R_{\lambda \infty} + \exp\left(-L_\lambda\, l\right)}{1 + R_{\lambda \rho}\, \psi_\lambda}\,(C_2 - \rho_\lambda\, C_1) - C_1\right]. \tag{3.49}$$

The thermal radiational characteristics of an optically infinitely thick layer $L\ell \to \infty$, with account taken of the reflection from the interface, are determined with the aid of Eqs. 3.47 and 3.49:

$$R'_{\lambda \infty}\left(\rho_{\lambda 0},\,\rho_\lambda\right) = \rho'_{\lambda 0} + \frac{\left(1 - \rho'_{\lambda 0}\right)\left(1 - \rho_\lambda\right)}{1 - \rho_\lambda R_{\lambda \infty}}\,[R_{\lambda \infty}\,(C_2 - \rho_\lambda\, C_1) - C_1]; \tag{3.50}$$

$$A'_{\lambda \infty}\left(\rho_{\lambda 0},\,\rho_\lambda\right) = 1 - \rho'_{\lambda 0} - \frac{\left(1 - \rho'_{\lambda 0}\right)\left(1 - \rho_\lambda\right)}{1 - \rho_\lambda R_{\lambda \infty}}\,[R_{\lambda \infty}\,(C_2 - \rho_\lambda\, C_1) - C_1]. \tag{3.51}$$

For diffuse irradiation of a plane layer the expressions for reflectance and transmittance, with account taken of the external and internal reflection, are determined with the aid of Eqs. 2.62 and 2.63 under the condition $E^*_{\lambda 2} \equiv 0$:

$$R_\lambda(\rho_{\lambda 0},\rho_\lambda) = \rho_{\lambda 0} + (1-\rho_\lambda)E^*_-(0)/E_\lambda = \rho_{\lambda 0} + [(1-\rho_{\lambda 0})(1-\rho_\lambda)R_{\lambda \infty}/(1-\rho_\lambda R_{\lambda \infty})]\cdot$$

$$[1-R_{\lambda \infty}\exp(-2L_\lambda \ell)]/(1-R^2_{\lambda \infty}\psi^2_\lambda)]; \tag{3.52}$$

$$T_\lambda(\rho_{\lambda 0},\rho_\lambda) = (1-\rho_\lambda)E^*_+(\ell)/E_\lambda = (1-\rho_{\lambda 0})(1-\rho_\lambda)/(1-\rho_\lambda R_{\lambda \infty})\cdot$$

$$[(1-R_{\lambda \rho}R^2_{\lambda \infty})\exp(-L_\lambda \ell)/(1-R^2_{\lambda \rho}\psi^2_\lambda)]. \tag{3.53}$$

The absorptance is determined from the following relationship:

$$A_{\lambda ef}(\rho_{\lambda 0},\rho_\lambda) = (1-\rho_{\lambda 0})-(1-1-\rho_{\lambda 0})(1-\rho_\lambda)[(R_{\lambda \infty}+\exp(L_\lambda \ell)]/$$

$$(1-\rho_\lambda R_{\lambda \infty})(1+R_{\lambda \rho}\psi_\lambda). \tag{3.54}$$

The thermal radiational characteristics of an optically infinitely thick layer, with account taken of the boundary reflection, are determined from Eqs. 3.52 and 3.54 under the condition $L_\lambda \ell \to \infty$:

$$R_{\lambda \infty}ef(\rho_{\lambda 0},\rho_\lambda) = 1-[(1-\rho_{\lambda 0})(1-R_{\lambda \infty})/(1-\rho_\lambda R_{\lambda \infty})]; \tag{3.55}$$

$$A_{\lambda\infty ef}(\rho_{\lambda 0},\rho\lambda)=(1-R_{\lambda\infty})(1-\rho_{\lambda 0})/(1-\rho_{\lambda}R_{\lambda\infty}). \tag{3.56}$$

Next we shall express the quantities $R_{\lambda\infty}$ and $A_{\lambda\infty}$, which determine the thermal radiational characteristics of a layer of substance without taking reflection into account, in terms of the thermal radiational characteristics of a layer $R_{\lambda\infty ef}(\rho_{\lambda 0},\rho_{\lambda})$ and $A_{\lambda\infty ef}(\rho_{\lambda 0},\rho_{\lambda})$, which are determined experimentally and which take reflection into account. It follows from Eqs. 3.55 and 3.56 that:

$$R_{\lambda\infty}=[R_{\lambda\infty ef}(\rho_{\lambda 0},\rho_{\lambda})-\rho_{\lambda 0}]/\{1-\rho_{\lambda 0}-\rho_{\lambda}[1-R_{\lambda\infty ef}(\rho_{\lambda 0},\rho_{\lambda})]\}; \tag{3.57}$$

$$A_{\lambda\infty}=A_{\lambda\beta ef}(\rho_{\lambda 0},\rho_{\lambda})(1-\rho_{\lambda}R_{\lambda\infty})/(1-\rho_{\lambda 0}). \tag{3.58}$$

As can be seen from these relationships, the boundary reflection causes a signifcant decrease in the reflectance and transmittance, with internal reflection playing a predominant role here.

The effect of reflection at the outer boundary $\rho_{\lambda 0}$ and of the integral internal reflection ρ_{λ} on thermal radiational characteristics of foodstuffs can be seen from the absorptance. In the case of a non-reflecting surface ($\rho_{\lambda 0}=\rho_{\lambda}=$ 0) Eq. 3.56 becomes Eq. 3.14. In that case the ratio of $A_{\lambda\infty ef}(\rho_{\lambda 0},\rho_{\lambda})$ to $A_{\lambda\infty}$ is equal to the correction of the absorptance $A_{\lambda\infty}$, which takes into account the boundary reflection, i.e.:

$$A_{\lambda\infty ef}(\rho_{\lambda 0},\rho_{\lambda})/A_{\lambda\infty}=(1-\rho_{\lambda 0})/(1-\rho_{\lambda}R_{\lambda\infty}). \tag{3.59}$$

This ratio depends on the refractive index n_{λ} and the absorption coefficient k_{λ}, which determine $\rho_{\lambda 0}$ and ρ_{λ}, respectively, and on the characteristics of the scattering medium $\chi_{\lambda}(\gamma)$ and Λ, which determine $R_{\lambda\infty}$. The magnitude of this ratio increases with an increase in n_{λ} and Λ. Thus, when $n_{\lambda}=1.5$ (e.g., for water near $\lambda=3.0$ and 18.0 μm, and for ice at $\lambda=3.5$ and 14.0 μm), with a spherical scattering indicatrix $\delta_f=0.5$, this ratio for weakly scattering media ($\Lambda<0.5$, $R_{\lambda\infty}<0.02$) lies in the range 0.9-1.2; for substances with an intermediate level of scattering ($\Lambda < 0.9$, $R_{\lambda\infty} < 0.4$) it lies at 1.2-1.4; and for strongly scattering substances ($\Lambda > 0.9$, $R_{\lambda\infty} > 0.4$) - at 1.4-2.5.

This means that the boundary reflection should be taken into account when investigating the optical properties of substances with a high moisture content.

3.5 Experimental-analytical methods for determining the optical properties

To determine the radiation field in foodstuffs with the aid of the basic equation for radiant energy transfer, which has been solved for different irradiation conditions, it is necessary to have the following data:

- in the case of diffuse irradiation: the averaged absorption coefficient $\bar{k}_\lambda$; the effective attenuation coefficient L_λ; and the bihemispherical reflectance of an optically infinitely thick layer $R_{\lambda\infty}$;

- in the case of directional irradiation: besides $\bar{k}_\lambda$, L_λ, and $R_{\lambda\infty}$, the absorption coefficient k_λ, the extinction coefficient ε_λ, and Duntley's parametrs C_1 and C_2 which take into account the incidence angle of the directional flux and the optical properties and the scattering indicatrix of the volume element.

The spectral diffuse-hemispherical reflectance $R_{\lambda\infty}$ [hemispherical reflectance ($\omega_R = 2\pi9$ as measured for diffuse radiation ($\omega \leq 2\pi$)] is an external parameter and can be determined experimentally according to one of the known methods described in Chapter 5.

The methods for determining the coefficients k_λ, ε_λ, $\bar{k}_\lambda$, L_λ, C_1 and C_2, which are all internal parameters, require special consideration. The methods proposed by the authors for determining k_λ, ε_λ, $\bar{k}_\lambda$, and L_λ are based on Eqs. 3.1, 3.2; 3.23, 3.24; and 3.34, 3.35, which interrelate the internal and external parameters, and on measured bihemispherical quantities R_λ, T_λ, and $R_{\lambda\infty}$ and measured directional-hemispherical quantities R'_λ, T'_λ, and $R'_{\lambda\infty}$.

To determine L_λ with the aid of Eqs. 3.1 and 3.2 it is necessary to have experimental data on the bihemispherical quantities R_λ and $R_{\lambda\infty}$ or T_λ and $R_{\lambda\infty}$.

One method for determining $L_\lambda(R_\lambda, R_{\lambda\infty})$ [18, 167] has been applied in determining L_λ for foodstuffs [44]:

$$L_\lambda(R_\lambda,R_{\lambda\infty})=(1/2\ell)\ln\{[R_{\lambda\infty}-(1-R_\lambda R_{\lambda\infty})/(R_{\lambda\infty}-R_\lambda)]\}. \tag{3.60}$$

For the same purpose another, less precise equation has been proposed [168, 169]:

$$L_\lambda(R_\lambda,R_{\lambda\infty})=(1/2\ell)\ln[R_{\lambda\infty}(1-R_{\lambda\infty}^2)/(R_{\lambda\infty}-R_\lambda)]. \tag{3.61}$$

A second method for determining $L_\lambda(T_\lambda,R_{\lambda\infty})$, which is based on measured T_λ and $R_{\lambda\infty}$, involves the following approximate relationship:

$$T_\lambda=(1-R_{\lambda\infty}^2)\exp(-L_\lambda\ell). \tag{3.62}$$

This relationship is obtained from Eq. 3.2 for thick samples $(L_\lambda\ell>3.0)$ or for weak scattering when the magnitude of $R_{\lambda\infty}^2\exp(-2L_\lambda\ell)$ in comparison with unity can be disregarded.

The method proposed by the authors is based on this relatioship. It yields the following simple equation for determining L_λ, which is applicable in the case of large $L_\lambda\ell$ values:

$$L_\lambda(T_\lambda,R_{\lambda\infty})=(1/\ell)\ln[(1-R_{\lambda\infty}^2)/T_\lambda]. \tag{3.63}$$

A third method for determining $L_\lambda(R_\lambda,T_\lambda,R_{\lambda\infty})$, which has been developed by the authors, is quite precise and highly sensitive to changes in the optical properties of substances [44]. The L_λ as a function of R_λ, T_λ, and $R_{\lambda\infty}$ is obtained by solving Eqs. 3.1 and 3.2 for L_λ:

$$L_\lambda(R_\lambda,T_\lambda,R_{\lambda\infty})=(1/\ell)\ln[(1-R_\lambda R_{\lambda\infty})/T_\lambda]. \tag{3.64}$$

A more complex relationship for determining L_λ on the basis of R_λ, T_λ, and $R_{\lambda\infty}$ has been described [168, 169]:

$$L_\lambda(R_\lambda,T_\lambda,R_{\lambda\infty})=(1/\ell)\{\ln R_{\lambda\infty}+\ln[T_\lambda/(R_{\lambda\infty}-R_\lambda)]\}. \tag{3.65}$$

A shortcoming of all three methods for determining L_λ is that it is impossible to avoid experimentally introduced errors and instrumental errors in determining R_λ, T_λ, $R_{\lambda\infty}$, and ℓ.

A fourth method for determining L_λ is based on the relative transmittance $T_{\lambda,1}/T_{\lambda,2}$ of two samples having different thicknesses and different values for $R_{\lambda\infty}$ [39]. To avoid instrumental error the bihemispherical T_λ should be measured for two thicknesses, ℓ_1 and ℓ_2. Then a system of algebraic equations of the type shown in Eq. 3.2 should be solved for L_λ and $R_{\lambda\infty}$. From Eq. 3.2 we can obtain the following relationship for L_λ on the basis of the relative transmittance of the samples having different thicknesses $(\ell_1>\ell_2)$:

$$L_\lambda(T_{\lambda,1},T_{\lambda,2})=[1/(\ell_1-\ell_2)][\ln(T_{\lambda,2}/T_{\lambda,1})-\Delta D_\lambda]; \tag{3.66}$$

where, ΔD_λ – correction factor which compensates for the losses due to reflection and which depends on the thickness of the sample and is determined from the equation:

$$\Delta D_\lambda=\ln\{[1-R_{\lambda\infty}^2\,\exp(-2L_\lambda\ell_1)]/[1-R_{\lambda\infty}^2\,\exp(-2L_\lambda\ell_2)]\}. \tag{3.67}$$

If the thickness of the samples differs N times $(N>1)$, then at $\ell_1>\ell_2$, by assuming that $L_\lambda\ell=\tau$, we can write:

$$L_\lambda(\ell_1-\ell_2)=L_\lambda\ell_2(N-1)=\tau_1-\tau_2=\Delta\tau. \tag{3.68}$$

And from this it follows:

$$\tau_2=L_\lambda\ell_2=\Delta\tau/(N-1),\quad \tau_1=N[\Delta\tau/(N-1)]. \tag{3.69}$$

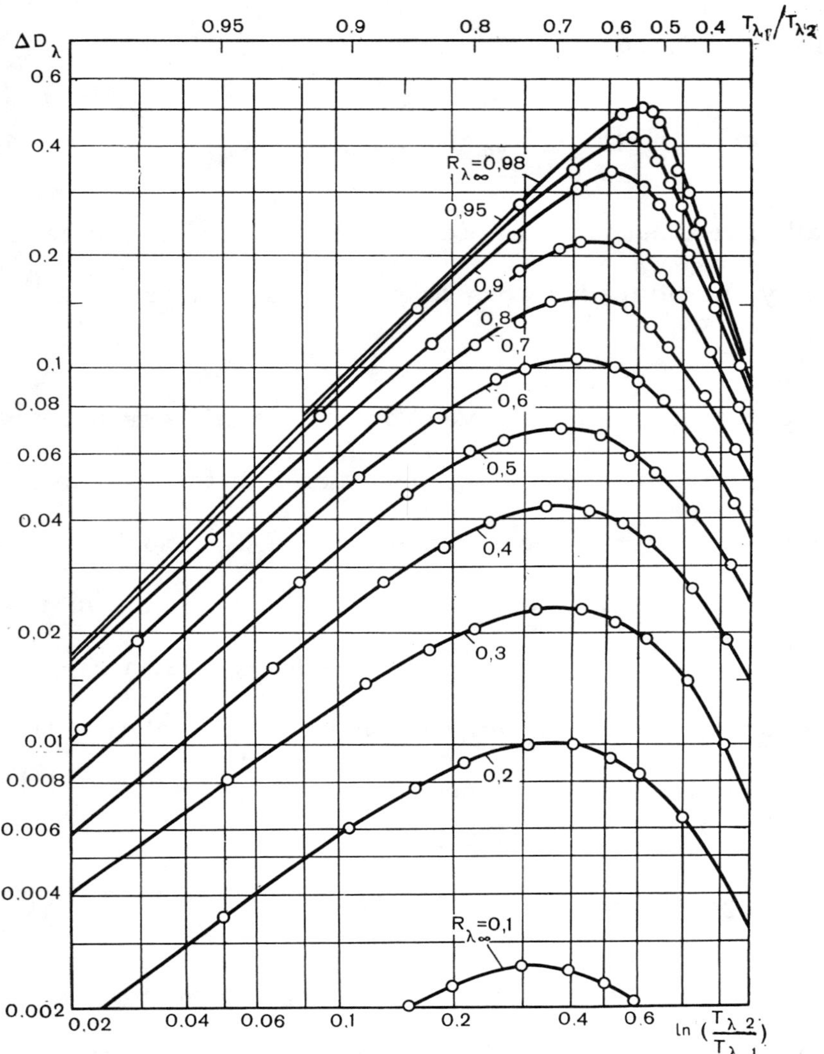

Figure 3.5 Correction ΔD_λ as a function of the observed relative transmittance of two samples at different values for $R_{\lambda\infty}$ [39].

By substituting these relationships into Eq. 3.67 for a general case we obtain the following equation describing the correction factor ΔD_λ:

$$\Delta D_\lambda = \ln\{[1-R_{\lambda\infty}^2 \exp(-2N\Delta\tau/(N-1))]/[1-R_{\lambda\infty}^2 \exp(-2\Delta\tau/(N-1))]\}. \qquad (3.70)$$

In Fig. 3.5 are shown the values of ΔD_λ as a function of $\ln(T_{\lambda,2}/T_{\lambda,1})$, which is determined with account taken of scattering in the case of diffuse irradiation. To put it differently, the plots represent the relative transmittance

$T_{\lambda,1}/T_{\lambda,2}$ for a series of values of $R_{\lambda\infty}$ when $\ell_1=2\ell_2(N=2)$. The approximate position of the maxima on such plots can be found with the aid of the equation:

$$\ln(T_{\lambda,2}/T_{\lambda,1})_{max}=\ln N/2. \tag{3.71}$$

This relationship makes it possible to select the thickness ratio of the sample. The correction ΔD_λ at $N=2$ is determined from the plots in Fig. 3.5.

To assess the precision of the methods considered here for determining the coefficient of effective attenuation L_λ, in Fig. 3.6 is shown the dependence of $L_\lambda \ell$ on ℓ in the case of starch. The least deviation from the average value of L_λ for a wide range of values of ℓ is observed when the third and fourth methods (Fig. 3.6, curves 1 and 4) are used. The first method for determining $L_\lambda(R_\lambda,$ $R_{\lambda\infty})$ (curve 3) can be used only if the samples are very thin (for starch at $\ell<0.2$ mm) and when the reflectance R_λ of the sample differs considerably from $R_{\lambda\infty}$ [at $(R_{\lambda\infty}-R_\lambda)>0.15$]. That this method is not quite precise is explained by the fact that the reflectance R_λ, as it has been shown [33], is less responsive to changes in the optical properties of the medium than T_λ. When ℓ is sufficiently large R_λ changes slightly with ℓ and approaches $R_{\lambda\infty}$. This sharply increases the error involved in the calculation of L_λ according to Eq. 3.60. Besides this error there might be instrumental errors involved in measuring L_λ and ℓ.

The second, approximate method for determining $L_\lambda(T_\lambda,R_{\lambda\infty})$ on the basis of Eq. 3.63 yields results which are close to precise values for thick samples (for starch, at $\ell>1.0$ mm, $T_\lambda<0.20$). This method is more sensitive to changes in the optical properties of the medium than the first method. Another advantage of this method lies in the fact that its precision increases with an increase in the thickness of the sample. For it is easier to prepare a thick sample than a thin one, and its thickness can be measured with a greater defree of precision.

Therefore, for practical purposes we recommend the third and fourth methods for determining L_λ.

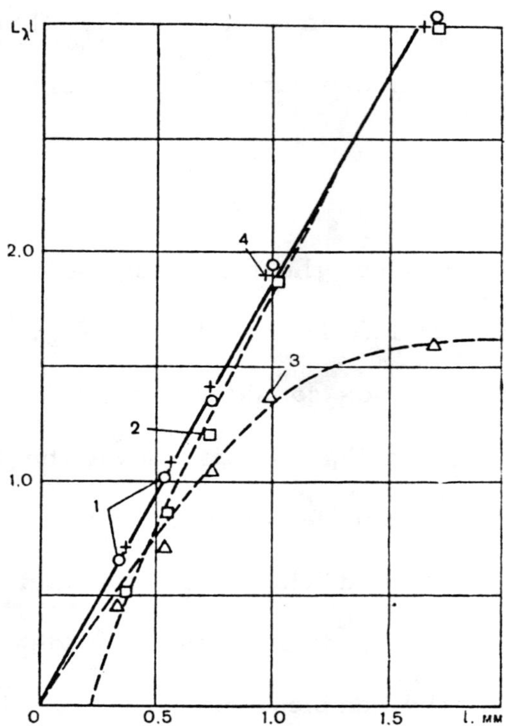

Figure 3.6 $L_\lambda \ell$ as a function of ℓ (mm) in the case of potato starch (W=11.8%) irradiated at λ=1.2 μm, as determined by different methods according to the equations [39, 44]:

    1 - Eq. 3.64;   3 - Eq. 3.60;
    2 - Eq. 3.63;   4 - Eq. 3.66.

As has been shown above, for a sufficiently thick layer of a scattering substance, when a direct radiant flux practically fails to pass through the layer $(T'_{\lambda B} \ll T'_\lambda)$, one can use Eqs. 3.26, 3.34, and 3.35. By subtracting term-by-term the left and right sides of Eq. 3.26 and 3.35 and by dividing Eq. 3.36 we obtain the known relationship [168, 169]:

$$(R_{\lambda\infty} - R_\lambda)/T_\lambda = (R'_{\lambda\infty} - R'_\lambda)/T'_\lambda. \tag{3.72}$$

For two samples having different thicknesses compensation for Eq. 3.35 yields:

$$T_{\lambda,1}/T_{\lambda,2} = T'_{\lambda,1}/T'_{\lambda,2}. \tag{3.73}$$

The obtained equations confirm the possibility of using the third and fourth methods for determining L_λ with the aid of directional-hemispherical R'_λ, T'_λ, and

$R'_{\lambda\infty}$.

By measuring the magnitudes of R'_λ and T'_λ for a sufficiently thick sample $(T'_{\lambda B} \ll T'_\lambda)$ and of $R'_{\lambda\infty}$ we can use Eq. 3.64 or Eq. 3.66 for determining the effective attenuation coefficient L_λ.

The absorption coefficient $\bar{k}_\lambda$ is usually calculated after L_λ has been estimated. It is calculated from L_λ and $R_{\lambda\infty}$ or from $R'_{\lambda\infty}$ according to the following simple relationships:

$$\bar{k}_\lambda = (1 - R_{\lambda\infty}) L_\lambda / (1 + R_{\lambda\infty}), \tag{3.74}$$

or

$$\bar{k}_\lambda = (1 - R'_{\lambda\infty}) L_\lambda / (1 + R'_{\lambda\infty}). \tag{3.75}$$

Duntley's parameters C_1 and C_2 can be determined analytically only if comprehensive data ara available on the bihemispherical R_λ and T_λ or $R_{\lambda\infty}$ and T_λ or on the directional-hemispherical R'_λ and T'_λ or $R'_{\lambda\infty}$ and T'_λ. C_1 and C_2 can then be determined from the following relationships derived from Eqs. 3.26, 3.34, and 3.35:

$$C_2 = T'_\lambda / T_\lambda, \tag{3.76}$$

$$C_1 = R_\lambda [(T'_\lambda / T_\lambda) - R'_\lambda], \quad C_1 = R_{\lambda\infty} [(T'_\lambda / T_\lambda) - R'_{\lambda\infty}]. \tag{3.77}$$

It is difficult to determine the extinction coefficient for absorbing-scattering substances, because it is difficult to determine experimentally the bidirectional transmittance $T'_{\lambda B}$ of a plane layer. When there is multiple scattering inside a layer it is hard to separate experimentally the straight line $T'_{\lambda B}$ from the diffuse one $(T'_\lambda - T'_{\lambda B})$, which are components of the flux passing through the layer. Therefore, the most precise calculated values for ε_λ can be obtained with the aid of Eqs. 3.39 and 3.40; this requires that we know five characteristics: C_1, C_2, T_λ, R_λ, and T'_λ or R'_λ of a thin sample for which the magnitude of $T'_{\lambda B}$ is sufficiently large. However, this method is rather complicated.

In practice the method of relative transmittance is still used for determining ε_λ [94]. The determination of the extinction coefficient according to the measured transmittance of samples having different thicknesses ($\ell_2 > \ell_1$) is based on the Bouguer-Lambert law:

$$\varepsilon_\lambda = 1/(\ell_2 - \ell_1)\ln(T'_{\lambda,1}/T'_{\lambda,2}). \qquad (3.78)$$

Here the term T'_λ stands for the bidirectional transmittance of the sample $T_\lambda(\theta,\theta_T)$ measured according to the usual method, with no account being taken of scattering. The error involved in measuring T'_λ with the aid of single-beam and double-beam spectrophotometers depends both on the scattering coefficient of the substance and on the thickness of the sample. Therefore, it is advisable to determine the real magnitude of ε_λ from Eqs. 3.39 and 3.40 in spite of their complexity.

The approximate value of the absorption coefficient k_λ for a radiation-scattering substance can be taken as one half of the value of $\bar{k}_\lambda$, which is determined from Eqs. 3.74 and 3.75. Indeed, the data in Table 1.1 indicate that for nonuniform diffuse irradiation:

$$k_\lambda = (1/m)\bar{k}_\lambda = (1/m)[(1-R_{\lambda\infty})L_\lambda/(1+R_{\lambda\infty})]; \qquad (3.79)$$

where, m – coefficient of the spatial distribution of the radiant flux as determined from Eq. 1.13. When studying the optical properties of foodstuffs subjected to diffuse irradiation it can be assumed that m=2, and the magnitude of k_λ can then be determined from Eq. 3.79. The magnitude of k_λ can also be determined from Eq. 3.15 if we know the backward scattering coefficient.

3.6 Graphic methods for determining the basic characteristics of the radiation field in foodstuffs

We have developed a simple graphic method, one that is convenient to use [44], which makes it possible to eliminate errors involved in the calculation and measurement of R_λ, T_λ, $R_{\Lambda\infty}$, and ℓ. According to this method it is necessary to measure R_λ or T_λ for two thicknesses, ℓ_1 and ℓ_2, of a layer and solve the algebraic equations of the type of Eq. 3.1 or Eq. 3.2 for $R_\lambda(L_\lambda \ell, R_{\lambda\infty})$ and $T_\lambda(L_\lambda \ell, R_{\lambda\infty})$ with respect to two unknowns, L_λ and $R_{\lambda\infty}$, by graphic means. The nomograms are constructed with the aid of Eqs. 3.1 and 3.2. Since the intersection angle of the curves $T_\lambda(\ell)$ and $R_\lambda(\ell)$ is very small, the error involved in determining $L_\lambda \ell$ and $R_{\lambda\infty}$ from measured T_λ and R_λ for samples with different thicknesses will be large. Therefore, we should also know the value of $R_{\lambda\infty}$. With the aid of nomograms we can determine $\bar{k}_\lambda$ and L_λ if T_λ and $R_{\lambda\infty}$ or R_λ and $R_{\lambda\infty}$ are known.

According to another method [168] we can determine L_λ and $R_{\lambda\infty}$ from R_λ', T_λ', and $R_{\lambda\infty}'$ measured under the conditions of directional irradiation. The method is based on the following relationship:

$$\ln[T_\lambda'/(R_{\lambda\infty}' - R_\lambda')] = L_\lambda \ell - \ln R_{\lambda\infty}. \qquad (3.80)$$

Eq. 3.80 is valid only at sufficiently large thicknesses of the layer when $T_{\lambda B}' \ll T_\lambda'$.

In Fig. 3.7 is shown a nomogram for the graphic determination of the characteristics L_λ and $R_{\lambda\infty}$ of the radiation field in foodstuffs as a function of their optical properties which change with time. For this one should make simultaneous measurements of the thermal radiational characteristics R_λ and T_λ of a layer having finite thickness at a given wavelength λ, and then determine L_λ and $R_{\lambda\infty}$ on the basis of the nomogram shown in Fig. 3.7.

Similar nomograms have also been constructed for determining the averaged absorption coefficient $\bar{k}_\lambda$ and the backward scattering coefficient s_λ (Fig. 3.8) [17, 39]. For this it is necessary to solve algebraic equations for $R_\lambda = f(\bar{k}_\lambda, \ell,$

$s_\lambda \ell$) or $T_\lambda = f(\bar{k}_\lambda \ell,\ s_\lambda \ell)$. On the basis of experimentally determined values of R_λ and T_λ we can determine on the nomogram the quantities L_λ, $R_{\lambda \infty}$, $\bar{k}_\lambda$, and s_λ. These quantities are needed for calculating $w_\lambda(x)$, $E_{0\lambda}(x)$, and $q_\lambda(x)$. The nomograms can also be used in determining L_λ and $\bar{k}_\lambda$ if T_λ and $R_{\lambda \infty}$ or R_λ and $R_{\lambda \infty}$ are known.

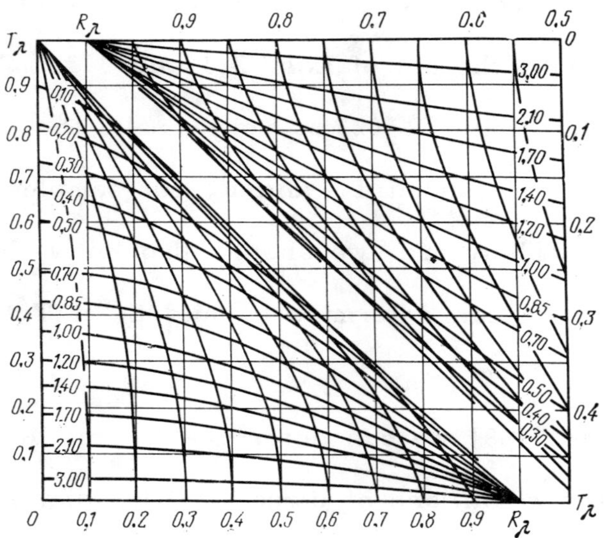

Figure 3.7 Nomogram for determining the basic characteristics L_λ and $R_{\lambda \infty}$ of the radiation field according to the simultaneously measured values of the reflectance and transmittance of a layer having finite thickness [17, 39].

Duntley has developed a graphic method for determining the averaged absorption coefficient $\bar{k}_\lambda$ and the backward scattering coefficient s_λ from measured values of R'_λ and T'_λ for two samples having different thicknesses ℓ_1 and ℓ_2 (Fig. 3.9). With our method we can also determine the parameters C_1 and C_2, because at sufficiently large thickness of the layer Eqs. 3.34 and 3.35 hold true. By using these equations we have developed a nomogram shown in Fig. 3.9. On the ordinate axis are plotted the values for $D_\lambda = \lg(1/T_\lambda)$ and on the abscissa - for R_λ; the curves are plotted at different $\bar{k}_\lambda \ell = \text{const}$ and $s_\lambda \ell = \text{const}$. The quantities R'_λ and T'_λ obtained for directional irradiation are used instead of R_λ and T_λ in the first approximation since in a majority of cases encountered in practice $C_1 \ll R_\lambda'$ and C_2 is close to unity. In a particular case where $C_1=0$ and $C_2=1$, $\bar{k}_\lambda \ell$ and $s_\lambda \ell$ are determined at once.

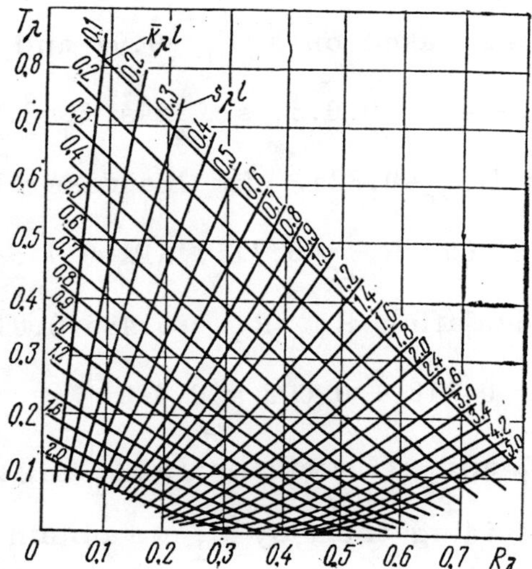

Figure 3.8 Nomogram for determining the averaged coefficients of absorption and backward scattering according to simultaneously measured values of R_λ and T_λ for a layer having finite thickness [17, 39].

In a general case where $C_1 \neq 0$ and $C_2 \neq 1$ the nomogram is used in the following way. First, R_λ' and T_λ' are measured for samples having $\ell_2 = N\ell_1$. Then the magnitudes of $\bar{k}_\lambda \ell_1$ and $s_\lambda \ell_2$ are obtained. These quantities are then used to determine $D_{\lambda,2c}$ from Fig. 3.9 (here, c – calculated value). The magnitude of $\Delta D_\lambda'$ is determined from the difference between $D_{\lambda,2}'$ and $D_{\lambda,2c}'$, while C_2 is calculated from the relationship $\lg C_2 = (\tfrac{1}{2})\Delta D_\lambda'$. The quantity $R_{\lambda 2c}'$ is determined from Fig. 3.9; then the difference $\Delta R_\lambda' = R_{\lambda 2c}' - R_{\lambda 2}'$ is calculated, and C_1 is obtained from $C_1 = (\tfrac{1}{2})\Delta R_\lambda' C_2$. The quantities R_λ and T_λ are determined from Eqs. 3.34 and 3.35:

$$R_\lambda = (R_\lambda' + C_1)/C_2; \tag{3.81}$$

$$D_\lambda = \lg(1/T_\lambda) = \lg(1/T_\lambda') + \lg C_2. \tag{3.82}$$

To illustrate the use of Duntley's nomogram in determining $\bar{k}_\lambda$ and s_λ for foodstuffs we shall use the values of R_λ' and T_λ' obtained for dough made of wheat flour (W=50.0%) at λ=1.1 µm. At ℓ_1=0.65 mm we have obtained the following: $R'_{\lambda 1}$=0.485, $T'_{\lambda 1}$=0.480, $D'_{\lambda 1}$=0.318; and at ℓ_2=2.60 mm and N=4: $R'_{\lambda 2}$=0.693, $T'_{\lambda 2}$=0.148, $D'_{\lambda 2}$=0.829. Calculations based on $\bar{k}_\lambda \ell_1$ =0.02 and $s_\lambda \ell_1$=1.13 from Fig. 3.9 yield the following: $\bar{k}_\lambda \ell_2$= $4\bar{k}\ell_1$=0.0.8, $s_\lambda \ell_2$=$4s_\lambda \ell_1$=3.92. Using these values we find from Fig. 3.9 that $D'_{\lambda 2c}$=0.825. In this case $\Delta D'_\lambda$=0.829-0.825= =0.004, $\lg C_2$=($\frac{1}{2}$)0.004=0.002, C_2=1.005, $R'_{\lambda 2c}$=0.715, $\Delta R'_\lambda$=0.715-0.693=0.022, C_1= =($\frac{1}{2}$)0.022·1.005=0.0111. After calculating $R_{\lambda 1} \cdot R_{\lambda 1}$=(0.485+0.011)/1.005=0.494 we obtain: $D_{\lambda 1}$=$\lg(1/T_\lambda)$=$D'_{\lambda 1}$ + $\lg C_2$= 0.318 + 0.002 = 0.320.

More precise values for $\bar{k}_\lambda \ell_1$ and $s_1 \ell_1$ will be: $\bar{k}_\lambda \ell_1$=0.03, $s_\lambda \ell_1$=1.25 at ℓ_1=0.65 mm and λ=1.1 µm. By dividing them by ℓ_1 we obtain the optical properties of the dough: $\bar{k}_\lambda$= 0.046 mm^{-1}, s_λ= 1.923 mm^{-1}.

Graphic methods make it possible to determine with a sufficient degree of precision for engineering purposes the optical properties of foodstuffs exposed to different infrared sources.

It can be seen from Eqs. 3.1, 3.2, 3.34, and 3.35 that, with the aid of the measured values of the directional-hemispherical thermal radiational characteristics R'_λ and T'_λ, we can in principle determine analytically the sought-for quantities L_λ and $R_{\lambda \infty}$. These quantities can be used to calculate also the other optical properties. For this purpose it is necessary to solve the above-mentioned system of transcendental equations (Eqs. 3.1, 3.2, 3.34, and 3.35) for L_λ, $R_{\lambda \infty}$, C_1, and C_2. The transcendental nature of Eqs. 3.1 and 3.2 makes their analytical solution difficult, and so many researchers sought various graphic methods for determining $\bar{k}_\lambda$, s_λ, L_λ, and $R_{\lambda \infty}$ [91, 114, 150, 168].

Nevertheless, an analytical solution of Eqs. 3.1 and 3.2 for L_λ and $R_{\lambda \infty}$, and then of the whole system of Eqs. 3.1, 3.2, 3.34, and 3.35 for C_1, C_2, R_λ, and T_λ is possible. And from the obtained values of $R_{\lambda \infty}$, C_1, and C_2 we can calculate $R_{\lambda \infty}$ from the known relationship (Eq. 3.26) [39, 44, 150].

An analytical solution of the transcendental equations (Eqs. 3.1 and 3.2) is possible through the following substitution [6]:

$$Z = \exp(-L_\lambda \ell) \tag{3.83}$$

In this case Eqs. 3.1 and 3.2 with regard to the unknown Z and $R_{\lambda\infty}$ acquire a simple form:

$$R_\lambda = R_{\lambda\infty}(1-Z^2)/(1-R_{\lambda\infty}^2 Z^2); \tag{3.84}$$

$$T_\lambda = (1-R_{\lambda\infty}^2)Z/(1-R_{\lambda\infty}^2 Z^2) \tag{3.85}$$

The absorbance is defined as follows:

$$A_\lambda = 1 - [(R_{\lambda\infty}+Z)/(1+R_{\lambda\infty}Z)]. \tag{3.86}$$

By using Eq. 3.86 we can express $R_{\lambda\infty}$ through R_λ, T_λ, and Z as follows:

$$R_{\lambda\infty} = (R_\lambda+T_\lambda-Z)/[1-(R_\lambda+T_\lambda)Z]. \tag{3.87}$$

The solution of Eqs. 3.1 and 3.2 for Z, with account taken of Eq. 3.87, yields an algebraic quadratic equation:

$$T_\lambda Z^2 - (1 + T_\lambda^2 - R_\lambda^2)Z + T_\lambda = 0. \tag{3.88}$$

Of the two solutions of this equation for Z one meets the condition of the problem and is in accord with its physical basis:

$$Z_\lambda = \{(1+T_\lambda^2-R_\lambda^2)-[(1+T_\lambda^2-R_\lambda^2)^2-4T_\lambda^2]^{\frac{1}{2}}\}/2T_\lambda. \tag{3.89}$$

The effective attenuation coefficient L_λ, with account taken of Eq. 3.82, becomes:

$$L_\lambda = -(1/\ell)\ln Z. \tag{3.90}$$

Thus, to calculate the magnitude of L_λ according to Eq. 3.90 and the magnitude of $R_{\lambda\infty}$ according to Eq. 3.87, with account taken of Eq. 3.89, it is necessary to know the bihemispherical thermal radiational characteristics R_λ and T_λ. Their experimental determination is difficult, especially in the region $\lambda > 1.4$ μm [39,48,118,150]. To overcome this difficulty we have developed an experimental-analytical method for determining these characteristics [6]. It is based on an analytical solution of Eqs. 3.1 and 3.2 and Eqs. 3.34 and 3.35, which relate four unknown quantities, R_λ, T_λ, C_1, and C_2, with the experimentally determinable R_λ'' and T_λ' for two samples having thicknesses ℓ_1 and ℓ_2. The analytical solution of Eqs. 3.34 and 3.35 is carried out with the aid of Duntley's method [150]. With this method R_λ' and T_λ', measured under the condition of directional irradiation of two samples having different thicknesses, $\ell_1 \neq \ell_2$, are used instead of R_λ and T_λ in the first approximation for calculating Z, L_λ, and $R_{\lambda\infty}$ according to Eqs. 3.87, 3.89, and 3.90. After that a more precise calculation is made.

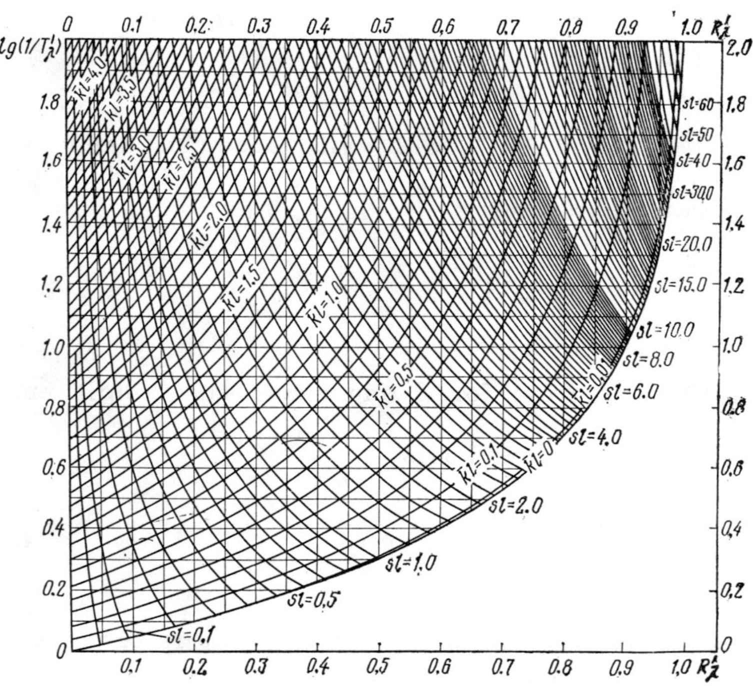

Figure 3.9 Nomogram for determining the optical properties $\overline{k}_\lambda \ell$ and $s_\lambda \ell$ from the experimentally obtained directional-hemispherical $R_\lambda(\theta; 2\pi)$ and $T_\lambda(\theta; 2\pi)$ [150].

The method of determining these characteristics consists of the following:

   - measurement of the directional-hemispherical thermal radiational character-
istics, $T_{\lambda 1}'$, $R_{\lambda 1}'$, $T_{\lambda 2}'$, and $R_{\lambda 2}'$, for samples having thicknesses ℓ_1 and ℓ_2;

- calculation of $Z_{\lambda 1c}$, $L_{\lambda 1c}$, and $R_{\lambda \infty c}$ with the aid of Eqs. 3.87, 3.89, and 3.90, respectively, according to the first approximation by using $T'_{\lambda 1}$ and $R'_{\lambda 1}$ instead of $T_{\lambda 1}$ and $R_{\lambda 1}$ in the first approximation for a sample having ℓ_1 thickness;

- calculation of the thermal radiational characteristics of the second layer (ℓ_2) by using the magnitudes of $L_{\lambda 1c}$ and $R_{\lambda \infty c}$:

$$T_{\lambda 2c} = (1 - R_{\lambda \infty c}^2) \exp(-L_{\lambda 1c}\ell_2) / [1 - R'_{\lambda \infty c} \exp(-2L_{\lambda 1c}\ell_2)], \qquad (3.91)$$

$$T_{\lambda 2c} = R_{\lambda \infty c} \{ [1 - \exp(-2L_{\lambda 1c}\ell_2)] / [1 - R_{\lambda \infty c} \exp(-2L_{\lambda 1c}\ell_2)] \}. \qquad (3.92)$$

- determination of the parameters C_1 and C_2 by comparing $T'_{\lambda 2}$ and $R'_{2\lambda}$ determined experimentally and $T'_{\lambda 2}$ and $R'_{\lambda 2}$ determined from Eqs. 3.91 and 3.92 according to Duntley's method [150]:

$$C_2 = (T'_{\lambda 2} / T_{\lambda 2c})^{\frac{1}{2}}, \qquad (3.93)$$

$$C_1 = \tfrac{1}{2}(R_{\lambda 2c} - R'_{\lambda 2})C_2; \qquad (3.94)$$

- use of Duntley's parameters C_1 and C_2, obtained from Eqs. 3.34 and 3.35, to determine the sought-for bihemispherical thermal radiational characteristics of a sample having ℓ_1 thickness or ℓ_2 thickness:

$$R_\lambda = (R'_\lambda + C_1)/C_2, \qquad (3.95)$$

$$T_\lambda = T'_\lambda / C_2; \qquad (3.96)$$

- calculation of the sought-for quantities Z_λ, L_λ, and $R_{\lambda \infty}$, which are necessary for calculating all the other characteristics, s_λ, $\bar{k}_\lambda$, etc, by substituting the obtained values of R_λ and T_λ into Eqs. 3.87, 3.89, and 3.90.

To verify the reliability of the results obtained we have compared the optical and bihemispherical thermal radiational characteristics of typical capillary-porous colloidal substances (bread dough, starch, the flesh of berries, grapes, etc) determined by our method and those determined by the graphic method and certain experimental methods [6, 39]. The results are in good agreement with one another.

Thus, for potato starch at $\lambda = 1.1$ µm, $L_\lambda = 979.2$ m^{-1} and $R_{\lambda\infty} = 0.85$, according to our method, which differ very little from values obtained experimentally: $L_\lambda = 952.8$ cm^{-1} (calculated from the experimental values of R_λ, T_λ, and $R_{\lambda\infty}$) and $R_{\lambda\infty} = 0.86$ (the relative error $\delta(R_{\lambda\infty}) = 1.1\%$, $\delta(L_\lambda) = 2.1\%$).

Table 3.7 Optical and bihemispherical thermal radiational characteristics of weakly scattering substances (the flesh of grapes) at $\lambda = 1.1$ µm obtained by the experimental-analytical method (1) and Duntley's graphic method (2) [6].

| Methods | Duntley's parameters | | L_λ (m^{-1}) | $R_{\lambda\infty}$ |
	$C_{1,\lambda}$	$C_{2,\lambda}$		
1	0.0361	1.0897	182.0	0.289
2	0.0335	1.0977	180.8	0.288

As can be seen from Table 3.7, there is good agreement between all the optical properties and thermal radiational characteristics as determined by the two methods. An advantage of the analytical method lies in the fact that it excludes errors that are inherent in the graphic methods.

Thus, the method proposed by the authors makes it possible to determine fairly precisely, with the aid of measured values of R_λ' and T_λ' of a layer, the bihemispherical thermal radiational characteristics R_λ, T_λ, and $R_{\lambda\infty}$ and the optical properties L_λ, $\bar{k}_\lambda$, and s_λ, as well as Duntley's parameters C_1 and C_2. We need to know all these characteristics in order to calculate the radiation field in light-scattering substances under different conditions of diffuse irradiation or directional irradiation at a certain angle.

PROPAGATION OF INFRARED RADIATION IN MULTILAYERED FOODSTUFFS WITH VARIABLE OPTICAL PROPERTIES

Infrared irradiation of foodstuffs affects their structure and intensifies the phase transformations and biochemical processes taking place inside them. As a result, variable time-dependent multilayered systems are formed with unstable optical properties.

4.1 Foodstuffs as multilayered absorbing-scattering systems

Owing to their different forms and multilayered structures foodstuffs can be regarded as absorbing-scattering substances. The penetration of infrared radiation in a substance affects not only its temperature fields but also the distribution of the moisture inside it. And this leads to changes in its optical properties. The presence of mobile boundaries where phase transformations take place causes changes in the multilayered structure. In the substance there can form "dry" and moisture-rich zones and zones of phase transformations. Thus, an isotropic layer of foodstuff can, when irradiated, become an optically multilayered system which selectively absorbs and scatters infrared radiation. The absorption of infrared radiation inside a layer is intensified by the movement of the evaporation zone.

Analogous physical processes take place in various foodstuffs when they are being dried, or when they become crystallized or frozen, or when they thaw and melt, etc. Therefore, it is necessary to know the principles governing the attenuation of radiation in multilayered systems. This requires a consideration of the boundary reflection and the selective absorption and scattering of radiation on optical inhomogeneities. These inhomogeneities strongly affect the energy distribution inside individual layers and the aggregate thermal radiational characteristics of each layer and of the system as a whole.

4.2 Propagation of radiation in multilayered foodstuffs subjected to diffuse irradiation

To solve problems of heat and mass transfer in foodstuffs exposed to infrared radiation it is necessary to determine the function $w(x, \tau)$ of heat distribution in each layer [20, 26, 34, 57, 77, 78, 82, 202]. This function should take into account the absorption of the total infrared radiation penetrating the substance. Other important factors involved here are: the conditions of irradiation; the selectivity of absorption and scattering; multiple scattering inside a layer of substance; and reflection of infrared radiation at the interface between different phases in the substance and at the outer boundaries of the layer.

The boundary layer, whose properties differ from those of the surrounding water vapor-air medium, should also be regarded as a separate layer in contact with the open surface of the substance. A model of such a multilayered substance exposed on both sides to radiation fluxes E_I and E_{II} is shown in Fig. 4.1.

Let us consider a general case of bilateral diffuse monochromatic irradiation of a multilayered system. The counterfluxes E_{i+} and E_{i-} inside the i-th layer can be determined by two different methods. One method involves compensation for reflection at the outer layer boundaries. In another method the boundaries are regarded as hypothetical nonabsorbing layers with nonsymmetrical coefficients of reflection: $\rho_{ij} \neq \rho_{ji}$ and of transmission:

$$t_{ij} = 1 - \rho_{ij} \neq t_{ji} = 1 - \rho_{ji}. \tag{4.1}$$

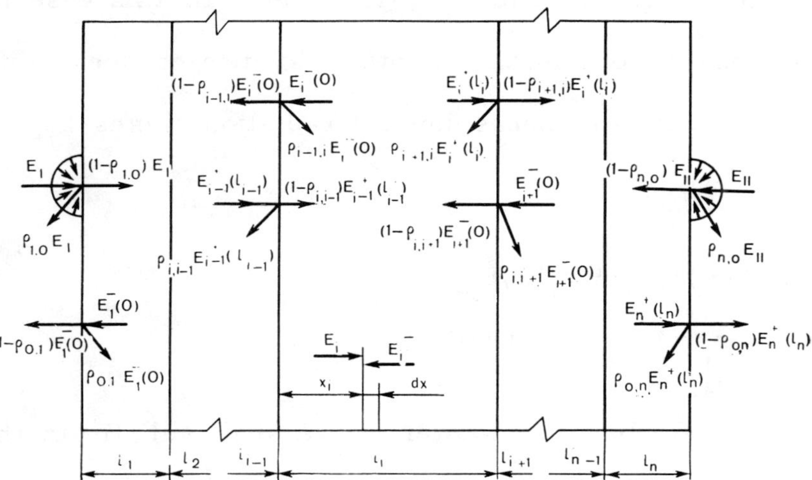

Figure 4.1 Bilateral diffuse irradiation of a multilayered system consisting of absorbing and scattering substances [47, 160].

The ρ_{ij} and t_{ij} are interrelated as follows [18]:

$$n_i^2 (1-\rho_{ij})=n_j^2 (1-\rho_{ji}); \qquad (4.2)$$

where, n_i and n_j – refractive index of the adjacent i-th and j-th layers, respectively. This equation can be used to calculate the coefficient of internal reflection ρ_{ji} of a layer if it forms an interface with air. For this purpose we need to know the experimental values of the external reflection coefficient ρ_{ij}. For example, for water in the infrared region at $\lambda=3.15$ µm $n_{\lambda j}=1.502$ and the reflection coefficient in this case of diffuse radiation coming from the air ($n_i \cong 1.0$) and incident upon the surface of the water is: $\rho_{ij}=0.09$. At $n=_{\lambda i} \cong 1.0$, $n_{\lambda j}=1.502$, and $\rho_{ij}=0.09$ the internal reflection coefficient in this case of diffuse radiation emerging from water and entering into the air is: $\rho_{ji}=0.596$. The magnitude of ρ_{ji} considerably exceeds that of ρ_{ij} owing to complete internal reflection. When the boundary reflection under boundary conditions is taken into account each i-th layer can be regarded as a separate layer if it is subjected to bilateral irradiation at external irradiance E_{iI} and E_{iII}, which are related through the boundary conditions to the counterfluxes E_i^+ and E_i^- (Fig. 4.1). In this case for any i-th layer having ℓ_i thickness we obtain the following system of linear differential equations describing the opposite hemispherical radiation fluxes E_{i+} and E_{i-}:

$$0 \leq x_i \leq \ell_i,$$

$$dE_{i+}/dx_i = [\overline{k}_i(x) + s_i(x)]E_{i+} + s_i(x)E_{i-}$$
$$-dE_{i-}/dx_i = -[\overline{k}_i(x) + s_i(x)]E_{i-} + s_i(x)E_{i+} \qquad (4.3)$$

Using the method described in Chapter 2, we shall substitute the effective optical depth τ_i for the variable x_i of the i-th layer and optical thickness $\tau_{\ell i}$ (as determined by Eqs. 2.12 and 2.13) for the layer's thickness ℓ_i, and introduce the Schuster number Λ_{efi} according to Eq. 2.14. Since Λ_{efi} is independent of $m_i(x)$ we shall assume that Λ_{efi} is a constant and average quantity for the entire i-th layer. In this case changes in the conditions of irradiation of a layer dx_i having x_i depth and the anisotropic nature of the scattering indicatrix are taken into account with the aid of the variables τ_i and $\tau_{\ell i}$, which depend on $m_i(x_i)$ and $\delta_s(x_i)$ according to Eqs. 2.11 and 2.12.

Eq. 4.3 can then be transformed into a new system of differential equations of the 1st order with respect to the dimensionless radiation fluxes $E_{i(+)}^*$ and $E_{i(-)}^*$ with a constant coefficient Λ_{efi}:

$$dE_{i(+)}^*/d\tau_i = -E_{i(+)}^* + \Lambda_{efi}E_{i(-)}^*$$
$$-dE_{i(-)}^*/d\tau_i = -E_{i(-)}^* + \Lambda_{efi}E_{i(+)}^*; \qquad (4.4)$$

$$\text{where,} \quad \tau_i = \int_0^{x_i} [\overline{k}_{\lambda i}(x_i) + s_{\lambda i}(x_i)]dx_i; \qquad (4.5)$$

$$\Lambda_{efi} = \delta_{si}\Lambda_i / [1-(1-\delta_{si})\Lambda_{\lambda i}] \qquad (4.6)$$

here, E_i^* – dimensionless radiation; $E_i^* = E_i/E_{i,I}$.

As the boundary conditions we shall use the relationships in which the denominate quantities E are replaced with dimensionless ones:

at $x_i=0$, $\tau_i=0$ $E^*_{i,I}=E^*_{(+),(i-1)}(\tau_{\ell,i-1})$,

$$E^*_{(+),i}(0)=(1-\rho_{i,(i-1)}\ E^*_{(+),(i-1)}(\tau_{\ell,i-1})+\rho_{i-1,}E^*_{(-),i}(0); \qquad (4.7)$$

at $x_i=\ell_i$, $\tau_i=\tau_{Pi}$ $E^*_{i,II}=E^*_{(-),i+1}(\tau_{\ell,i+1}=0)$,

$$E^*_{(-),i}(\tau_{\ell i})=\rho_{i+1,i}E^*_{(+),i}(\tau_{\ell i})+(1-\rho_{i,i+1})E^*_{(-),i+1}(0); \qquad 4.8)$$

where, $E^*_{iI}=E_{i,I}/\delta_0 T^4_I$, $E^*_{i,II}=E_{i,II}/\delta_0 T^4_{II}$ – the dimensionless radiant fluxes relative to the radiation flux emitted by a blackbody at the temperature of infrared generators I and II to which a multilayered system is exposed on both sides.

The radiation fluxes $E_{i,I}$ and $E_{i,II}$ should be determined with account taken of multiple reflections in the operating chamber and inside the multilayered system.

The radiation fluxes $E^*_{i,I}$ and $E^*_{i,II}$ at the boundary of the i-th $(1<i<n)$ layer, with account taken of multiple reflection, are expressed through external radiation fluxes E^*_I and E^*_{II} as follows:

$$E^*_{i,I}=[E^*_I T_{1\ldots(i-1),0}+E^*_{II}T_{n\ldots i,0}R_{(i-1)\ldots1,i}]\cdot$$
$$[1-R_{(i-1)\ldots1,i}R_{i\ldots n(i-1)}]^{-1}; \qquad (4.9)$$

$$E^*_{i,II}=[E^*_{II}T_{n\ldots(i+1),0}+E^*_I T_{1\ldots i,0}R_{(i+1)\ldots n,i}]\cdot$$
$$[1-R_{(i+1)\ldots n,i}R_{i\ldots1,(i+1)}]^{-1}. \qquad (4.10)$$

The radiation fluxes incident upon the 1st and the n-th layers, with account taken of Eqs. 4.9 and 4.10, are $E^*_{1,I}=E^*_I$, $E^*_{n,II}=E^*_{II}$:

$$E^*_{1,II}=[E^*_{II}T_{n\ldots2,0}+E^*_I T_{10}R_{2\ldots n,1}](1-R_{2\ldots n,1}R_{1,2})^{-1} \qquad (4.11)$$

$$E^*_{n,I}=[E^*_I T_{1\ldots(n-1),0}+E^*_{II}T_{n,0}R_{(n-1)\ldots1,n}]\cdot$$
$$[1-R_{(n-1)\ldots1,n}R_{n,(n-1)}]^{-1}. \qquad (4.12)$$

In Eqs. 4.9–4.12 the following designations are used: $T_{1\ldots(i-1),0}$ and $T_{1\ldots i,0}$ – the transmittance of a multilayered system from the 1-st to the (i-1) layer and from i-th layer, respectively, when the 1-st layer is exposed to an external radiation flux E_I; $T_{n\ldots i,0}$ and $T_{n\ldots(i,e)0}$ – the transmittance of a multilayered system from the n-th to the i-th layer and from the (i+1)-th layer, respectively, when the n-th layer is exposed to an external radiation flux E_{II}; $R_{(i-1)\ldots1,i}$, $R_{i\ldots1,i}$, $R_{i\ldots1,i}$, $R_{i\ldots n,i}$, $R_{(i+1)\ldots n,i}$ – the reflectance inside a stack of layers exposed to internal radiation at the corresponding boundary of the i-th layer.

The magnitudes of the above-mentioned characteristics for a multilayered system can be determined by taking into account the boundary reflection with the aid of the known relationships for a system consisting of two layers: j and i [18. 44]:

$$R_{i+j}=R_i+T_i^2 R_j/(1-R_i R_j);\tag{4.13}$$

$$T_{i+j}=T_i T_j/(1-R_i R_j).\tag{4.14}$$

By using these equations we can obtain the relationships for a system consisting of n layers with the aid of relationships for T_i, R_i, T_j, and R_j, which have been determined with account taken of the boundary reflection from Eqs. 3.45–3.48. For example, the transmittance of a system consisting of n layers can be represented as a system consisting of (n-1) layers and a n-th layer and, with account taken of Eq. 4.14, is:

$$T_{1\ldots n}=T_{(1\ldots n-1)+n}=[T_{1\ldots(n-1),0}T_{n,n-1}]/[1-R_{(n-1)\ldots1n} \cdot R_{n,(n-1)}];\tag{4.15}$$

where, $T_{n,(n-1)}$ – transmittance of the n-th layer when it is irradiated from the direction of the (n-1) layer; $T_{1\ldots(n-1),0}$ – a system consisting of a 1-st layer and (n-1) layers, when the 1-st layer is irradiated from the air; $R_{(n-1)\ldots1,n}$ – reflectance of a multilayered system from the (n-1) to the 1-st layer when exposed to internal radiation from the direction of the n-th layer;

$R_{n,(n-1)}$ - reflection of n-th layer when exposed to internal radiation from the direction of the (n-1)-th layer.

A general solution of Eq. 4.4 for counterfluxes under the boundary conditions defined by Eqs 4.7 and 4.8, with account taken of Eqs. 4.9-4.12, yields:

$$E_{(+)i}^{*}(x_i)=E_{i,I}^{*}C_{i,i-1}/[1-B_{i-1,i}B_{i+1,i}\Psi_i^2][exp(-L_i^{*}\tau_i)-B_{i+1,i}\Psi_i^2exp(L_i^{*}\tau_i)]+$$

$$E_{i,II}C_{i,i+1}\Psi_i/[1-B_{i-1,i}B_{i+1,i}\Psi_i^2][exp(L_i^{*}\tau_i)-B_{i-1,i}exp(-L_i^{*}\tau_i)]; \qquad (4.16)$$

$$E_{(-)i}^{*}(x_i)=E_{i,I}^{*}C_{i,i-1}\Psi_i/[1-B_{i-1,i}B_{i+1,i}\Psi_i^2]\{exp[L_i^{*}(\tau_{\ell i}-\tau_i)]-$$

$$B_{i-1,i}exp[-L_i^{*}(\tau_{\ell i}-\tau_i)]\}+E_{i,II}^{*}C_{i,i+1}/[1-B_{i-1,i}B_{i+1,i}\Psi_i^2] \cdot$$

$$\{exp[-L_i^{*}(\tau_{\ell i}-\tau_i)]-B_{i-1,i}\Psi_i^2exp[L_i^{*}(\tau_{\ell i}-\tau_i)]\}. \qquad (4.17)$$

The quantities Ψ_i, $R_{i\infty}$, and L_i^{*} in Eqs. 4.16 and 4.17 are determined from the following relationships:

$$\Psi_i=R_{i\infty}exp(-L_i^{*}\tau_i), \qquad (4.18)$$

$$R_{i\infty}=(1/\Lambda_{efi})[1-(1-\Lambda_{efi}^2)^{\frac{1}{2}}], \qquad (4.19)$$

$$L_i^{*}=(1-\Lambda_{efi}^2)^{\frac{1}{2}}=L_i\varepsilon_{efi}^{-1}. \qquad (4.20)$$

The coefficients $C_{i,j}$ and $B_{i,j}$, which take into account the reflection at the boundaries of the i-th layer, are defined as:

$$C_{i,j}=(1-\rho_{i,j})/(1-\rho_{j,i}R_{i\infty}), \qquad (4.21)$$

$$B_{j,i}=(R_{i\infty}-\rho_{j,i})/[R_{i\infty}(1-\rho_{j,i}R_{i\infty})]. \qquad (4.22)$$

The amount of the radiant energy absorbed per unit time at depth x_i by a volume element is determined from the equation of the conservation of energy:

$$w_i^{*}(x_i)=(1-\Lambda_{efi})[E_{(+)i}^{*}(x_i)+E_{(-)i}^{*}(x_i)]=(1-\Lambda_{efi})E_{i,0}^{*}(x_i); \qquad (4.23)$$

where, $E^*_{i,0}(x_i)=E^*_{(+)i} + E^*_{(-)i}$ — spatial irradiance at depth x_i;

$$w^*_i(x_i)=(1-\Lambda_{efi})E^*_{i,I}(1+R_{i\infty})C_{i,i-1}/(1-B_{i-1,i}\Psi^2_i)\{\exp(-L^*_i\tau_i)-$$

$$B_{i+1,i}\Psi^2_i/R_{i\infty}[\exp(L^*_i\tau_i]\}+(1-\Lambda_{efi})E^*_{i,II}(1+R_{i\infty})C_{i,i+1}/(1-B_{i-1,i}B_{i+1,i}\Psi^2_i)\cdot$$

$$\{\exp[-L^*_i(\tau_{\ell i}-\tau_i)]-B_{i-1,i}(\Psi^2/R_{i\infty})\exp[L^*_i(\tau_{\ell i}-\tau_i)]\}. \tag{4.24}$$

Thus, the problem of determining the radiation field in a multilayered system is reduced to one of calculating the transmittance and reflectance of the system, with account taken of the boundary reflection, by adding up the layers according to Eqs. 4.13-4.15.

The effect of boundary reflection on the distribution of radiant energy in a multilayered system and on its thermal radiational characteristics can be determined by the method according to which the boundaries are regarded as hypothetical nonabsorbing layers; these layers are characterized by nonsymmetrical coefficients of reflection, $\rho_{i,j}\neq\rho_{j,i}$, and of transmittance, $t_{i,j}\neq t_{j,i}$, which are interrelated through Eqs. 4.1 and 4.2. It is assumed that the isolated i-th layer is a combination of three layers: two boundaries of the layer, i,i-1 and i,i+1, and the i-th layer itself. The i-th layer is characterized by the reflectance R_i and transmittance T_i, which can be determined, with no account being taken of the boundary reflection, from Eqs. 3.1 and 3.2:

$$T_i=[(1-R^2_{i\infty})/(1-\Psi^2_i)]\exp(-L^*_i\tau_{\ell i}); \tag{4.25}$$

$$R_i=R_{i\infty}[1-\exp(-2L^*_i\tau_{\ell i})]/(1-\Psi^2_i). \tag{4.26}$$

In this case we can assume that the radiation fluxes $E^*_{i,1}=E^*_{(+),i}(0)$ and $E^*_{i,2}=E^*_{(-),i}(\tau_{\ell i})$ fall upon both sides of the i-th layer isolated within the system. These radiation fluxes are related to the fluxes $E^*_{i\,I}=E^*_{(+)i-1}(\tau_{\ell i-1})$ and $E^*_{i\,II}=E^*_{(-)i+1}(0)$ incident upon the boundaries of this layer through the following conditions:

with $x_i=0$ and $\tau_i=0$

$$E^*_{i,1}=E^*_{(+)i}(0)=(1-\rho_{i,i-1})E^*_{i,I} + \rho_{i-1,i}E^*_{(-)i}(0); \qquad (4.27)$$

with $x_i=\ell_i$ and $\tau_i=\tau_{\ell i}$

$$E^*_{i,2}=E^*_{(-)i}(\tau_{\ell i})=(1-\rho_{i,i+1})E^*_{i,II} + \rho_{i+1,i}E^*_{(+)i}(\tau_{\ell i}). \qquad (4.28)$$

Under these boundary conditions Eq. 4.4 yields:

$$E^*_{(+)i}(\tau_i)=[E^*_{i,1}/(1-\Psi^2_i)][\exp(-L^*_i\tau_i)-\Psi^2_i\exp(L^*_i\tau_i)]+[E^*_{i,2}\Psi_i/(1-\Psi^2_i)]\cdot$$

$$[\exp(L^*_i\tau_i)-\exp(-L^*_i\tau_i)]; \qquad (4.29)$$

$$E^*_{(-)i}(\tau_i)=[E^*_{i,2}/(1-\Psi^2_i)]\{\exp[-L^*_i(\tau_{\ell i}-\tau_i)]-\Psi^2_i\exp[L^*_i(\tau_{\ell i}-\tau_i)]\}+$$

$$[E^*_{i,1}\Psi_i/(1-\Psi^2_i)]\{\exp[L^*_i(\tau_{\ell i}-\tau_i)]-\exp[-L^*_i(\tau_{\ell i}-\tau_i)]\}. \qquad (4.30)$$

Eqs. 4.27 and 4.25, with account taken of Eqs. 4.29 and 4.30, yield the

following relationships which make it possible to determine the effect of the bound-

ary reflection on the distribution of radiation inside a multilayered system:

$$E^*_{i,1}=[C_{i,i-1}E^*_{i,I}/(1-D_{i+1,i}D_{i-1,i})]+[C_{i,i+1}D_{i-1,i}E^*_{i,II}/(1-D_{i+1,i}D_{i-1,i})]; \quad (4.31)$$

$$E^*_{i,2}=[C_{i,i+1}/(1-D_{i+1,i}D_{i-1,i})]+[C_{i,i-1}D_{i+1,i}E^*_{i,I}/(1-D_{i+1,i}D_{i-1,i})]; \qquad (4.32)$$

where, $C_{i,j}=(1-\rho_{i,j})/(1-\rho_{j,i}R_i),$ $\qquad (4.33)$

$$D_{j,i}=\rho_{j,i}T_i/(1-\rho_{j,i}R_i). \qquad (4.34)$$

For the i-th layer the function $w^*_i(\tau_i)$ is defined by Eq. 4.23, with account tak-

en of Eqs. 4.29 and 4.30:

$$w^*_i(\tau)=(1-\Lambda_{efi})E^*_{i,j}(1+R_{i\infty})/(1-\Psi^2_i)[\exp(-L^*_i\tau_i)-\Psi^2_i\exp(L^*_i\tau_i)/R_{i\infty}]+$$

$$(1-\Lambda_{efi})E^*_{i,2}(1+R_{i\infty})/(1-\Psi^2_i)\{\exp[-L^*_i(\tau_{\ell i}-\tau_i)]-\Psi^2_i\exp[L^*_i(\tau_{\ell i}-\tau_i)]/R_{i\infty}\}. \quad (4.35)$$

A comparison of Eqs. 4.35 and 4.33 shows that the method which represents the boundaries as reflecting and transmitting layers provides a simpler solution. Here the total number of layers increases and becomes $(2n+1)$, and this makes it more difficult to calculate $E^*_{i,I}$ and $E^*_{i,II}$ for a multilayered system by adding up the layers in succession. However, the method makes it possible to establish the relationship between the thermal radiational characteristics of the i-th layer which can be calculated with or without consideration of the boundary reflection.

The internal boundary reflection strongly affects the distribution of the absorbed energy and that of the radiant fluxes throughout the thickness of the i-th layer inside a multilayered system. This is taken into account by the coefficients $C_{i,j}$, $B_{i,j}$, $C'_{i,j}$, and $D_{i,j}$ which depend on $\rho_{i,j}$, $R_{i\infty}$, R_i, and T_i.

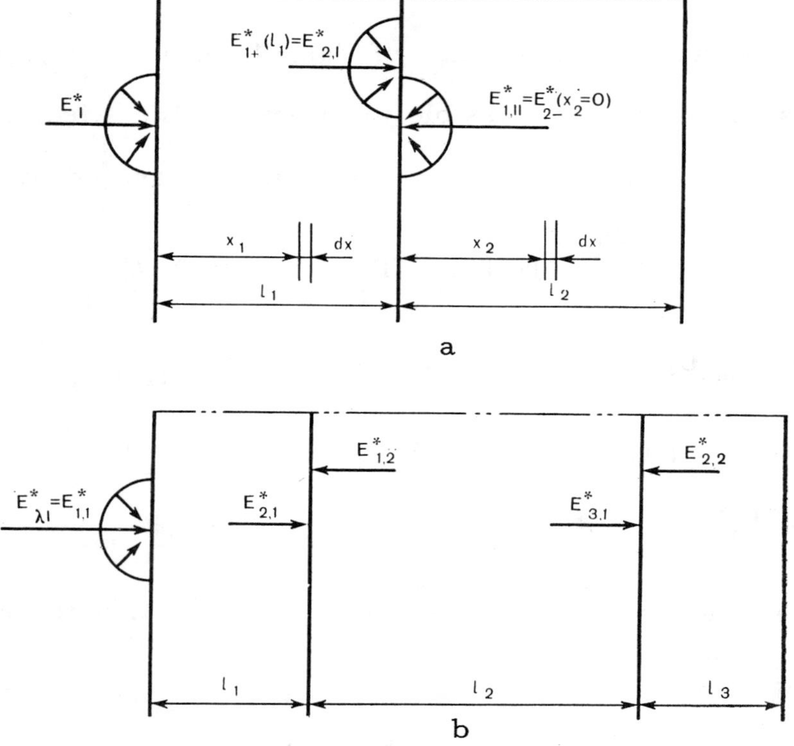

Figure 4.2   Radiation fluxes in a two-layer system (a) and a three-layer system (b) in the case of unilateral diffuse irradiation [23].

To illustrate how the developed equations are used in calculating the radiation fields in multilayered foodstuffs we shall consider a simple case of unilateral irradiation of a two-layer system. Let us assume that the first layer is an optically thin one and the second, an optically infinitely thick one ($L_2 \ell_2 \to \infty$), and that the two layers do not have continuous interfaces (Fig. 4.2,a). Such a system includes confectioneries powdered with sugar and bread and flour products which do not form a clearly defined interface with the surrounding medium. In this case n=2 and it is not necessary to take into consideration the boundary reflection ($\rho_{i,j}=0$). In the case of unilateral irradiation ($E_{2,II}=0$) Eqs. 4.16–4.35 are considerably simplified. For the first, optically thin layer the counterfluxes $E^*_{(+)\ell}(x_1)$ and $E^*_{(-)\ell}(x_1)$ and the function of the energy absorbed at depth x_1 per unit time $w^*_1(x_1)$ are determined according to Eqs. 4.29, 4.30, and 4.35 under the following conditions:

- the dimensionless quantity $L_\ell \tau_1$ is replaced by $L_1 x_1$, $L_\ell \tau_{\ell 1}=L_1 \ell_1$;

- the radiation fluxes incident upon the first layer are: $E^*_{\ell,1}=E^*_{\ell,I}$, $E^*_{\ell,2}=E^*_{\ell,II}$; and by considering that $T_2=0$ and $R_2=R_{2\infty}$, Eqs. 4.31 and 4.32 yield $E^*_{\ell,1}=E^*_{\ell,I}=E^*$, $E^*_{2,II}=0$,

$$E^*_{1,2}=E^*_{1,II}=E^*_{(-),2}(x_2=0)=E^*_f T_1 R_{2\infty}/(1-R_1 R_{2\infty}); \qquad (4.36)$$

$$E^*_{2,I}=E^*_{(+),1}(\ell_1)=E^*_f T_1/(1-R_1 R_{2\infty}). \qquad (4.37)$$

The quantities T_1 and R_1, which are determined from Eqs. 4.25 and 4.26 under the condition $L^*_i \tau_{\ell i}=L_1 \ell_1$, should be substituted into Eqs. 4.36 and 4.37.

For the second layer, under the condition that $L_2 \ell_2 \to \infty$ and $\Psi_2=0$, Eqs. 4.29, 4.30, and 4.35 are transformed into a simpler form:

$$E^*_{(+),2}(x_2)=E^*_{2,I}\exp(-L_2 x_2); \qquad (4.38)$$

$$E^*_{(-),2}(x_2)=E^*_{2,I}R_{2\infty}\exp(-L_2 x_2); \qquad (4.39)$$

$$w^*_2(x_2)=(1-\Lambda_{ef2})E^*_{2,I}(1+R_{2\infty})\exp(-L_2 x_2). \qquad (4.40)$$

Fig. 4.3 shows the distribution functions of spatial irradiance $E_{\lambda 0i}^{*}=E_{+i}^{*}+E_{-i}^{*}$ and of the energy absorbed per unit time $w_{\lambda i}^{*}$ in the case of a sugar-powdered layer of fruit candy subjected to monochromatic irradiation at $\lambda=1.1$ μm. Calculations have been carried out according to Eqs. 4.29–4.35 for a sugar-powdered layer ($\ell_1=0.2$ mm) and for a layer of fruit candy ($\ell_2>7$ mm) which practically does not transmit radiation ($T_{\lambda 2}<0.01$, $R_{\lambda 2}\cong R_{\lambda 2\infty}$). It can be seen that at the interface, when $x_1=\ell_1=0.2$ mm and $x_2=0$, $E_{\lambda 01}^{*}=E_{\lambda 02}^{*}$. This is the condition for the conjugation of radiation fields in multilayered systems. At the same time the amount of absorbed radiant energy $w_{\lambda 2}^{*}$ in a layer of fruit candy at $x_2=0$ is eight times more than $w_{\lambda 1}^{*}$ in a sugar-powdered layer at $x_1=\ell_1$; in other words, at the point of contact of the two layers the function w_{λ}^{*} changes abruptly, in proportion to the ratio between the probabilities of the absorption of radiation in the neighboring layers:

$$w_{\lambda 2}^{*}/w_{\lambda 1}^{*}=(1-\Lambda_{ef2})/(1-\Lambda_{ef1})=0.016/0/002=8.$$

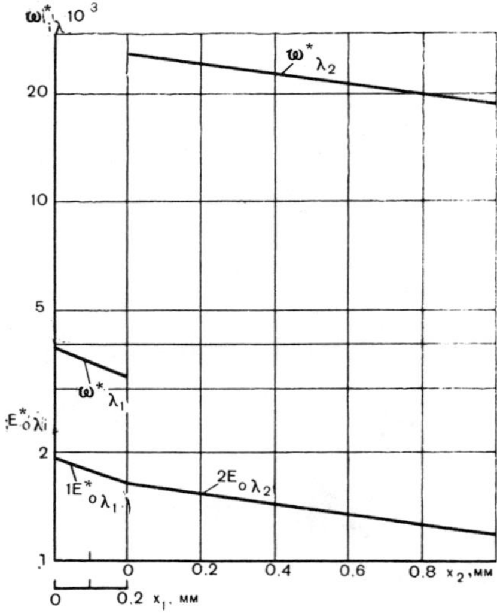

Figure 4.3 Distribution of spatial irradiance $E_{\lambda 0}^{*}$ and of the absorbed radiant energy w_{λ}^{*} at $\lambda=1.1$ μm in a case of unilateral irradiation of a two-layer system: sugar-powdered fruit candy [39].

Therefore, more heat is emitted by the layer of fruit candy owing to the absorption of the penetrating infrared radiation. Since the coefficient $L_{1\lambda}$ for powdered sugar at $\lambda=1.1$ μm ($L_{1\lambda}=0.53$ mm^{-1}) is greater than for fruit candy ($L_{2\lambda}=0.31$ mm^{-1}), the spatial irradiance $E_{0\lambda}^{*}$ and the w_{λ}^{*} for powdered sugar decrease with x_1 more rapidly (Fig. 4.3). This is shown by the larger inclination of the w_{λ}^{*} curves.

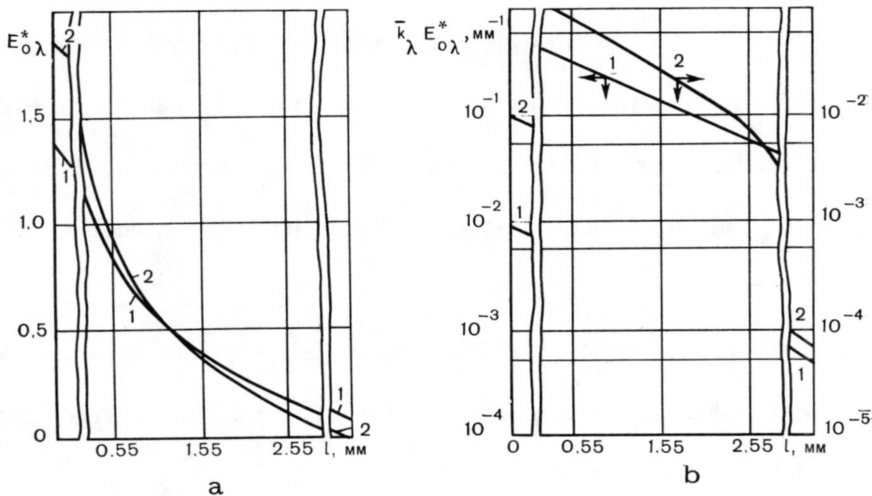

Figure 4.4 Distribution of the spatial irradiance $E_{\lambda 0}^{*}$ (a) and of the absorbed energy $\bar{k}_{\lambda}E_{0\lambda}^{*}$ (b) at $\lambda=1.1$ μm for a case of unilateral irradiation of a three-layer system: wheat grains of different structures [23]:
    1 - vitreous wheat grain;
    2 - farinaceous wheat grain.

When investigating the consistency of grains or during their disinsectization it is important to know the distribution of the absorbed energy of integral or monochromatic radiation inside the grains subjected to infrared irradiation.

To simplify the problem we can consider a grain to be a three-layer absorbing-scattering system: the first layer is the upper layer facing the infrared source; the second layer is the endosperm; and the third layer is the lower shell (Fig. 4.2,b). For unilateral irradiation with E_{I}^{*} flux from above the first and second layers will be irradiated by diffuse fluxes not only from the top but also from below owing to reflection at the interfaces between the first and the second and between the second and the third layers. This reflection is due to the different structures of the separate component parts of a grain. Such a model has been verified by experimental investigations [23]. Indded, the reflectance of the shell is always less than that of the whole grain. For the spectral reflectance of a whole grain is augmented by the reflection of the endosperm-lower shell system; in other words, the upper layer is also irradiated from below.

By using Eq. 4.35 for each layer separately we can determine the distribution of the energy of monochromatic radiation per unit time by a volume element absorbed throughout the thickness of the grain. With the aid of experimental data [23] on $\overline{k}$, L, and R_∞ we have calculated and constructed profiles of radiation field inside the husk and the endosperm of a grain.

In Fig. 4.4 are shown the distribution functions $E_{\lambda 0}^*$ of the transferred energy with respect to the thickness of a grain of wheat. For a farinaceous grain $E_{\lambda 0}^*$ decreases rapidly with an increase in the thickness of the grain. This is due to the large coefficient of effective attenuation of farinaceous grain as compared to that of vitreous grain. At wavelength 1.1 μm the value of L_λ for farinaceous and vitreous grain is 0.96 and 0.69 mm^{-1}, respectively. The value of $E_{\lambda 0}^*$ for the upper layer of farinaceous grain is greater than that for the upper layer of vitreous grain. This is due to the large reflectance of the farinaceous endosperm.

The absorption coefficient of the shell for farinaceous endosperm is different from that for vitreous endosperm (at $\lambda=1.1$ μm, $\overline{k}_{\lambda\ \text{shell}}=0.005\ \text{mm}^{-1}$; for vitreous endosperm $\overline{k}_\lambda=0.30\ \text{mm}^{-1}$, and for farinaceous endosperm $\overline{k}_\lambda=0.08\ \text{mm}^{-1}$). Therefore, the slope of the curve $\overline{k}_\lambda E_{\lambda 0}^*(x)$, which shows the distribution of the absorbed energy throughout the thickness of the grain, will differ from the slope of the curve $\bar{k}_\lambda E_{\lambda 0}^*(x)$ for the shell. More radiant energy is absorbed per unit time by a volume element in the endosperm than in the shell owing to the high absorbtance of the endosperm.

Thus, the separate component parts of a grain, owing to their different structures, have different absorption coefficients. And this accounts for the anomalous distribution of the absorbed radiant energy inside such a complex substance as wheat grain.

Next we shall discuss ways of calculating energy transfer in grapes subjected to diffuse and directional irradiation at a certain angle. For this purpose a grape can be represented by a model consisting of three absorbing–scattering layers: skin–flesh–skin. Because of the effect of this boundary reflection on the distribution of radiant energy and on the thermal radiational characteristics, calculations here are very complicated. To simplify them we can assume that the skin, which is very thin, is the boundary which reflects, absorbs and transmits radiation, and that it is penetrated by radiation fluxes $E_{1\lambda}$, T_λ, and $E_{2\lambda}$.

For bilateral irradiation at a certain angle $\theta = \text{arc } \cos\mu$ with radiation fluxes $E_{\lambda I}$ and $E_{\lambda II}$, the energy absorbed by a unit volume per unit time at depth x is defined as follows [6, 39]:

$$w_\lambda(x) = \bar{k}_\lambda E_{\lambda 1} T_\lambda (1+R_{\infty\lambda})/(1-\Psi_\lambda^2)\{C_2[\exp(-L_\lambda x)-\Psi_\lambda^2 \exp(L_\lambda x)/R_{\infty\lambda}]+$$

$$C_1 \exp(-\varepsilon_\lambda \ell/\mu)[\exp[-L_\lambda(\ell-x)]-\Psi_\lambda^2 \exp[L_\lambda(\ell-x)]/R_{\beta\lambda}]\}-[\bar{k}_\lambda(C_1+C_2)-k_\lambda] \cdot$$

$$E_{\lambda 1} T_\lambda \exp(-\varepsilon_\lambda x/\mu)+\bar{k}_\lambda E_{\lambda 2} T_\lambda (1+R_{\infty\lambda})/(1-\Psi_\lambda^2)\{C_2[\exp[-L_\lambda(\ell-x)]-\Psi_\lambda^2 \exp[L_\lambda \cdot$$

$$(\ell-x)]/R_{\infty\lambda}]+C_1 \exp(-\varepsilon_\lambda \ell/\mu)[\exp(-L_\lambda x)-\Psi_\lambda^2 \exp(L_\lambda x)/R_{\infty\lambda}])-[\bar{k}_\lambda(C_1+C_2)-$$

$$k_\lambda]]E_{\lambda 2} T_\lambda \exp[-\varepsilon_\lambda(\ell-x)/\mu]; \qquad (4.41)$$

where, $\Psi_\lambda = R_{\infty\lambda} \exp(-L\ell)$. $\qquad\qquad (4.42)$

For bilateral irradiation with a diffuse flux the absorbed energy for a separate layer in a multilayered system is:

$$w_{\lambda i}(x_i) = L_{\lambda i} E_{i,1\lambda}(1-R_{\infty\lambda i})/(1-\Psi_{i\lambda}^2)[\exp(-L_{\lambda i}x_i)-\Psi_{\lambda i}^2\exp(L_{\lambda i}x_i)/R_{\infty i\lambda}]+$$

$$L_{\lambda i}E_{\lambda i,2}(1-R_{\infty i\lambda})/(1-\Psi_{i\lambda}^2)\{\exp[-L_{\lambda i}(\ell_i-x_i)]-\Psi_{i\lambda}^2\exp[L_{i\lambda}(\ell_i-x_i)]/R_{\infty i\lambda}\}. \quad (4.43)$$

The radiation flux incident upon the i-th layer is determined under the following boundary conditions:

for the first layer:

$$x_1=0, \quad E_{1,1\lambda}=E_{I\lambda}, \qquad\qquad (4.44)$$

$$x_1=\ell_1, \quad E_{1,2\lambda}=(E_{II\lambda}T_{\lambda,3+2}+E_{I\lambda}T_{\lambda1}R_{\lambda,2+3}; \qquad\qquad (4.45)$$

for the second layer:

$$x_2=0, \quad E_{2,1\lambda}=(E_{I\lambda}T_{1\lambda}+E_{\lambda II}T_{\lambda,3+2}R_{\lambda1})M_{1,2+3}, \qquad\qquad (4.46)$$

$$x_2=\ell_2, \quad E_{2,2\lambda}=(E_{\lambda II}T_{\lambda3}+E_{\lambda,1+2}R_{\lambda3})M_{3,2+1}, \qquad\qquad (4.47)$$

for the third layer:

$$x_3=0, \quad E_{\lambda3,1}=(E_{\lambda I}T_{\lambda,1+2}+E_{\lambda II}T_{\lambda3}R_{\lambda,2+1})M_{3,2+1}, \qquad\qquad (4.48)$$

$$x_3=\ell_3, \quad E_{\lambda,3,2}=E_{\lambda II}; \qquad\qquad (4.49)$$

where,

$$M_{i,j}=1/(1-R_{\lambda i}R_{\lambda j}). \qquad\qquad (4.50)$$

The problem of determining the radiation field according to Eq. 4.64 under the boundary conditions defined in Eqs. 4.44-4.50 in a three-layer system is reduced to one of determining, with account taken of the boundary reflection, the transmittance and reflectance of the system by the addition method with the aid of Eqs. 4.13 and 4.14.

As an example of the practical application of the obtained relationships describing energy transfer inside a layer of a typical scattering substances we have investigated the spectral properties R_λ and T_λ of the skin and pulp of different kinds of grapes in the visible and near infrared spectral regions from 0.4 to 5.0

μm. The obtained optical properties and thermal radiational characteristics of a grape subjected to unilateral and bilateral nonsymmetrical irradiation (E_I/E_{II}= 1.5) are used in calculating the radiation fields at wavelengths corresponding to the maxima of the spectra of solar radiation (λ=0.55 μm) and of the infrared lamp KG-220-1000 (λ=1.1 μm), and the absorption of water (λ=1.2 μm). It has been found (Fig. 4.5) that in the case of bilateral diffuse or directional irradiation grapes are heated more uniformly than in the case of unilateral irradiation.

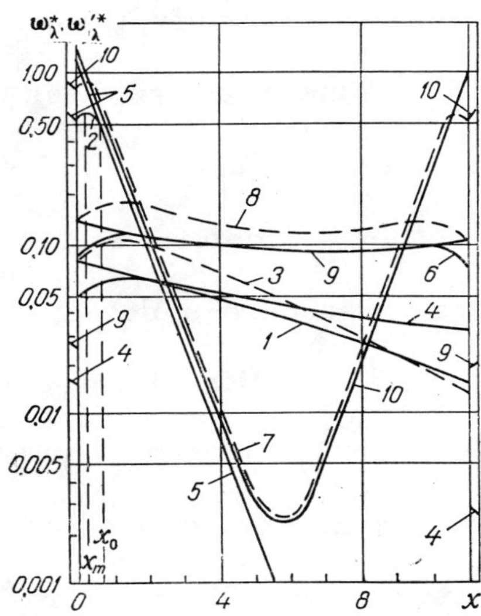

Figure 4.5 Distribution of the absorbed radiant energy in a grape under different irradiation conditions and at different wavelengths [6, 42]:

    1, 2, 3 - unilateral-directional irradiation;
    4, 5   - unilateral-diffuse irradiation;
    6, 7, 8 - bilateral-directional irradiation;
    9, 10   - bilateral-diffuse irradiation;
    1,4,6,9 - λ=1.1 μm;
    2,5,7,10 - λ=0.55 μm;
    3, 8   - λ=1.2 μm;
    $w_\lambda^*, w_\lambda'^*$ - in mm^{-1};
    x - in mm.

In the case of directional irradiation the amount of absorbed energy per unit time reaches a maximum at a certain depth x_m; and in the case of difffuse irradiation the maximum absorption of energy occurs in the surface layer at x=0. When x is greater than x_0, the nature of the dependence of w and that of w

on x under different irradiation conditions are approximately the same (Fig. 4.5, curves 5 and 2). But when x is smaller than x_0, these dependences differ in character. The presence of a maximum in the amount of energy absorbed in the case of directional irradiation is due to the fact that the absorption coefficient for directional radiation flux is different from that for scattered radiation flux: $\bar{k}$ is greater than k m times.

As another example let us consider the method of calculating the transfer of spectral radiant energy in stone fruits. To simplify the calculations for the purposes of engineering design we shall assume that the fleshy mesocarp of a fruit (plum, apricot, cherry, peach) is the main layer, considering that the epicarp is very thin. According to our method we determine the radiation fluxes $E_{\lambda I}$ and $E_{\lambda II}$ at the mesocarp-epicarp and mesocarp-endocarp interfaces, with account taken of the thermal radiational characteristics of the component parts of the fruit and of multiple reflection which occurs between layers in any multilayered system [39, 47]. In this case the problem is reduced to one of bilateral irradiation of a plane layer. A solution to this problem with respect to the counterfluxes E_+ and E_- is described in Chapter 2. The difference here lies in the non-symmetrical conditions of irradiation.

The energy transfer in a plane single-layer selectively absorbing-scattering substance subjected to unilateral and bilateral diffuse and directional irradiation at a certain angle with $E'_{\lambda I}$ and $E_{\lambda II}$ radiation fluxes is calculated with the aid of the basic equation for energy transfer [39, 47]. Its solution yields the magnitude of counterfluxes E'_+, E'_-, and E_+, E_-.

On passing through the epicarp the directional radiation flux $E'_{\lambda I}$ undergoes attenuation and is reflected from the mesocarp-endocarp interface whose reflectance is $R_{\lambda(2+3)}$. Owing to multiple reflection between layers 1 and (2+3) the magnitude of the radiation flux passing through the epicarp, $E'_{\lambda I}$, will be:

$$E'_{\lambda I} = E'_{\lambda 1} T_{\lambda 1} / (1 - R_{\lambda 1} R_{\lambda (2+3)}). \tag{4.51}$$

On passing through a system of layers (1+2) (epicarp and mesocarp) the directional radiation flux $E'_{\lambda I}$ changes into a diffuse one. It is reflected from the endocarp, and as a result the mesocarp is irradiated by the radiation flux $E_{\lambda II}$. The magnitude of $E_{\lambda II}$ is determined with account taken of multiple reflection between layers 3 and (2+1):

$$E_{\lambda II} = E'_{\lambda 1} T_{(1+2)} R_{\lambda 3} / (1 - R_{\lambda (2+1)} R_{\lambda 3}). \tag{4.52}$$

In Eqs. 4.51 and 4.52 the thermal radiational characteristics of a two-layer system are determined according to a known method [39, 47]:

for the (1+2) system (epicarp+mesocarp):

$$T_{\lambda (1+2)} = T_{\lambda 1} T_{\lambda 2} / (1 - R_{\lambda 1} R_{\lambda 2}), \tag{4.53}$$

for the (2+3) system (mesocarp+endocarp):

$$R_{\lambda (2+3)} = R_{\lambda 2} + T_{\lambda 2}^2 R_{\lambda 3} / (1 - R_{\lambda 2} R_{\lambda 3}), \tag{4.54}$$

for the (2+1) system (mesocarp+epicarp):

$$R_{\lambda (2+1)} = R_{\lambda 2} + T_{\lambda 2}^2 R_{\lambda} / (1 - R_{\lambda 2} R_{\lambda 1}). \tag{4.55}$$

The spatial irradiance equals the sum of the directional and the scattering components [39, 47]:

$$E_{\lambda 0}(x) = E_{\lambda 0s} + E'_{\lambda 0B}. \tag{4.56}$$

In our case of nonsymmetrical conditions of irradiation, with account taken of the above-mentioned results and of Eqs. 4.51 and 4.52 for directional and diffuse radiation fluxes, we find that the directional component of the irradiance $E'_{\lambda 0B}$ is:

$$E'_{\lambda 0B} = [E'_{\lambda 1} T_{\lambda 1} / (1 - R_{\lambda 1} R_{\lambda(2+3)})] \exp(-\varepsilon_\lambda x) \tag{4.57}$$

and that the diffuse (scattered) component of spatial irradiance, with account taken of the reflection from the endocarp, is:

$$E_{\lambda 0s} = E'_{\lambda 1} T_{\lambda 1} [1 - R_{\lambda 1} R_{\lambda(+3)}]^{-1} (1 + R_{\lambda\infty})(1 - \Psi_\lambda^2)^{-1} \{ C_2 [\exp(-L_\lambda x) - \Psi_\lambda^2 R_{\lambda\infty}^{-1} \cdot$$

$$\exp(L_\lambda x)] + C_1 \exp(-\varepsilon_\lambda \ell) [\exp[-L_\lambda(\ell-x)] - \Psi_\lambda^2 R_{\lambda\infty}^{-1} \exp[L_\lambda(\ell-x)] \} -$$

$$E'_{\lambda 1} T_{\lambda 1} (C_1 + C_2) [1 - R_{\lambda 1} R_{\lambda(2+3)}]^{-1} \exp(-\varepsilon_\lambda x) + E'_{\lambda 1} T_{\lambda(1+2)} R_{\lambda 3} [1 - R_{\lambda(1+2)} \cdot$$

$$R_{\lambda 3}]^{-1} (1 + R_{\lambda\infty})(1 - R_{\lambda ef} \Psi_\lambda^2)^{-1} [\exp[-L_\lambda(\ell-x)] - R_{\lambda ef} \Psi_\lambda^2 R_{\lambda\infty}^{-1} \exp[L_\lambda(\ell-x)]]; \tag{4.58}$$

where, $R_{\lambda ef} = (R_{\lambda\infty} - R_{\lambda 1}) / R_{\lambda\infty}(1 - R_{\lambda\infty} R_{\lambda 1})$, $\qquad$ (4.59)

$$\Psi_\lambda = R_{\lambda\infty} \exp(-L_\lambda \ell). \tag{4.60}$$

The radiant energy absorbed per unit time and per unit volume is as follows:

$$w'_\lambda(x) = \bar{k}_\lambda E_{\lambda 0s} + k_\lambda E'_{\lambda 0B} = w_{\lambda s} + w'_{\lambda B}. \tag{4.61}$$

The quantity $E_{\lambda 0s}$ in Eq. 4.61, determined with the aid of Eq. 4.58, takes into account the nonsymmetrical conditions of irradiation to which the mesocarp is subjected, as well as the reflection of radiation from the endocarp. Herein lies the difference between Eqs. 4.58 and 4.61, on the one hand, and those described in literature, on the other [39, 47].

Using Eqs. 4.56–4.61 and Eqs. 4.51–4.55 we have calculated with the aid of a computer the radiation fields for apricot in the visible and near infrared spectral regions ($\lambda = 0.4 - 2.5$ μm) of solar radiation at atmospheric masses m=1 ans m=2. The obtained complex functions of the distribution of spectral spatial irradiance $E_{\lambda 0}(x)$ and of the absorbed energy flux $w_\lambda(x)$ of solar radiation in a mesocarp layer of apricot at two wavelengths, $\lambda = 1.0$ μm and $\lambda = 1.2$ μm, are $E_\lambda = 740 \cdot 10^{-3}$ watt/m^{2}μm and $E_\lambda = 430 \cdot 10^{-3}$ watt/m^{2}μm, respectively. The effect of the endocarp (the stone in apricot) noticeably affects the distribution of w_λ and $E_{\lambda 0}$ at depth

x=6-10 mm. The presence of a maximum in w_λ in the case of directional irradiation is due to the difference between the absorption coefficient k_λ of directional
radiation and the absorption coefficient $\bar{k}_\lambda$ of diffuse scattered irradiation, and
to the presence of a maximum in $E_{\lambda 0s}$ for scattered radiation at x=0.4-0.6 mm.
The total irradiance $E_{\lambda 0}=E_{\lambda os} + E_{\lambda 0B}$ is characterized by the absence of a maximum.

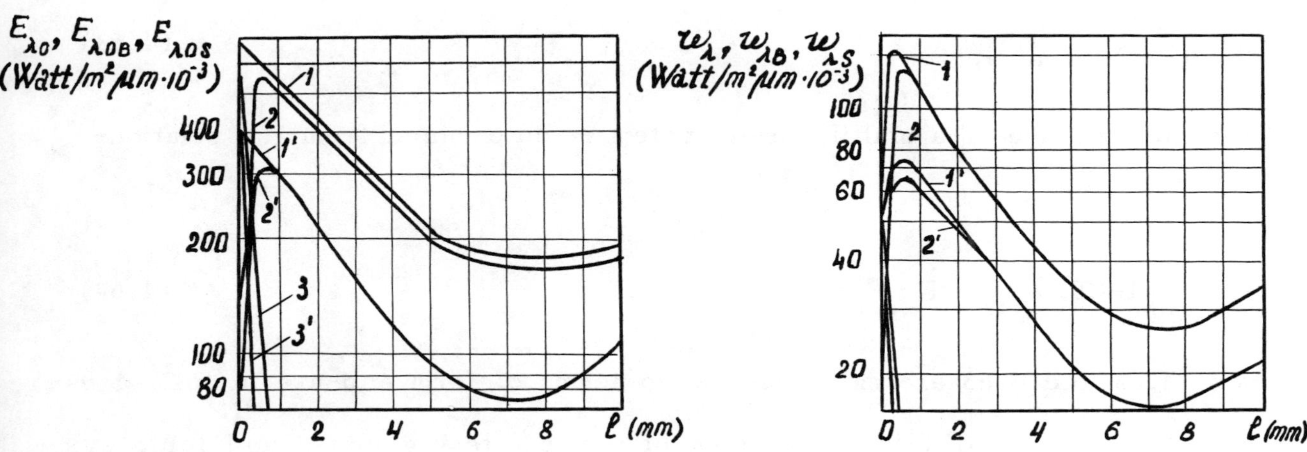

Figure 4.6   Distribution of the spectral spatial irradiance and of the absorbed
energy flux of solar radiation in a mesocarp layer of apricot at $\lambda=1.0$ μm (1,2,
3) and $\lambda=1.2$ μm (1',2',3') [39, 47]:

    1,1' - $E_{0\lambda}, w_\lambda$;
    2,2' - $E_{\lambda 0s}, w_{\lambda s}$;
    3,3' - $E_{\lambda 0B}, w_{\lambda B}$.

4.3   Calculating the thermal radiational characteristics of multilayered foodstuffs

The thermal radiational characteristics of a multilayered system, with compensation made for absorption, multiple scattering, and boundary reflection, are
determined by the summation method for each layer with the aid of Eqs. 4.16 and
4.17, with account taken of Eqs. 4.11 and 4.12. When $E_{II}=0$ Eq. 4.16, with account taken of Eq. 4.11, yields:

$$R_{1\ldots n} = \rho_{1,0} + (1-\rho_{0,1})E^{*}_{(-)1}(0)(E_{II}=0)/E^{*}_{I} = \rho_{1,0} + R_{1,0} +$$

$$(R_{2\ldots n,1}T_{1,0}T_{1,2})/(1-R_{2\ldots n,1}R_{1,2}). \tag{4.62}$$

The transmittance of a multilayered system is determined from Eq. 4.17, with account taken of Eq. 4.12, under the condition $E_{II}=0$:

$$T_{1\ldots n} = (1-\rho_{0,n})E^{*}_{n}(\ell_{n})(E_{II}=0)/E_{I} = T_{1\ldots(n-1),0}T_{n,n-1}/$$

$$(1-R_{(n-1)\ldots,n}R_{n,,(n-1)}). \tag{4.63}$$

The absorptance of a multilayered system is determined from the relationship:

$$A_{1\ldots n} = 1-(T_{1\ldots n} + R_{1\ldots n}). \tag{4.64}$$

Eqs. 4.62 and 4.63 are more general in terms of form and are in full agreement with Eqs. 4.13 and 4.14 for a stack of non-scattering plates and for a system of scattering layers with no account being taken of boundary reflection.

Eqs. 4.16 and 4.17 can be used to derive general relationships describing the thermal radiational characteristics of a layer adjacent to different layers (media) inside a multilayered system. The transmittance and reflectance of the i-th layer are determined from Eqs. 4.16 and 4.17 under the condition $E_{i,II}=0$:

$$T_{i} = (1-\rho_{i+1i})E^{*}_{i}(\ell_{i})(E_{i,II}=0)/E_{i,I} = (1-\rho_{i+1,i})\cdot$$

$$C_{i,i-1}(1-B_{i+1,i}R^{2}_{i\,\infty})\exp(-L_{i}\ell_{i})/(1-B_{i-1,i}B_{i+1,i}\psi^{2}_{i}); \tag{4.65}$$

$$R_{i} = \rho_{i,i-1} + (1-\rho_{i-1,i})E^{-}_{i}(0)(E_{i,II}=0)/E_{i,I} = \rho_{i,i-1} +$$

$$(1-\rho_{i-1,i})C_{i,i-1}R_{i\infty}(1-B_{i+1,i}\exp(-2L_{i}\ell_{i})/(1-B_{i-1,i}B_{i+1,i}\psi^{2}_{i}); \tag{4.66}$$

where, $C_{i,j}$ and $B_{i,j}$ - coefficients determined by Eqs. 4.21 and 4.22, respectively. For a particular case involving a single layer in the air (or in vacuo)

$\rho_{i,i-1}=\rho_{1,0}$, $\rho_{i+1,i}=\rho_{0,1}$, and Eqs. 4.65 and 4.66 fully accord with Eqs. 3.52 and 3.53.

Eqs. 4.62 and 4.63, with account taken of Eqs. 4.65 and 4.66, yield relationships for calculating the thermal radiational characteristics of a two-layer system, with compensation made for boundary reflection:

$$R_{1+2}=\rho_{1,0}+R_{1,0}+T_{1,0}T_{1,2}R_{2,1}/(1-R_{1,2}R_{2,1}); \tag{4.67}$$

$$T_{1+2}=T_{1,0}T_{2,1}/(1-R_{1,2}R_{2,1}). \tag{4.68}$$

In these equations the reflectance of the first layer $R_{1,0}$ subjected to external irradiation, and the transmittance of the first and second layers $T_{1,2}$ and $T_{2,1}$ subjected to internal irradiation are defined as follows:

$$R_{1,0}=R_{1,\infty}C_{1,0}[1-B_{2,1}\exp(-2L_1\ell_1)]/(1-B_{0,1}B_{2,1}\psi_1^2); \tag{4.69}$$

$$T_{1,2}=(1-\rho_{0,1})C_{1,2}(1-B_{0,1}R_{1,\infty}^2)\exp(-L_1\ell_1)/(1-B_{0,1}B_{2,1}\psi_1^2); \tag{4.70}$$

$$T_{2,1}=(1-\rho_{0,2})C_{1,2}(1-B_{0,2}R_{2,\infty}^2)\exp(-L_2\ell_2)/(1-B_{0,2}B_{1,2}\psi_2^2). \tag{4.71}$$

The method for representing the boundaries in the form of reflecting and transmitting layers makes it possible to establish the relationship between the thermal radiational characteristics of the i-th layer of a system, which are calculated with and without compensation being made for boundary reflection.

Eqs. 4.29 and 4.30, with account taken of Eqs. 4.31, 4.32, 4.25, and 4.26, yield general relationships for the reflectance and transmittance of the i-th layer inside the system, with the boundary reflection being taken into consideration:

$$T_i(\rho_{i,j},\rho_{j,i})=C_{i,i-1}T_i[(1+D_{i+1,i}R_i)/T_i]/(1-D_{i+1,i}D_{i-1,i}); \tag{4.72}$$

$$R_i(\rho_{i,j},\rho_{j,i})=C_{i,i-1}R_i[(1+D_{i+1,i}T_i/R_i]/(1-D_{i+1,i}D_{i-1,i}); \tag{4.73}$$

where, R_i and T_i are determined from Eqs. 4.25 and 4.26.

The reflection coefficient at the interface $\rho_{i,j}$ for the external flux incident upon the i-th layer is different from the reflection coefficient at the interface $\rho_{j,i}$ for the flux emerging from the layer. These coefficients depend on the incidence angle θ, the refractive index n_λ, and the absorption coefficient k_λ, and are defined by Fresnel's formulas (Eqs. 2.58, 2.59).

The effect of the boundary reflection on the absorptance can be deduced from a comparison of the two equations:

$$A_i = 1 - (R_i + T_i); \tag{4.74}$$

$$A_i(\rho_{i,0}, \ \rho_{0,i}) = [1 - (R_i + T_i)](1 - \rho_{i,0})\rho_{0,i}T_i / [(1 - \rho_{0,i}R_i)^2 -$$

$$\rho_{0,i}^2 T_i^2]. \tag{4.75}$$

For a strongly scattering medium at $\Lambda_{ef} = 0.9$, $\varepsilon\ell = 1.0$, $\rho_{i,0} \cong 0.1$, and $\rho_{0,i} \cong 0.4$, according to Eq. 4.75, with the boundary reflection being taken into account, is: $A_i(\rho_{i,0}, \ \rho_{0,i}) = 0.75$, or 8.35 times greater (error >800%) than $A_i = 0.9$ as calculated with the aid of Eq. 4.74, with no account being taken of the boundary reflection. With an increase in n_λ and Λ_{ef} the radiance coefficient $\rho_{0,1}$ and the radiant emittance R_i increase, which results in a bigger error when determining the absorptance of a multilayered system with no account being taken of the boundary reflection.

In Table 4.1 are shown the $R_{\lambda(1+2)}$ values for a two-layer system as calculated with the aid of Eq. 4.67 at $\rho_{1,0} = 0$ and as measured experimentally. The satisfactory agreement between the experimentally obtained and the calculated values of $R_{\lambda(1+2)}$ for a two-layer system shows that Eqs. 4.62-4.75 can be used for determining the thermal radiational characteristics of multilayered foodstuffs. When the optical properties $L_{\lambda i}$, $R_{\lambda\infty i}$, and $\rho_{\lambda i}$ are known, these equations can be used, without resorting to measurements, for calculating the thermal radiational characteristics R_λ, T_λ, and A_λ of a multilayered system, at different thicknesses of the component layers.

Table 4.1 Thermal radiational characteristics of a two-layer system (cacao beans)

[39]

λ (μm)	$R_{\lambda1,0}$	$T_{\lambda1,0}$	$R_{\lambda1,2}$	$T_{\lambda1,2}$	$R_{\lambda2,1}$ cacao kernel without shell	$R_{\lambda(1+2)}$ for a two-layer system exper-al	calc.(Eq.4.67)
1.0	0.605	0.194	0.655	0.178	0.596	0.645	0.639
1.1	0.634	0.212	0.678	0.202	0.670	0.698	0.687
1.2	0.629	0.204	0.665	0.195	0.482	0.657	0.657

4.4 Propagation of radiation in foodstuffs with variable optical properties

The analytical relationships given in Chapter 2 are derived with variables $\bar{k}_\lambda(x)$ and $s_\lambda(x)$ on the assumption that the absorption coefficient k_λ and the scattering coefficient σ_λ are constant; i.e., that they are independent of coordinate x and of time. Now, when foodstuffs are subjected to infrared processing, their optical properties change, depending on the exposure period and the depth of the penetration of infrared radiation. These changes result from changes in the physico-chemical properties, the structure and the distribution of density in foodstuffs, and the moisture content throughout the thickness of the layer of substance.

We shall consider a simple case where the optical properties of a substance are variable only with respect to the coordinate. This enables us to obtain sufficiently simple solutions which hold true for finite intervals of time in which the optical properties which vary along the coordinate can be considered to be approximately constant.

The obtained simple solutions are important for yet another reason. During infrared irradiation of a substance with an integral flux under technological conditions the spectral composition of the radiation changes with respect to the coordinate owing to multiple scattering on the optical inhomogeneities of the substance. And this leads to changes in the average integral coefficients of absorption and scattering.

The optical properties of foodstuffs exposed to solar and infrared radiation are independent of the flux density of incident radiation and are determined by by the physico-chemical properties and the changes in them during the irradiation. And both the coefficients of absorption and scattering, which characterize the optical properties of foodstuffs capable of absorbing and scattering radiation, change with the thickness of the given layer.

Let us consider a general case of bilateral irradiation of a plane layer (Fig. 2.4) with diffuse radiation fluxes $E_{1\lambda}$ and $E_{2\lambda}$. The optical properties of the layer, depending on coordinate x, are defined by the coefficients of absorption and scattering averaged with respect to the hemisphere:

$$\bar{k}_\lambda(x)=m(x)k_\lambda(x); \quad s_\lambda(x)=m(x)\delta_s(x)\sigma_\lambda(x). \tag{4.76}$$

The coefficients m and δ_s, as determined by Eqs. 1.13 and 1.21, also depend on x. They take into account the conditions of irradiation of a layer element with dx thickness and the form of the scattering indicatrix. The coefficients can have the following values: $2<m>2$, $1<m<\infty$, $0<\delta_s\leq1$. The dependence of the optical properties on x can be due to several factors: a change in the spectral coefficient as a function of the coordinate, $k_\lambda(x)$; a change in the spectral scattering coefficient as a function of the coordinate, $\sigma_\lambda(x)$; a change in the indicatrix of scattering as a function of the coordinate, $\chi_\lambda(\gamma)$; and changes in the spatial distribution of the radiation flux and its spectral composition.

In this case we obtain a system of linear differential equations of the first order, similar to Eq. 4.3, with two variable coefficients, $\bar{k}_\lambda(x)$ and $s_\lambda(x)$, relative to the counterfluxes $E_{\lambda+}$ and $E_{\lambda-}$.

We can obtain a new system of linear differential equations by means of a few simple transformations. These involve the substitution of the optical depth for x:

$$\tau_\lambda = \int_0^x [\bar{k}_\lambda(x) + s_\lambda(x)]dx, \qquad (4.77)$$

and of the optical thickness τ_ℓ, determined from Eq. 4.77 at x=ℓ, for the thickness of the layer ℓ; and the introduction of the Schuster number $\Lambda_{ef\lambda}(x)$, which is variable with respect to x. The Schuster number represents the effective survival probability of the photon during the elementary act of scattering at depth x:

$$\Lambda_{ef\lambda}(x)=s_\lambda(x)/[\bar{k}_\lambda(x) + s_\lambda(x)]=\delta_s(x)\Lambda_\lambda(x)/\{1-[1-\delta_s(x)]\Lambda_\lambda(x)\}. \qquad (4.78)$$

This new system of linear differential equations contains one variable coeffficient $\Lambda_{ef\lambda}(\tau_\lambda)$ with respect to the dimensionless values of the net radiant flux $q_\lambda^* = E_{+\lambda}^* - E_{-\lambda}^*$ and spatial irradiance $E_{0\lambda}^* = E_{+\lambda}^* + E_{-\lambda}^*$:

$$dq_\lambda^*/d\tau_\lambda =-[1-\Lambda_{ef\lambda}(\tau_\lambda)]E_{0\lambda}^*; \quad dE_{0\lambda}^*/d\tau_\lambda =-[1+\Lambda_{ef\lambda}(\tau_\lambda)]q_\lambda^*. \qquad (4.79)$$

The boundary conditions here are:

when $\tau_\lambda = 0$
$$E_{0\lambda}^*(0)=(1+R_\lambda)E_{1\lambda}^*+T_\lambda E_{2\lambda}^*,$$
$$q_\lambda^*(0)=(1-R_\lambda)E_{1\lambda}^*-T_\lambda E_{2\lambda}^*; \qquad (4.80)$$

when $\tau_\lambda = \tau_{\ell\lambda}$
$$E_{0\lambda}^*(\tau_{\ell\lambda})=T_\lambda E_{1\lambda}^*+(1+R_\lambda)E_{2\lambda}^*,$$
$$q_\lambda^*(\tau_{\ell\lambda})=T_\lambda E_{1\lambda}^*-(1-R_\lambda)E_{2\lambda}^*. \qquad (4.81)$$

The solution of Eq. 4.79 for q_λ^* or $E_{0\lambda}^*$ with an arbitrary function $\Lambda_{ef\lambda}(\tau_\lambda)$

yields a self-adjoint differential equation of the second order:

$$\{d[(dq^*_\lambda/(1-\Lambda_{ef\lambda}(\tau))/d\tau]/d\tau\}-[1+\Lambda_{ef\lambda}(\tau)]q^*_\lambda=0. \qquad (4.82)$$

Its solution can be obtained in an explicit form only when the function $\Lambda_{ef\lambda}(\tau)$ is known.

In many real processes involving the transfer of monochromatic radiant energy in the case of unilateral irradiation the dependence of the average effective survival probability of the photon, $\Lambda_{ef\lambda}$, on the optical depth τ_λ can be approximated by the following function:

$$\Lambda_{ef\lambda}(\tau)=1 - [a/(\tau + b)]. \qquad (4.83)$$

The constants a and b stand for the dimensionless value of the averaged absorption coefficient $a=\bar{k}^*_\lambda(0)$ and that of the averaged extinction coefficient $b=\varepsilon^*_{ef\lambda}(0)$, respectively, for a layer of unit thickness at $\tau=0$:

$$\Lambda_{ef\lambda}(0)=1 - [\bar{k}^*_\lambda(0)/\varepsilon^*_{ef\,\lambda}(0)]=1 - [\bar{k}^*_\lambda(0)/\bar{k}^*_\lambda(0) + s^*_\lambda(0)]. \qquad (4.84)$$

Such an increase in the magnitude of the average effective survival probability of the photon is due to the following circumstance. The density of the surface layer of foodstuffs of plant and animal origin decreases rapidly with the depth. Meanwhile, as shown by experimental data [26, 39], the absorption and scattering coefficients decrease in direct proportion to the density. Furthermore, when capillary-porous substances are subjected to infrared irradiation a crust is formed whose density decreases with the depth. Therefore, with an increase in τ the magnitude of $\Lambda_{ef\lambda}$ increases since the absorption coefficient $\bar{k}_\lambda$ decreases.

In the process of irradiation with directional and imperfectly diffuse fluxes the magnitudes of m_λ and $\delta_{s\lambda}$ inside a layer also change with depth. These indicators take into account the conditions of irradiation and the form of the scattering indicatrix. They also change relatively rapidly inside the surface layer [33]. This is due to the fact that at large τ an "effect of depth" is created in absorbing-scattering substances when the conditions of irradiation of the layer element are close to ideal diffuse irradiation and m and δ_s are constant.

In a general case of bilateral irradiation of a plane layer the complex relationship between $\Lambda_{ef\lambda}$ and τ can be described as follows:

$$\Lambda_{ef\lambda}(\tau)=1 - [a/(\tau + b)] - [a/(c - \tau)]; \tag{4.85}$$

where, a, b, c - constants which depend on the optical properties of the substance and on the spatial distribution of monochromatic incident radiation fluxes $E_{\lambda 1}$ and $E_{\lambda 2}$.

By taking into consideration Eq. 4.85 and substituting a new variable ξ $(\tau, \Lambda_{ef\lambda})$ for τ, Eq. 4.82 can be transformed into an equation of Bessel's type [77]:

$$(d^2q_\lambda/d\xi^2) + [(1 - 2\chi)dq_\lambda/\xi d\xi] + [(\beta^2\gamma^2\xi^{2\gamma-2} + (\alpha^2 - \nu^2\gamma^2)/\xi^2]/q_\lambda=0. \tag{4.86}$$

The solution of this equation yields a cylindrical function [77] of the type:

$$q_\lambda(\xi)=\xi^\chi Z_\nu(\beta\xi^\gamma). \tag{4.87}$$

This function is expressed through modified Bessel's functions of the first order $J_\nu(Z)$ and the second order $K_\nu(Z)$.

A general solution of Eq. 4.87 under conditions of bilateral irradiation, with account taken of Eqs. 4.79 and 4.85, yields:

$$q_\lambda(\tau)=Z_\nu(\beta\xi^\gamma)=Z_{2ai}\{2[2a/(b+c)]^{\frac{1}{2}}i[(\tau+b)(c-\tau)]^{\frac{1}{2}}\}. \tag{4.88}$$

This relationship can be expressed through Bessel's functions $J_\nu(Z)$ and $K_\nu(Z)$.

For strongly absorbing substances ($R_{\infty\lambda}>0.5$; $\Lambda_{ef\lambda}>0.8$) belonging to the third and fourth group according to the classification based on optical properties (Chapter 6), when the values of Z are large, it suffices to use in Eq. 4.88

only the first terms of the expansion of Bessel's functions. Then, a general solution under the conditions of bilateral irradiation assumes a simpler form:

$$q_\lambda(\tau)=\{C_1\exp[-2(2a\xi/(b+c))^{\frac{1}{2}}]+C_2\exp[2a\xi/(b+c))^{\frac{1}{2}}]\}\xi^{-1/4}; \qquad (4.89)$$

where, $\xi=(\tau+b)(c-\tau)$. Integration constants C_1 and C_2 in this equation are determined from the boundary conditions (Eqs. 4.80 and 4.81).

The amount of radiant energy absorbed per unit time at depth x by a layer element is determined on the basis of the law on the conservation of energy, with account taken of Eq. 4.79:

$$w_\lambda^*(\tau)=-dq_\lambda^*/d\tau=[1-\Lambda_{ef\lambda}(\tau)]E_{0\lambda}^*(\tau). \qquad (4.90)$$

In the case of unilateral irradiation ($E_{\lambda 2}=0$) of a layer having finite thickness τ_ℓ and variable optical properties the quantities $q_\lambda(\tau)$ and $E_{0\lambda}(\tau)$ for total radiation are determined as follows [51]:

$$q_\lambda(\tau)=[E_{1\lambda}(1-R_{\infty\lambda})/(1-B_1\Psi_\lambda^2)]\{\exp[-2(2ab)^{\frac{1}{2}}(\xi^{\frac{1}{2}}-1)]+$$

$$\Psi_\lambda^2\exp[2(2ab)^{\frac{1}{2}}(\xi^{\frac{1}{2}}-1)]\}\xi^{-1/4}; \qquad (4.91)$$

$$E_{0\lambda}(\tau)=\{[E_{1\lambda}(1+R_{\infty\lambda}9/(1-B_1\Psi_\lambda^2)][(1+4(2ab\xi)^{\frac{1}{2}})/4a\}\{\exp[-2(ab)^{\frac{1}{2}}(\xi^{\frac{1}{2}}-1)]-$$

$$[\Psi_\lambda^2(4(2ab\xi)^{\frac{1}{2}}-1)/4(2ab\xi)^{\frac{1}{2}}+1)]\exp[2(2ab)^{\frac{1}{2}}(\xi^{\frac{1}{2}}-1)]\}\xi^{-1/4}. \qquad (4.92)$$

In these equations the following designations are used:

$$R_{\infty\lambda}=\{4[(2ab)^{\frac{1}{2}}-a]+1\}/\{4[(2ab)^{\frac{1}{2}}+a]+1\}; \qquad (4.93)$$

$$\psi_\lambda^2=\{[4(2ab\xi_\ell)^{\frac{1}{2}}-a)+1]/[4(2ab\xi_\ell)^{\frac{1}{2}}+a)-1]\}\exp[-4(2ab)^{\frac{1}{2}}(\xi_\ell^{\frac{1}{2}}-1)]; \qquad (4.94)$$

$$B_1=\{4[(2ab)^{\frac{1}{2}}-a)-1]/\{4[(2ab)^{\frac{1}{2}}+a)+1]\}; \qquad (4.95)$$

$$\xi=1+\tau/b, \quad \xi_\ell=1+\tau_\ell/b. \qquad (4.96)$$

The thermal radiational properties of a layer having finite thickness and variable optical properties are determined from Eq. 4.91, with account taken of Eqs. 4.80 and 4.81, under conditions of unilateral irradiation $E_{2\lambda}$:

$$R_\lambda = 1 - q_\lambda(0)/E_{1\lambda} = [R_{\infty\lambda} - (1 + B_1 - R_{\infty\lambda})\psi_\lambda^2]/(1 - B_1\psi_\lambda^2); \qquad (4.97)$$

$$T_\lambda = q_\lambda(\tau_\ell)/E_{1\lambda} = [(1-R_{\infty\lambda})/(1-B_1\psi_\lambda^2)][8(ab\xi)^{\frac{1}{2}}\exp[-2(ab)^{\frac{1}{2}}(\xi_\ell^2-1)]/$$

$$4[(2ab\xi_\ell)^{\frac{1}{2}}+a)-1]. \qquad (4.98)$$

For an optically infinitely thick layer the net radiant flux and spatial irradiance are determined from Eqs. 4.91 and 4.92 at $\tau_\ell \to \infty$:

$$q_\lambda(\tau)=E_\lambda(1-R_{\infty\lambda})(1+\tau/b)^{-1/4}\exp[-2(ab)^{\frac{1}{2}}(1+\tau/b)^{\frac{1}{2}}-1]; \qquad (4.99)$$

$$E_{0\lambda}(\tau)=E_\lambda(1+R_{\infty\lambda})\{1+4[2ab(1+\tau/b)]^{\frac{1}{2}}/4a \cdot \exp[-2(ab)^{\frac{1}{2}}[(1+\tau/b)^{\frac{1}{2}}-1] \cdot$$

$$(1+\tau/b)^{-1/4}\}. \qquad (4.100)$$

The obtained general relationships (Eqs. 4.88-4.100) for solving Eqs. 4.78-4.81 describing the transfer of monochromatic radiant energy in substances with optical properties that are variable with respect to the coordinate under conditions of uni- and bilateral irradiation are also applicable in the case where the substances are exposed to total radiant flux. Here the spectral optical characteristic $\Lambda_{ef\lambda}(\tau)$ should be replaced by an integral characteristic $\Lambda_{ef}(\tau)$.

Using Eqs. 4.88-4.100 we can calculate the total radiation field in selectively absorbing-scattering substances without integrating the analytical expressions for $q_\lambda(x)$ and $E_{0\lambda}(x)$, Here we should also compensate for changes in the average integral optical characteristics relative to the coordinate with the aid of the function $\Lambda_{ef}(\tau)$ and the optical depth τ. The functions $w^*(x)$ and $q(x)$ shown in Fig. 4.9 are calculated by this method and by the method of integration with respect to the spectrum of the emitter. The results show that the proposed method is sufficiently precise and can be used in engineering calculations.

4.5 Methods for calculating the transfer of total radiation in foodstuffs having variable optical properties

In practice the thermal processing of foodstuffs is carried out by integral irradiation. Experimental investigations show that most substances subjected to thermal processing scatter infrared radiation to a considerable extent and exhibit selective optical properties. They are not ideal "gray" substances, and their spectral characteristics, which indicate their absorbing and scattering properties, depend on the wavelengths of the radiation source.

During the propagation of total radiation in the irradiated layer multiple scattering takes place. Each act of scattering is accompanied by a change in the spectral composition of the incident flux of total radiation as it penetrates the substance. The spectral region with a large spectral absorption coefficient k_λ is the first to disappear from the general spectral composition of the radiation entering the layer. And the radiation corresponding to those spectral regions for which k_λ is the smallest and the scattering coefficient σ_λ is the largest penetrates more deeply into the sample.

The effect of multiple reflection of radiation is also due to changes in the spectral composition of the total radiation incident upon the outer surface of the substance. Each act of reflection is accompanied by the disappearance from the original spectral composition first of all of the spectral region with the larger spectral absorptance A_λ of the substance.

All this means that the calculations should be carried out simultaneously under external and internal conditions of irradiation. This makes it more difficult to determine the total radiation inside the layer and the heat transfer between the layers of the substance and the radiation source. The problem is further complicated by the selective spectral emissive and thermal radiationl characteristics of the substance and the infrared generator.

To determine the specific characteristics of the transfer of total radiant energy in selectively absorbing-scattering foodstuffs we shall consider the propagation of spectral radiation in a semi-infinite layer of two typical strongly scattering substances: potato starch (w=11.8%) and bread crumbs (w=42.8%), subjected to integral irradiation with an infrared KGT-220-1000 lamp.

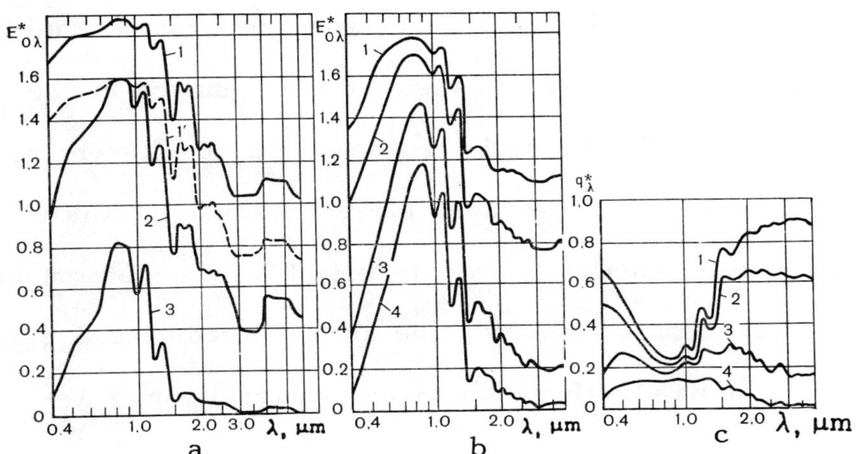

Figure 4.7 Changes in the spectral composition of the spatial irradiance $E^*_{o\lambda}$ (a, b,) and net radiation flux q^*_λ (c) in a semi-infinite layer of potato starch (a) and bread crumbs (b, c) at different depth x (mm) [16,39]: 1 - x=0; 2 - x=0.2; 3 - x=1.0; 4 - x=2.0; (curve 1' is obtained by superimposing with curve 2).

We shall rewrite Eqs. 2.31, 2.32, and 2.40 in a form that is convenient for analysis in which the dimensionless quantities q^*_λ, w^*_λ, and $E^*_{\lambda o}$ are independent of the incident flux E_λ:

$$E^*_{\lambda 0}(x)=E_{\lambda 0}/E_\lambda=(1+R_{\lambda\infty})\exp(-L_\lambda x), \qquad (4.101)$$

$$q^*_\lambda(x)=q_\lambda/E_\lambda=(1-R_{\lambda\infty})\exp(-L_\lambda x), \qquad (4.102)$$

$$w^*_\lambda(x)=w_\lambda/k_\lambda E_\lambda=(1-\Lambda_\lambda)(1+R_{\lambda\infty})\exp(-L_\lambda x). \qquad (4.103)$$

Fig. 4.7(c) shows that strong scattering and weak absorption take place in the spectral region 0.7-1.4 μm.　As a result the net radiation flux q_λ^* changes very little in this region with a change in x from x=0 mm to x=2 mm, and the spatial irradiance $E_{\lambda 0}^*$ remains high (Fig. 4.7 a,b).　A noticeable decrease in q_λ^* and $E_{\lambda 0}^*$ in this spectral region is observed only near the absorption bands at λ = 1.0, 1.2., and 1.45 μm.　In the range from 0.4-0.6 μm and 1.5-6.0 μm, where the values of the absorption coefficient are larger, there is a more pronounced change in the spectral composition of $E_{\lambda 0}^*$ and q_λ^* with a change in the depth x.　This is especially evident for q_λ^* (Fig. 4.7,c).

The selectivity of the optical properties and the effect of multiple scattering on the optical inhomogeneities of a substance cause changes in the average integral absorption coefficient $\bar{k}$, the backward scattering coefficient s, and the effective attenuation coefficient L of the layer element having dx thickness at depth x. These coefficients depend on the coordinate x and can be determined by averaging with respect to the spectral composition the spatial irradiance E_{x0}^* (λ) and the radiation flux $q_x^*(\lambda)$ at a fixed value of x in the spectral range $\lambda_1 - \lambda_2$ of the incident radiation flux E_λ^*.

The average integral absorption coefficient $\bar{k}(x)$, acting at depth x, is determined by averaging with respect to the spectrum the spatial irradiance $E_{\lambda 0}(x)$:

$$\bar{k}(x) = \left[\int_{\lambda_1}^{\lambda_2} E_{\lambda 0}(x) \bar{k}_\lambda \, d\lambda \right] / \left[\int_{\lambda_1}^{\lambda} E_{\lambda 0}(x) \, d\lambda \right]. \tag{4.104}$$

This relationship is derived with the aid of Eqs. 2.39 and 2.40, which interrelate $w_\lambda(x)$ and $E_{\lambda 0}(x)$.　By integrating these equations with respect to λ within the limits $\lambda_1 - \lambda_2$ of the incident radiation flux E we establish the relationship between the integral values of the absorbed energy at depth x per unit time and the spatial irradiance $E_0(x)$:

$$w(x) = \int_{\lambda_1}^{\lambda_2} w_\lambda(x) \, d\lambda = \int_{\lambda_1}^{\lambda_2} \bar{k}_\lambda E_{\lambda 0}(x) \, d\lambda; \tag{4.105}$$

$$E_0(x) = \int_{\lambda_1}^{\lambda_2} E_{\lambda 0}(x) \, d\lambda. \tag{4.106}$$

By multiplying and dividing the right side of Eq. 4.105 by $E_0(x)=\int_{\lambda_1}^{\lambda_2} E_{\lambda 0}(x)d\lambda$, we obtain the following equation which is convenient to use in engineering design:

$$w(x)=\overline{k}(x)E_0(x); \qquad (4.107)$$

where, $\overline{k}(x)$ - average integral absorption coefficient of radiation at depth x, which depends on the coordinate x and is defined by Eq. 4.104.

In an analogous way we can derive an equation for the average integral co-efficient of effective attenuation L on the basis of Eqs. 2.41 and 2.32:

$$w(x)=\int_{\lambda_1}^{\lambda_2} L_\lambda q_\lambda(x)d\lambda=L(x)q(x); \qquad (4.108)$$

where, $L(x)$ - average integral coefficient of effective attenuation at depth x, which depends on the spectral composition of the net radiation flux $q_\lambda(x)$ and is defined as follows:

$$L(x)=\int_{\lambda_1}^{\lambda_2} q_\lambda(x)L_\lambda d\lambda \;/\!\!\int_{\lambda_1}^{\lambda_2} q_\lambda(x)d\lambda. \qquad (4.109)$$

The average integral coefficient of backward scattering $s(x)$ can be deter-mined from the analytical relationship (Eq. 3.17) which contains the coefficients L and k according to the relationship:

$$s(x)=[L^2(x)-\overline{k}^2(x)]/2\overline{k}(x). \qquad (4.110)$$

The coefficient $s(x)$ at depth x for the opposite integral radiation fluxes E_+ and E_- can be determined from the relationship:

$$s_i(x)=[\int_{\lambda_1}^{\lambda_2} E_{\lambda i}(x)s_{\lambda i}d\lambda]/[\int_{\lambda_1}^{\lambda_2} E_{\lambda i}(x)d\lambda]. \qquad (4.111)$$

The radiation which penetrates deeply into a substance corresponds to those spectral regions for which the absorption coefficient k_λ is the smallest and the scattering coefficient σ_λ is the largest. Therefore, with an increase in x the mag-nitude of $\overline{k}(x)$ decreases and that of $s(x)$ increases. The attenuation of the integral radiant energy in the case of a semi-infinite medium $(L_\lambda l \to \infty)$ proceeds according to a rule which differs from the exponential law since the attenuation coefficient $L(x)$ varies with the coordinate x:

$$L(x) = \{\overline{k}(x)[\overline{k}(x)+2s(x)]\}^{\frac{1}{2}}. \tag{4.112}$$

It can be assumed that owing to the opposite nature of the dependences of $\overline{k}(x)$ and $s(x)$ on x, $L(x)$ depends only slightly on x. This makes it possible, for purposes of simplification, to introduce the average value of L with respect to the layer and to use all the relationships obtained above for the spectral quantities q_λ, w_λ, R_λ, T_λ, and A_λ in calculating the integral quantities q, w, R, T, and A by substituting the average integral quantity L for the spectral quantity L_λ.

The limits of the spectral interval $\lambda_1 \div \lambda_2$ for the integral incident radiation E, used in Eqs. 4.104–4.112, need to be substantiated.

The limits of the spectral range of the radiation emitted by the blackbody, if we disregard that part of the spectrum for which the minimal values of spectral radiance do not exceed 0.1% of the maximal value, can be determined from the following relationships [12, 39]:

$$0.25\lambda_{max} < \lambda < 13\lambda_{max};$$
$$917/T < \lambda < 47684/T. \tag{4.113}$$

For practical calculations we only need to limit the spectral region corresponding to the greater part of the energy ($\sim$95%) coming from the infrared generator:

$$0.4\lambda_{max} \varsigma \lambda < 4.0\lambda_{max}. \tag{4.114}$$

Thus, for a Ni-Cr spiral at $T=1270$ K ($\lambda_{max}=1.9$ μm) such limits lie at $\lambda_1=0.76$ and $\lambda_2=7.6$ μm; and for a KGT-220-1000 lamp at $T=2250$ K ($\lambda_{max}=1.16$ μm) – at $\lambda_1=.46$ and $\lambda_2=4.6$ μm [8, 102, 109].

For an emitter with a tungsten filament a general equation describing the maxmum of the spectral emissive power λ (in μm) in relation to temperature has the following form [68, 69]:

$$\lambda_{max}=0.66+0.509\exp[(2100-T)9.64\cdot10^{-4}]. \tag{4.115}$$

The application of this equation is limited to 1200 K$\leq$T$<$3200 K. Fairly precise calculations needed in engineering design can be carried out according to the spectral composition of 90% of the total energy emitted by the infrared source. Under such conditions the following empirical equations hold true, with account taken of Eq. 4.115 [68, 69]:

$$\lambda_1=0.52\lambda_{max};$$
$$\lambda_2=(4.27\div6.2)^{10}10^{-4}T. \tag{4.116}$$

Next we shall consider how Eqs. 4.91-4.100 can be used for calculating the integral radiation field in the case of typical capillary-porous foodstuffs: potato starch and bread crumbs. Let us assume that an optically infinitely thick plane layer $(L\ell\to\infty)$ is subjected to unilateral irradiation with a diffuse flux.

With the aid of Eqs. 4.104-4.109 we can determine the magnitudes of $\bar{k}(x)$ and $L(x)$, which can then be used to calculate the effective coefficient of extinction $\varepsilon_{ef}(x)$, the backward scattering coefficient $s(x)$, and the average effective survival probability of the photon $\Lambda_{ef}(x)$ (Schuster number) as a function of x (Fig. 4.8). It can be seen that Λ_{ef} increases with an increase in depth, and at large values of x approaches unity; i.e., pure scattering begins. In an adjacent zone $\ell<1.0$ mm the average absorption coefficient $\bar{k}(x)$ decreases with a decrease in the coordinate x (the maximum of $\bar{k}(x)$ lies at x=0).

The analogous dependences of the integral coefficients $\bar{k}(x)$, $L(x)$, $s(x)$, $\varepsilon_{ef}(x)$ on x are obtained for a whole number of strongly scattering foodstuffs ($\Lambda_{ef}>$$>0.8$) both of animal and plant origin subjected to high-temperature infrared and solar irradiation [39, 44].

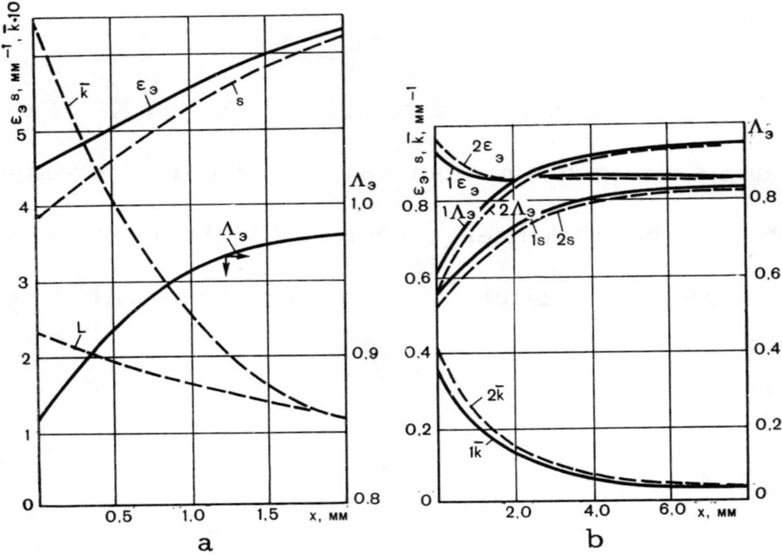

Figure 4.8 Integral optical properties of potato starch (a) and bread crumbs (b) subjected to infrared irradiation as a function of depth x at different temperatures of the filament [16, 39]: 1 – T=2400 K; 2 – T=2600 K.

These results can be used to average the process of the propagation of integral high-temperature radiation in foodstuffs and to develop sufficiently precise analytical relationships for calculating $E(x)$ and $q(x)$, with account taken of the changes in the optical properties of a substance as a function of the coordinate x.

Let us compare the following three methods for calculating the transfer of infrared radiant energy in foodstuffs:

– method of integration with respect to λ of the equations for w_λ, $E_{\lambda 0}$, and q_λ;

– method based on the variable optical properties which take into account changes in the average integral optical properties and the spectral composition of $E_{\lambda 0}$, w_λ, and q_λ with a change in the coordinate;

– method based on the optical and thermal radiational characteristics averaged according to the spectrum of the infrared source within the limits λ_1-λ_2: these characteristics are used in spectral equations in place of w_λ, $E_{\lambda 0}$, and q_λ.

Of these three methods the most precise and also the most time-consuming is the integration method. As the first approximation, which simplifies the integration of the complex functions w_λ, $E_{\lambda 0}$, and q_λ according to the spectrum, we replace integration by summation with a final spectral interval $\Delta\lambda$. In this case it suffices to consider $\Delta\lambda$ to be equal to 0.1 μm. Narrowing $\Delta\lambda$ to 0.05 μm does not improve the results of the calculation. This is due to the fact that the linear dispersion of the monochromators in spectrophotometers in the infrard spectral region 1.0-15.0 μm is 0.100-2.00 μm/mm, while in the investigation of optically scattering media the width of the slit used is 0.5-2.0 mm.

The simplest, but least precise method is that based on the averaged optical and thermal radiational characteristics. According to this method, in the case of integral irradiation of a semi-infinite layer $w(x)$, $E_0(x)m$ and $q(x)$ are determined with the aid of Eqs. 2.39, 2.40, and 2.32, respectively. To do this the spectral quantities $\bar{k}_\lambda$, L_λ, and $R_{\lambda\infty}$ are replaced with the average integral characteristics $\bar{k}$, L, and R_∞, which are independent of x.

The optical integral characteristics $\bar{k}$, s, and L are determined from Eqs. 3.10-3.12 in which the integral thermal radiational characteristics T and R of a layer of finite thickness and R_∞ are used instead of R_λ, T_λ, and $R_{\lambda\infty}$. These characteristics are determined by averaging the magnitudes of R_λ, T_λ, and $R_{\lambda\infty}$ with respect to the spectrum of the incident radiation flux E from the chosen infrared emitter. Owing to the selective nature of the spectral thermal radiational characteristics of foodstuffs and the emissive characteristics of the infrared source the integral hemispherical thermal radiational properties of one and the same substance will be different with different emitters. These properties should be calculated according to the following approximation:

$$A = \left[\int_{\lambda_1}^{\lambda_2} E_\lambda A_\lambda \, d\lambda\right] / \left[\int_{\lambda_1}^{\lambda_2} E_\lambda \, d\lambda \cong \left[\sum_{\lambda_1}^{\lambda_2} E_\lambda A_\lambda \Delta\lambda\right] / \left[\sum_{\lambda_1}^{\lambda_2} E_\lambda \Delta\lambda\right]; \qquad (4.117)$$

where, A_λ - spectral hemispherical quantity (A_λ, R_λ, T_λ, $R_{\lambda\infty}$) except for the emissivity factor ε_T which is averaged with respect to the spectral composition of the incident radiation flux from the chosen emitter; E_λ - spectral radiation flux which depends on the properties of the chosen infrared emitter and the irradiation conditions.

The method based on averaged characteristics is an approximate one since it does not take into account the changes in the spectral composition of the radiation resulting from multiple selective scattering and from the absorption of radiation inside the layer. The assumption that L, $\bar{k}$, and s are constant throughout the thickness of the layer is not confirmed by experimental results (see Fig. 4.8).

The method based on the variable optical properties, proposed by the authors, makes it possible with the aid of analytical relationships, Eqs. 4.101-4.103, to calculate without integration and with a sufficient degree of precision the integral radiation fields for selectively absorbing substances.

According to this method the optical properties of the layer, depending on the coordinate x, are characterized by coefficients $\bar{k}(x)$ and $s(x)$ (which are variable with respect to x) averaged according to the spectral composition and the hemisphere. These coefficients contain the overall information on the optical distribution of energy and the spectral composition of the integral radiation at depth x.

By substituting the optical depth τ for the variable x according to Eq. 4.77, and the optical thickness τ_ℓ (determined from Eq. 4.77) for the thickness of the layer ℓ at $x=\ell$, and by introducing the Schuster number $\Lambda_{ef}(\tau)$, we obtain a system of linear differential equations containing one variable coefficient $\Lambda_{ef}(\tau)$. These equations can be used in calculating the dimensionless net radiation flux $q^* = E_+^* - E_-^*$ and the spatial irradiance $E_0^* = E_+^* + E_-^*$ for integral radiation of the type defined by Eq. 4.79 under the boundary conditions expressed by Eqs. 4.80 and 4.81.

In the case of integral radiation under these boundary conditions it is necessary to substitute for R_λ and T_λ the integral reflectance and transmittance of a layer averaged with respect to the spectrum of the corresponding integral flux E_i incident upon the layer.

The dependence of the average effective survival probability of the photon $\Lambda_{ef}(\tau)$ on the optical depth τ for substances subjected to unilateral irradiation with a high-temperature infrared source is approximated by Eq. 4.83. Constants a and b in this equation stand for the dimensionless absorption coefficient $a=\overline{k}^*(0)$ and the dimensionless extinction coefficient $b=\overline{k}^*(0)+s^*(0)$. These coefficients are determined for a layer of unit thickness by averaging the spectral coefficients $\overline{k}_\lambda$ and s_λ with respect to the spectral composition of $E_0^*(0)$ and $q^*(0)$ at the layer's boundary at $\tau = 0$.

In a general case of bilateral irradiation of a plane layer the dependence of $\Lambda_{ef}(\tau)$ can be described by Eq. 4.85. Here, a, b, and c are constants which depend on the spatial distribution and spectral composition of the incident integral radiation fluxes E_1^* and E_2^*.

A general solution of Eq. 4.79 with $\Lambda_{ef}(\tau)$ functions of the type defined by Eqs. 4.83 and 4.85 in the case of unilateral and bilateral irradiation is given in Eqs. 4.88-4.96. Therefore, the equations for $w_\lambda(x)$, $E_{\lambda 0}(x)$, and $q_\lambda(x)$ can be used in calculating $w(x)$, $E_0(x)$, and $q(x)$ in the case of integral radiation.

The change in the spectral composition of the radiation with a change in the coordinate is calculated with the aid of the optical depth τ and the variable average integral optical characteristics $\overline{k}(x)$ and $\varepsilon_{ef}(x)$, which determine the type of $\Lambda_{ef}(\tau)$ function. For this it is necessary to determine at x=0 the constants $a=\overline{k}^*(0)$ and $b=\varepsilon_{ef}(0)$ for a layer of unit thickness. This is done by averaging $\overline{k}_\lambda$ and L_λ with respect to the spectral composition of $E_0^*(0)$ and $q^*(0)$, which at x=0 are determined from Eq. 4.80 with the aid of data on R and T. These two quantities can be calculated from Eqs. 4.97 and 4.98 or experimentally, with account taken of scattering (see Chapter 5).

In practice the thickness of a layer of irradiated foodstuffs is usually such that the layer is not transparent, and the incident flux is partially reflected and absorbed by the substance. In this case for an optically infinitely thick layer the net radiation flux q, the spatial irradiance E_0, and the energy absorbed per unit time, w, are determined according to Eqs. 4.99 and 4.100 at $\tau_\ell \to \infty$ as follows:

$$q^*(\tau)=E^*(1-R_\infty)(1+\tau/\varepsilon^*)^{-1/4}\exp[-2(2\overline{k}^*\varepsilon^*)^{\frac{1}{2}}(1+(\tau/\varepsilon^*))^{\frac{1}{2}}-1]; \qquad (4.118)$$

$$E_0^*(\tau)=E^*(1+R_\infty)\{1+4[2\overline{k}^*\varepsilon^*(1+\tau/\varepsilon^*)]^{\frac{1}{2}}\}/4\overline{k}^*(1+\tau/\varepsilon^*)^{1/4}.$$

$$\exp[-2(\overline{k}^*\varepsilon^*)^{\frac{1}{2}}(1+(\tau/\varepsilon^*)^{\frac{1}{2}}-1)]\}; \qquad (4.119)$$

$$w^*(\tau)=[1-\Lambda(\tau)]E_0^*(\tau); \qquad (4.120)$$

where, R_∞ – reflectance of an optically infinitely thick layer:

$$R_\infty=\{4[(2\overline{k}^*\varepsilon^*)^{\frac{1}{2}}-\overline{k}^*]+1\}/\{4[(2\overline{k}^*\varepsilon^*)^{\frac{1}{2}}+\overline{k}^*]+1\}. \qquad (4.121)$$

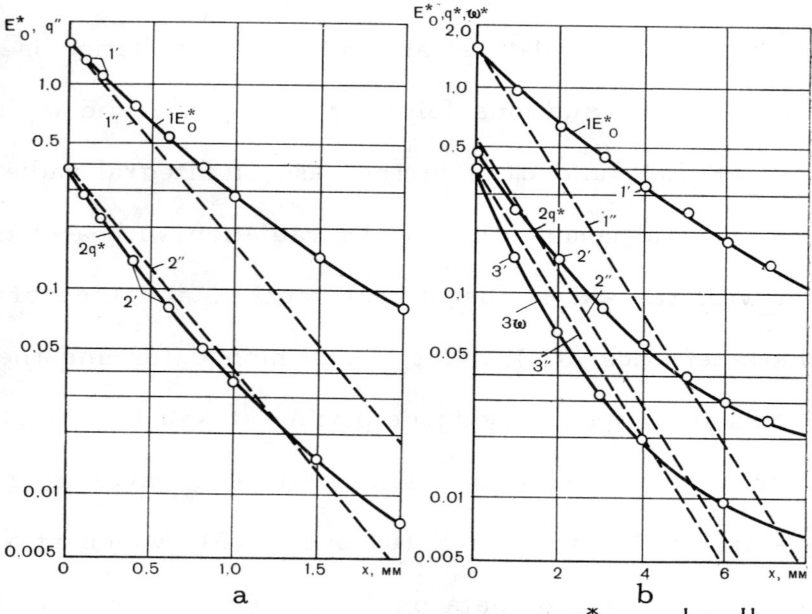

Figure 4.9 Changes in the spatial irradiance $E_0^*(1, 1', 1'')$, the net radiation flux $q^*(2, 2', 2'')$, and the absorbed energy $w^*(3, 3', 3'')$ of integral radiation in potato starch (a) and bread crumbs (b) [39, 44]: 1,2,3 – calculated according

to the method based on variable optical properties with the aid of Eqs. 4.118-4.120; $1',2',3'$ - calculated according to the numerical integration method; $1'',2'',3''$ - calculated according to the method based on averaged optical and thermal radiational characteristics.

Eqs. 4.118 and 4.119 make it possible to calculate the integral radiation field in absorbing-scattering substances without integrating the analytical spectral functions $q_\lambda(x)$ and $E_{0\lambda}(x)$ with respect to x. In these calculations the changes in the average integral optical properties are accounted for with the aid of the function $\Lambda_{ef}(\tau)$ and the optical depth τ. It is sufficient to determine only the quantities $a=\bar{k}_o^*$ and $b=\bar{k}_o^*+s_o^*$ by averaging the spectral coefficients $\bar{k}_\lambda$ and s_λ for a given substance according to the spectral composition $q(0)$ and $E_0(0)$ at x=0. The constants a and b determined in this way at x=0 can be used to calculate from Eqs. 4.93, 4.97, and 4.98 the integral thermal radiational characteristics R_∞, R, and T for a substance having variable absorption and scattering coefficients, as well as to determine the functions $w(\tau)$ and $E_0(\tau)$, $q(\tau)$ with the aid of Eqs. 4.90-4.100.

To assess the precision of the proposed method and of the analytical equations which characterize the distribution of the integral infrared flux in selectively absorbing-scattering substances calculations were carried out of R_∞, $E_0(x)$, and $q(x)$ for potato starch and bread crumbs (Fig. 4.9 and Table 4.2).

By using experimental data [39, 44] for potato starch subjected to irradiation with a KGT-220-1000 lamp at x=0, $k^*(0)=0.641$, and $s^*(0)=3.862$, the integral reflectance $R_\infty=0.612$ is obtained from Eq. 4.121, which differs by 0.003 only (error $\sim 0.5\%$) from the integral reflectance $R_\infty=0.615$ obtained by integrating the spectral reflectance $R_{\lambda\infty}$ for the entire spectrum of the infrared lamp. The precise values for $E_0^*(x)$ and $q^*(x)$ (curves $1'$ and $2'$) were calculated by integrating the analytical expressions for the dimensionless spectral net radiant flux $q_\lambda^*(x)$ and the spectral dimensionless spatial irradiance $E_{0\lambda}^*(x)$ for the entire spectrum of the infrared lamp at fixed values of the coordinate x.

From Fig. 4.9 it can be seen that at the layer's boundary x=0 and at different values of x, the values of E_0^* and q^* obtained by integration with respect to the spectrum differ on the average by 1.0% from those calculated with the aid of Eqs. 4.118 and 4.119. This shows that the proposed method is highly precise for solving problems of energy transfer in absorbing-scattering substances with variable optical properties.

Table 4.2   Integral reflectance R_λ of foodstuffs subjected to irradiation with a KGT-220-1000 lamp [16, 39, 44]

Foodstuffs	Temperature of emitter T(K)	R_∞ (Eq.4.121)	R_∞ obtained by integrating the entire spectrum
Potato starch	2500	0.612	0.615
Bread crumbs	2400	0.526	0.519
Bread crumbs	2600	0.537	0.536

The approximate values of E_0^* and q^* were determined from Eqs. 2.39 and 2.32 with the aid of the optical properties $\bar{k}$, s, and L and the thermal radiational characteristics T, R, and R_∞ averaged with respect to the spectrum of the KGT-220-1000 lamp.

For a layer $\ell=1.0$ mm of potato starch (w≅11.8%) irradiated with an NIK lamp (T=2250 K) R=0.595 and T=0.066; and at $\ell=10.0$ mm (an optically infinitely thick layer) $R_\infty=615$. The optical integral characteristics determined from Eqs. 3.10-3.12 are: L=2.264 mm^{-1}. $\bar{k}=0.54$ mm^{-1}, s=4.73 mm^{-1}. These data can be used in calculating the approximate energy distribution in potato starch.

A comparison of the approximate and precise values of E_0^*, w^*, and q^* for potato starch (Fig. 4.9 a) shows that the former are determined with an error of only 10-20% when the values w^* and q^* are increased by two orders of magnitude (from 0.01 to 1.00). For the spatial irradiance E^* the error is large, reaching 45.5% at x=1.0 mm and >500% at x=0.2 mm. The nature of the dependence of E_0^*

on x differs considerably from the exponential law.

For bread crumbs the values of $E_0(x)$, $w(x)$, and $q(x)$ calculated according to the approximate method (Fig. 4.9 b, dashed lines) differ from those determined by integration: at $x=0$. $\Delta E=8.9\%$, $\Delta w=9.7\%$, and $\Delta q=3.7\%$; and at $x=2.0$ mm, $\Delta E=51.6\%$, $\Delta w=31.0\%$, and $\Delta q=15.2\%$. Thus, the nature of the dependence $E_0(x)$ and of $q(x)$ differs considerably from the exponential dependence. And the function of distribution of the absorbed energy $w(x)$ at a depth up to 4.5 mm is satisfactorily approximated by the exponential dependence (error $<31\%$). This indicates that the approximate method for determining $w(x)$ can be used in calculating radiant heat transfer when designing drying units. However, the approximate methods for calculating $E_0(x)$ and $q(x)$ with the integral characteristics L and k are not sufficiently precise (the error should not exceed 5-10%) for engineering design purposes.

In the case of bread crumbs the method based on variable optical properties is considerably simplified since $\varepsilon_{ef}(x)$ is only slightly dependent on x. An analysis of the dependence $\varepsilon_{ef}(x) = f(x)$ (Fig. 4.9 b) shows that the integration of $\tau = \int_0^x \varepsilon_{ef}(x)dx$ with an error $<0.5\%$ can be replaced by the product $\tau = \varepsilon_{av}x$. Then, for the purpose of calculating the integral radiation field in a layer of bread crumbs irradiated by a quartz lamp Eqs. 4.118 and 4.119 can be rewritten as follows:

$$E_0(\tau)=[E(1+\bar{R}_\infty)]\{[1+4(2\bar{k}\varepsilon_{av}(1+x)^{\frac{1}{2}}]/4\bar{k}(1+x)^{1/4}\}\cdot$$

$$\exp[-2(2\bar{k}\varepsilon_{av})^{\frac{1}{2}}(1+x)^{\frac{1}{2}}-1]; \tag{4.122}$$

$$q(\tau)=[(1-\bar{R}_\infty)/4\bar{k}(1+x)^{1/4}\exp[-2(2\bar{k}\varepsilon_{av})^{\frac{1}{2}}(1+x)^{\frac{1}{2}}-1]. \tag{4.123}$$

Here, $\bar{k}$, ε_{av}, and x - dimensionless quantities. The results of the calculations based on Eqs. 4.122 and 4.123 accord within 1% with the values of $E_0(x)$ and $q(x)$ obtained by integrating the spectral functions $E_{\lambda 0}(x)$ and $q_\lambda(x)$ according to the spectrum of the radiation source.

We may conclude, therefore, that the method based on the averaged optical and thermal radiational characteristics can be used only for purposes of approximate estimations in engineering design work. To obtain more precise data it is advisable to use the method of integration or the method based on variable optical properties which take into account changes in the average integral optical properties with a change in the coordinate.

4.6 Calculating the integral thermal radiational characteristics of foodstuffs

The integral hemispherical thermal radiational characteristics of one and the same substance vary, depending on the emitter used. This is due to the selectivity of the spectral thermal radiational characteristics of the substance and the emissive characteristics of the infrared source. They also depend on the irradiation conditions and are determined with the aid of an approximate relationship (Eq. 4.117).

The results of the calculations based on this equation accord well with experimental data. Thus, for a layer of bread 11.0 mm thick irradiated with a KGT-220-1000 lamp ($T=2500$ K) the transmittance according to Eq. 4.117 is 0.81% as compared with 0.79% obtained experimentally by irradiating the layer with a high-temperature emitter.

The emissivity ε of a plane layer having finite thickness for a medium with selective optical properties does not equal the absorptance A of the layer averaged according to Eq. 4.117. Indeed, A is the absorptance of the layer with respect to the incident radiation flux emitted by a selected infrared lamp at temperature T_1, while $\varepsilon(T_2)$ is the emissivity of the layer. This emissivity equals numerically the layer's absorptance in relation to the blackbody radiation at a temperature which is the same as the surface temperature of the layer, T_2. Therefore, the emissivity can be determined from the following equation:

$$\varepsilon(T_2) = [\int_{\lambda_1}^{\lambda_2} R_{\lambda a}(T_2) A_\lambda d\lambda] / [\int_{\lambda_1}^{\lambda_2} R_{\lambda a}(T_2) d\lambda] \cong [\sum_{\lambda_1}^{\lambda_2} R_{\lambda a}(T_2) A_\lambda \Delta\lambda] /$$

$$\sum_{\lambda_1}^{\lambda_2} R_{\lambda a}(T_2) \Delta\lambda] . \tag{4.124}$$

As pointed out above, and as it can be seen from Eqs. 4.117 and 4.124, the emissivity of a substance depends on its temperature and optical properties, while its absorptance depends also on the spectral composition of the incident radiation flux.

The spectral radiation coefficient r_λ and the spectral absorption coefficient k_λ represent the internal characteristics of the optical properties of a substance, and therefore they depend on the structure and the state of the substance. As a result, the independence of the equalities $r_\lambda/k_\lambda=r_{\lambda a}$ and $R_\lambda/A_\lambda=R_{\lambda a}$ (here, b - blackbody) of the optical properties of the substance itself, an independence established under the condition of equilibrium between the radiation and the substance, holds true even in the absence of such an eqilibrium provided that the equilibrium has no significant effect on the state of the substance.

For the integral or averaged emissivity $\varepsilon(T)$ and the absorptance A of a layer of substance the ratio $\varepsilon(T)/A$ does not remain constant and depends on the spectral thermal radiational characteristics of the layer and on the temperature of the layer and the emitter. We shall consider the dependence on the temperature of the emitter by looking at a concrete example.

Starch with 11.8% moisture content at T=293 K has the following properties:

- absorptance $A_1=0.385$, averaged according to the spectrum of a KGT-220-1000 lamp at $T_1= 2500$ K;

- absorptance $A_2=0.912$, averaged according to the spectrum of a metallic plate emitter at $T_2=590$ K;

- emissivity $\varepsilon(T_1)=0.94$, obtained by averaging according to Eq. 4.124 from the spectrum of blackbody radiation at $T_2=293$ K.

It can be seen that $A_1 \neq A_2 \neq \varepsilon$, and that the ratio $\varepsilon/A_1=2.44$ (i.e. it is not unity). This means that the application of Kirchhoff's law to the integral characteristics must be well-founded. Thus, if the temperature of the emitter is close to that of a blackbody; i.e., if it is close to the state of thermodynamic equilibrium, the ratio ε/A approaches unity. But if $T_1 \neq T_2$. $\varepsilon/A \neq 1$. Therefore, one must exercise care when using published data on the emissivity of capillary-porous colloidal substances.

The integral (averaged) thermal radiational characteristics obtained with the aid of Eqs. 4.117 and 4.124 can be used only in estimating the magnitudes of the incident radiation flux E, the net radiation flux E_n and the effective radiation flux E_{ef}. This is due to the fact that when calculating heat transfer between substances having selective radiation properties it is necessary to take into account changes in the absorptance during multiple reflection. Each act of the reflection of energy from the surface of the emitter, from the walls of the infrared unit and from the irradiated substance is accompanied by changes in the spectral composition of the incident radiation. And this leads to changes in the total absorptance of the substance.

Table 4.3 Total absorptance of potato starch with compensation (A_1) and without compensation (A_2) for reflection [39].

Infrared unit	A_1	A_2	$\delta(\%)$
KGT-220-1000 with reflector (T_1=2500K, R_1=0.91)	0.294	0.385	31.0
Ni-Cr spiral with reflector (T_1=1033K, R_1=0.5)	0.802	0.820	2.2
Metallic plate (T_1=570 K, R_1=0.2)	0.900	0.912	1.3

Estimations of E_n with and without compensation for the selectivity of the thermal radiational characteristics of substances [26, 39, 44] show that the error can be as high as 20-30%. For substances with strongly selective optical properties the error can be even greater.

A precise determination of E_n and E_{ef} requires the integration of the spectral quantities $E_{\lambda n}$ and $E_{\lambda ef}$ throughout the spectrum, which are calculated with account taken of multiple reflection for each wavelength. However, this makes the calculations too complicated.

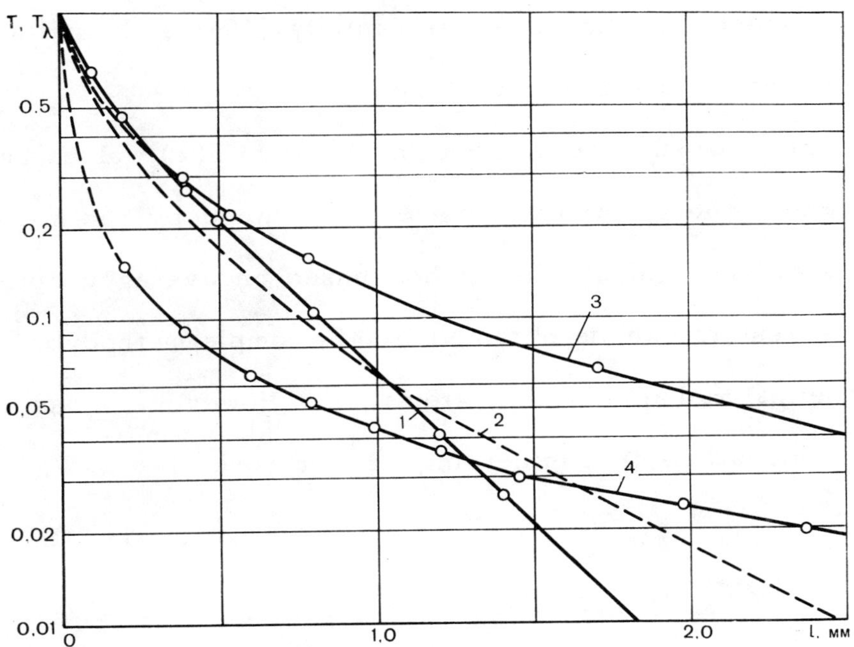

Figure 4.10 Dependence of the total transmittance $T(\theta;2\pi)$ and spectral transmittance $T_\lambda(\theta;2\pi)$ of potato starch (w=11.8%) on the thickness of a layer ℓ (mm) irradiated by different high-temperature emitters [39]: 1,2 - KGT-220-1000 lamp (1 - calculated by the method based on averaged characteristics; 2 - calculated by numerical integration); 3 - at λ=1.1 µm; 4 - 250 w Philips lamps [184].

The results of the calculations of the total absorptance of potato starch are summarized in Table 4.3. It can be seen that the absorptance depends largely on the spectral composition of the incident flux and on the irradiation conditions which determine the extent of multiple reflection.

The simple method based on the averaged characteristics L, $\overline{k}$, and R_∞ can be used (Eq. 3.2) to calculate the dependence of the total transmittance T on the thickness of the layer. The results of such calculations for potato starch (Fig. 4.10) show that this method can be used in approximate calculations of the integral quantities R, T, and A for foodstuffs.

Experimental results [184] (Fig. 4.10, curve 4), obtained by using a 250 w lamp with no account taken of the scattering of radiation by the sample, are smaller than the calculated values for a thin layer ($\ell < 1.6$ mm). The nature of the $T(\ell)$ dependence in this instance is similar to the $T_\lambda(\ell)$ dependence at $\lambda = 1.1$ μm (Fig. 4.10, curve 3), which is obtained experimentally [184] and with account taken of the scattering of radiation by the sample.

More precise results, which practically coincide with $T(\ell)$ values calculated by integration throughout the spectrum, have been obtained by the method based on variable optical properties and by the method based on averaged optical properties. Less precise results can be obtained by the simple method based on averaged optical and thermal radiational characteristics. However, it can be used only for approximate estimations in engineering design work.

SPECTRAL INSTRUMENTS FOR MEASURING THE THERMAL RADIATIONAL CHARACTERISTICS OF FOODSTUFFS

As it follows from the physical principles discussed above for measuring the thermal radiational characteristics of foodstuffs, the reflectance, transmittance, and absorptance of foodstuffs are largely determined by their optical properties: and, depending on the irradiation conditions, their magnitudes can vary by more than 20% (Tables 3.1-3.4, 3.6).

In order to describe the thermal radiational characteristics in relation to the irradiation conditions a dual terminology for R_λ and T_λ is introduced. It characterizes the irradiation conditions and the optical properties of the sample, which depend on the conditions employed for measuring them. The first part of the term stands for the irradiation conditions, and the second - for the conditions used in carrying out the measurements. As can be seen from the data in Table 1.2 and Fig. 1.7, the reflectance and transmittance of light-scattering substances can in principle be measured by nine different methods. A classification of these methods is given in Table 5.1 in which are shown the measured values of R_λ and T_λ and their designation according to Table 1.2.

Under this classification there are three main ways of irradiating a substance: by directional $F(\theta')$, diffuse $F(\omega')$, and hemispherical $F(2\pi)$ radiant fluxes. And as there are three different ways of measuring these fluxes one can speak of nine possible methods for measuring R_λ and as many methods for measuring T_λ.

5.1 Specific features of measuring the thermal radiational characteristics of foodstuffs

In order to measure the thermal radiational characteristics of foodstuffs it is necessary to take into account their specific physico-chemical and biological properties. Such properties are the basic indicators of their quality and set limits on the choice of the irradiation parameters. In practice consideration is not always given to the thermolability of foodstuffs, the continuity and irreversibility of biochemical processes, the comparability of large optical inhomogeneities with the cross section of incident irradiation, and the strong scattering of infrared radiation.

Table 5.1 Classification of methods for measuring of the reflectance and transmittance of foodstuffs [39, 54, 163].

Incident radiation	Methods for measuring reflected and transmitted:		
	directional radiation	diffuse radiation	hemispherical radiation
Directional	I. Bidirectional reflectance $R_\lambda(\theta,\phi; \theta_R, \phi_R)$ or transmittance $T_\lambda(\theta, \phi; \theta_T, \phi_T)$.	II. Directional-diffuse reflectance $R_\lambda(\theta, \phi: \omega_R)$ or transmittance $T_\lambda(\theta, \phi; \omega_T)$.	III. Directional-diffuse reflectance $R_\lambda(\theta,\phi; 2\pi)$ or transmittance $T_\lambda(\theta, \phi; 2\pi)$
Diffuse	IV. Diffuse-directional reflectance $R_\lambda(\omega; \theta_R, \phi_R)$ or transmittance $T_\lambda(\omega; \theta_T, \phi_T)$.	V. Bidiffuse reflectance $R_\lambda(\omega; \omega_R)$ or transmittance $T_\lambda(\omega; \omega_T)$.	VI. Diffuse-hemispherical reflectance $R_\lambda(\omega; 2\pi)$ or transmittance $T_\lambda(\omega; 2\pi)$.
Hemispherical	VII. Hemispherical-directional reflectance $R_\lambda(2\pi; \theta_R, \phi_R)$ or transmittance $T_\lambda(2\pi; \theta_T, \phi_T)$.	VIII. Hemispherical-diffuse reflectance $R_\lambda(2\pi; \omega_R)$ or transmittance $T_\lambda(2\pi; \omega_T)$.	IX. Bihemispherical reflectance $R_\lambda(2\pi; 2\pi)$ or transmittance $T_\lambda(2\pi; 2\pi)$.

The data obtained in experimental investigations of the degree of scattering and spreading of the cross section of directional radiation in foodstuffs are of paramount importance. They are necessary for selecting the optico-geometric parameters of the measuring system and of the radiation detector in various attachments for spectrophotometers. As can be seen from data in Figs. 2.14 and 2.15, the cross section of a narrow directional radiation flux in foodstuffs becomes strongly diffused. The effect of such flux spreading should be taken into account when deciding on the dimensions of the samples of foodstuffs in order to prevent energy losses due to the spreading of the radiation cross section penetrating the sample.

Thus, the following properties of foodstuffs should be considered when working out methods for measuring their thermal radiational characteristics:

- thermolability, which rules out the use of intensive integral radiation flux during measurements;

- irreversibility and continuity of biochemical processes, which require that measurements of all the thermal radiational characteristics should be carried out in a single experiment and simultaneously;

- strong scattering of radiation and the spreading of the cross section of directional radiation flux, which set limits to the minimal dimensions of the sample, the radiation detector, and the dimensions of the hemispherical and spherical optical components in measuring devices;

- large optical inhomogeneities, which must be comparable with the cross section of the incident radiation flux; this requires that wide fluxes be used or that a sample of extended length be moved relative to the incident flux;

- dependence of the thermal radiational characteristics on the irradiation conditions; this requires that the measurements be made under different irradiation conditions corresponding to those used in industrial infrared units. In designing thermal radiational units we must know the hemispherical characteristics $R_\lambda(2\pi)$

and $T_\lambda(2\pi)$ of foodstuffs under different irradiation conditions $(\theta\,;\,\omega;2\pi)$. An effective method for carrying out numerous measurements of R_λ and T_λ is one which meets the following conditions:

- the sample should be exposed to a monochromatic flux; this reduces to a minimum the heat-up of the sample and radiation-induced changes in the physico-chemical properties of foodstuffs;

- the incident radiation should be modulated so that the radiation coming from the sample will have no effect on the measurements;

- the radiation detector should register most ($>80\%$) of the total radiant energy reflected or transmitted by the layer;

- the method should permit direct measurements of diffuse reflection or transmission without the use of any standards; the data obtained should be easy to process.

To develop new methods for measuring the spectral thermal radiational characteristics of light-scattering substances subjected to directional irradiation we should analyze the distribution of a narrow parallel radiation flux in these substances. A characteristic feature of the distribution is the rapid and nearly complete transformation of the narrow radiation flux into diffuse radiation due to multiple scattering (Fig. 2.14). This is accompanied by an extensive spreading of the cross section of the radiation flux and a considerable increase in the cross section of the flux leaving the sample. Multiple scattering also causes an increase in the cross section of the flux reflected by the sample (Fig. 2.15).

A special device has been developed [118] for investigating the radial dependence of a radiation flux which is reflected and transmitted by the sample. It consists of a variable iris diaphragm and a photometric sphere (Fig. 5.1).

The iris diaphragm (2), equipped with blackened metallic blades 0-2 mm thick, is used to vary the diameter of the cross section of the reflected or transmitted radiation flux from 3 to 40 mm. After passing through the diaphragm the flux en-

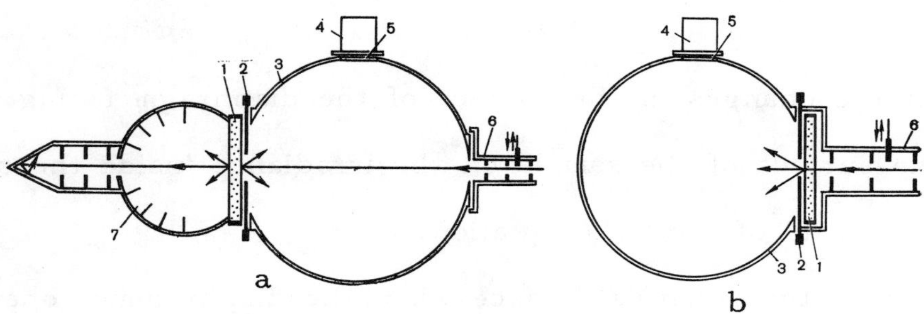

Figure 5.1 Schematic diagrams of a setup for investigating the radial dependence of a radiation flux reflected (a) and transmitted (b) by a light-scattering medium [39, 50, 118].

ters the photometric sphere (3) which eliminates zonal nonuniformity of the sensitivity of the radiation detector (4). To provide diffuse scattering inside the photometric sphere (200 mm in diameter) it is coated with $BaSO_4$. There are three openings in the sphere. In one opening, 54 mm in diameter, is a holder (1) for samples, 30 mm in diameter and 10 mm thick. In a second opening, 30 mm in diameter, is a radiation detector (4) – photometer FEU-62 – operating in the range of 0.4–1.2 μm. The opening in front of the detector is closed with semitransparent glass (5) which provides diffuse scattering. The intensity of the signal from the detector is measured with the aid of a UF-206 reading device, which can record currents up to 1 μA.

The unit can operate with any radiation source which provides a narrow directional flux. The effective cross section of the reflectance and transmittance fluxes is measured with the aid of monochromators in conjunction with spectrophotometers VSU-2P and SF-4A, as well as an optical quantum generator LG-56. This generator gives a narrow directional monochromatic flux with a cross section 2 mm in diameter at wavelength 0.63 μm and with a very low angular divergence. The radiant flux after passing through the system of diaphragms (6) 3 mm in diameter, the iris diaphragm (2), and the sample, is absorbed by a blackbody (7). To measure the radial dependence of the radiant flux reflected and transmitted by the sample the setup is arranged according to diagrams a and b in Fig. 5.1, respectively.

The effect of the changes in the surface of the diaphragm facing the sphere's interior and in the surface of the sample on the irradiance inside the sphere can be estimated with the aid of Taylor's equation.

The irradiance of the sphere's surface when the diaphragm is open, E_1, and when the diaphragm is closed, E_2, can be described as follows:

$$E_1=[R/(1-\phi_1 R)](F/S; \quad E_2=[R/(1-\phi_2 R)]F/S; \tag{5.1}$$

where, $\phi_1=1-(1/S)[(1-R)S+(1-R_D)S_D+(1-R_0)S_0++(1-R)S];$ (5.2)

$$\phi_2=1-(1/S[(1-R)S+(1-R_D)S_D'+(1-R_0)S_0'+(1-R)S]; \tag{5.3}$$

here, S_D, S_D' – surface area of the opened and closed diaphragm, respectively.

The effect of the changes in the area of the diaphragm on the irradiance inside the sphere is expressed by the equation:

$$E_1/E_2=(1-\phi_2 R)/(1-\phi_1 R). \tag{5.4}$$

The results obtained with the aid of Eq. 5.4 accord well with experimental data. They both show that the irradiance inside the sphere varies by only 0.8% during a change in the reflection of the MS-14 standard and a maximal change in the diameter of the diaphragm from 3 to 30 mm. This is due to the small magnitude of the ratios S_d/S and $S_0/S\ll 1$. The irradiance increases until the area of the diaphragm opening equals that of the effective cross section of the radiation flux transmitted by the sample. The radiant flux is measured at fixed openings of the diaphragm at intervals of 2 mm from 3 to 30 mm.

The relative flux density $j_i(R,T)$ of the radiation which is reflected by the ring-shaped element of the sample's surface $\Delta S_i=S_i-S_{i-1}$ contained between the radii $r_i>r_{i-1}$, $\Delta R_\lambda(\theta; 2\pi)$, is calculated according to the following equation:

$$j_i^*(R,T) = (1/j_0)\{[N_i - N_{(i-1)}]/[S_i - S_{(i-1)}]\};\qquad (5.5)$$

where, N_i, $N_{(i-1)}$ – signals of the radiation detector when the areas of the diaphragm opening are S_i and $S_{(i-1)}$, respectively, which correspond to the radii r_1 and $r_{(i-1)}$, respectively; $j_0 = N_0/S_0$ – relative flux density of the radiation passing through the diaphragm with a radius $r_0 = 3$ mm which corresponds to the area S_0 of the opening.

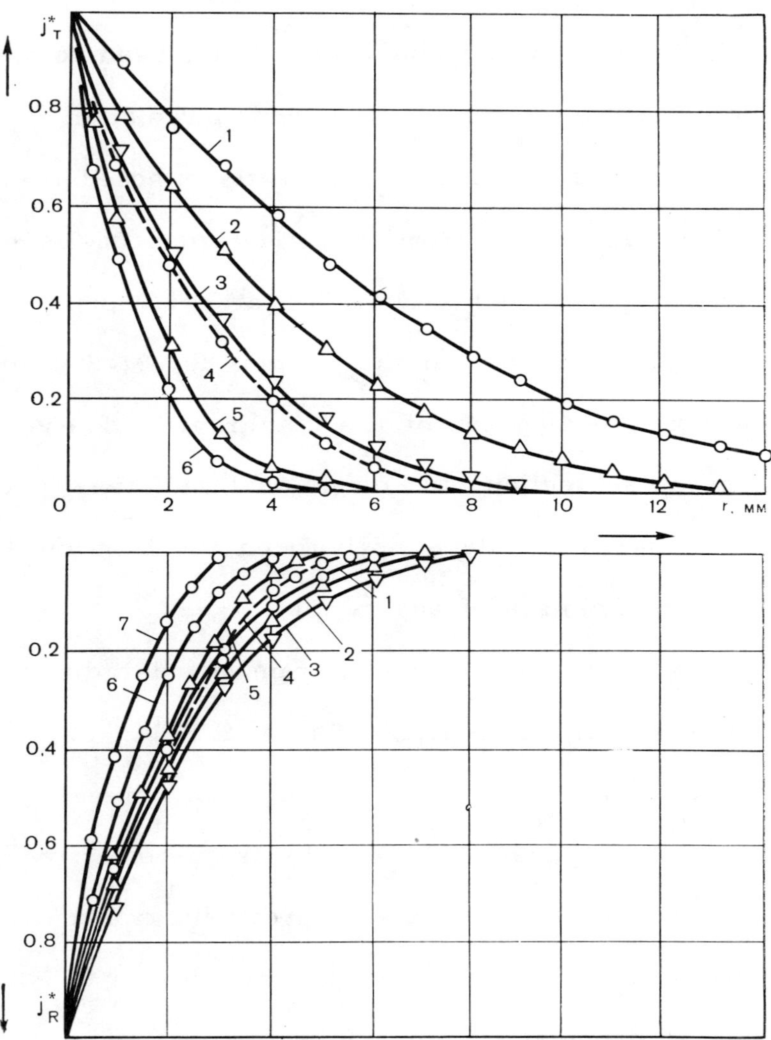

Figure 5.2 Flux density of transmitted radiation j_T^* and reflected radiation j_R^* at $\lambda = 0.63$ µm as a function of the radius of the effective cross section for a reference reflector [39, 50, 118]: 1-4 – polytetrafluoroethylene-4 (ℓ: 1 – 9.0, 2 – 5.0, 3 – 3.1, 4 – 2.2 mm); 5 – enamel BM-548 ($\ell = 0.25$ mm); 6 – paper ($\ell = 0.09$ mm, 75g/m^2); 7 – opal glass MS-14.

To work out the method and check the reproducibility of the results we have taken several typical scattering materials: reflection standards (opal glass MS-14), polytetrafluoroethylene-4, enamel BM-548, and 0.09 mm-thick paper (75g per m^2), whose optical properties can be considered constant durin the experiment. A comparison of the data in Fig. 2.15 and 5.2 shows that when the dependences of j on r are of the same nature the effective cross sections of the reflected and transmitted radiation fluxes ($\lambda=0.63$ μm) for the investigated materials are different.

The methods and devices described above have been used to measure the diameters of the effective cross section (D_{ef}) of reflected radiation $R_\lambda(\theta; 2\pi)$ and transmitted radiation $T_\lambda(\theta; 2\pi)$ for different foodstuffs exposed to a narrow (D= 2 mm) directional radiation flux coming from an optical quantum generator LG-56 operating at $\lambda=0.63$ μm and to a monochromatic flux at $\lambda=1.0$ μm.

As can be seen from Fig. 5.3, for all the samples $R_\lambda(\theta; 2\pi)$ and $T_\lambda(\theta; 2\pi)$ increase with an increase in the diameter of the diaphragm. These results confirm that in evey instance the cross section of a directional radiation flux undergoes spreading. The degree of spreading depends both on the type of the substance under investigation and the thickness of the sample.

The effective diameter of the reflected and transmitted radiation flux is determined on the basis of the changes in $R_\lambda(\theta; 2\pi)$ and $T_\lambda(\theta; 2\pi)$. We consider the diameter at which $R_\lambda(\theta; 2\pi)$ and $T_\lambda(\theta; 2\pi)$ cease to increase despite an increase in the diameter of the diaphragm ($D=D_{ef}$) to be the effective diameter. For example for a layer (3.5 mm) of the soft part of bread its transmittance continues to increase until the diameter of the diaphragm reaches 24 mm (Fig. 5.3, curve 1). Any further increase in the diameter of the diaphragm does not lead to an increase in $T_\lambda(\theta; 2\pi)$. Thus, the effective diameter corresponding to $T_\lambda(\theta; 2\pi)$ for the soft part of bread ($\ell=3.5$ mm) is 24 mm.

When investigating the thermal radiational characteristics with the aid of spectrophotometers (SF-4A), IKS-12, IKS-14, IKS-21), the sample is irradiated with a flux which diverges at a small angle. By using the effective diameter of the spreading of a narrow beam we can determine the dimensions of the effective cross sections of an incident flux whose cross section has an arbitrary configuration and arbitrary dimensions. The following method has been developed for determining the dimensions of the sample and the radiation detector [39, 118].

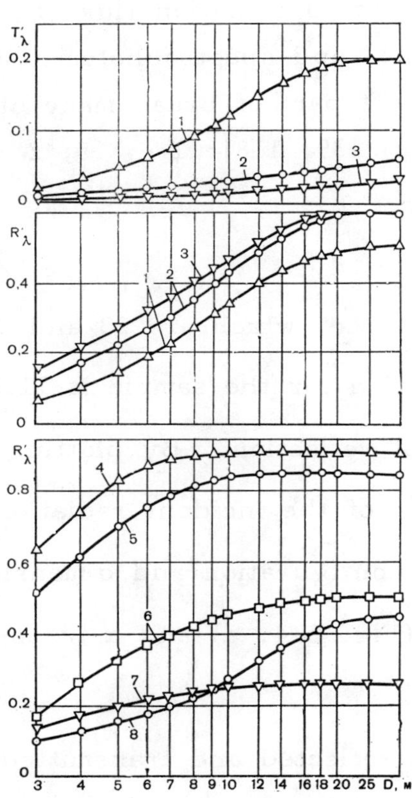

Figure 5.3 Spectral directional-hemispherical reflectance $R_\lambda(\theta; 2\pi)$ and transmittance $T_\lambda(\theta; 2\pi)$ as a function of the diameter D (mm) of the diaphragm at $\lambda=0.63$ µm for various substances [39, 118]: 1,2,3 – soft part of bread made of first-grade wheat flour (w=11.8%); (ℓ: 1 – 3.5, 2 – 5.5, 3 – 10.0 mm); 4 – potato starch, $\ell=5.0$ mm; 5 – paper (75 g/m^2), $\ell=0.09$ mm; 6 – bread crumbs (w=11.8%), $\ell=10.0$ mm; 7 – bread crust together with soft part of bread (w= 11.8%), $\ell=8.5$ mm; 8 – potato (w=75.5%), $\ell=5.0$ mm.

A sample of the soft part of bread (6x10 mm^2) is irradiated inside an SF-4A spectrometer. The effective cross section of the flux as it propagates through the sample ($\ell=3.5$ mm) is determined as follows. It can be seen from Fig. 5.3 that the effective diameter of the cross section of the radiation flux is 25 mm for the transmitted flux and 22 mm for the reflected flux when the sample is irradiated with a flux 2 mm in diameter. By subtracting from the magnitudes of these effective diameters the value for the diameter of the cross section of the incident

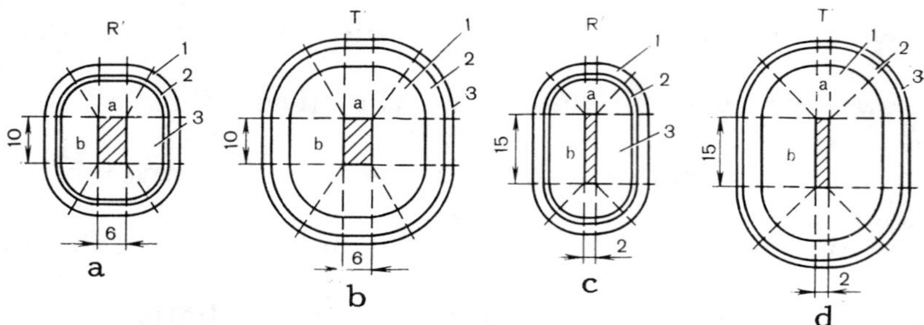

Figure 5.4 Characteristic configurations and dimensions of the effective cross sections of a reflected radiation flux (a, c) and a transmitted radiation flux (b, d), measured with the aid of a spectrometer SF-4A (a, b), and spectrophotometers IKS-12, IKS-14, and IKS-21 (c, d), for samples of the soft part of bread made of first-grade flour (w=12.3%) at different thicknesses ℓ (mm) [39, 118]: 1 - 3.5; 2 - 5.5; 3 - 10.0.

flux we obtain the effective diameters for a point source, which are 23 and 20 mm, respectively. The effective radius of the cross section for the sample is 11.5 mm in the case of transmission and 10 mm in the case of reflection. By plotting these radii from different points of the cross section of the incident radiation flux and by drawing an enveloping curve we obtain the configuration and dimensions of the effective cross sections for a reflected flux (Fig. 5.4, a) and a transmitted flux (Fig. 5.4, b).

The areas of the effective cross sections for a reflected and transmitted flux are shown in Fig. 5.4 c and d, respectively for the same sample, but in this case they are measured with the aid of an IKS-14 spectrophotometer (the cross section of the incident radiation flux in this case is 2x15 mm^2).

As can be seen, the areas of the effective cross section for transmitted radiation are somewhat larger than those for reflected radiation. This is due to a greater deviation of the directional flux during transmission. Furthermore, with an increase in the thickness of the sample the cross section corresponding to transmission also increases, while that of reflection decreases.

So that the radiation transmitted and reflected by a sample can be measured by the radiation detector the receiving area of the detector must not be smaller than the area of the scattered flux and must have the same configuration. The radiation-receiving area in existing detectors is much smaller: 12 mm in diameter in photoresistors 6AN; 3×10 mm^2 in RTE; 6.5×12.5 mm^2 in bolometers, etc. This means that only a portion of the reflected and transmitted radiation is registered.

Diagrams showing the distribution of radiation flux in a sample of the soft part of bread ($\ell = 10$ mm) irradiated with the aid of spectrophotometers IKS-14 and SF-4A are given in Fig. 5.5 (the cross sections of the incident flux - 2×15 and 6×10 mm^2, respectively). The radiation-receiving area of the detector and the dimensions of the sample in this case should not be less than 38×51 mm^2. Under these conditions the transmittance and reflectance cannot be determined without error. The magnitude of the error will depend on the degree of the spreading of the cross section of the directional radiation flux, the dimensions of the sample, and the radiation-registering area of the detector. The dimensions of the sample should be such that there will be no radiation losses at the edges of the sample or radiation losses due to the shielding of the transmitted or reflected radiation by the clamps holding the sample.

To estimate how precise the measurements are of the transmittance and reflectance of a sample with SF-4A we shall use the diagram shown in Fig. 5.5. The area under the curves is proportional to the relative magnitude of the flux transmitted by the sample. The error in measuring T_λ is zero if the detector receives all the transmitted radiation, or 10% if it registers 90% of the transmitted radiation.

The area under the curves has been calculated by means of graphic integration. The crosshatched area makes up 90% of the total area. If the former is known, we can determine the minimal dimension of one side of the detector (error in measuring T_λ - 10%). The minimal dimension of the first and the second side of the detector is 22 and 24 mm, respectively. Thus, in order to determine the transmittance of a layer of the soft part of bread ($\ell = 10$ mm, $w = 42\%$) with the

aid of a SF-4A spectrometer (the cross section of the incident radiation - 6x10
mm) with an error of less than 10%, the dimensions of the radiation detector and
of the sample should be not less than 22x24 mm. These requirements are met by
FESS-U10 photocells (whose total registring area is 32 mm in diameter). In measurements of T_λ made with the aid of all other devices (with the exception of
photometers with an integrating sphere) the error is more than 10%.

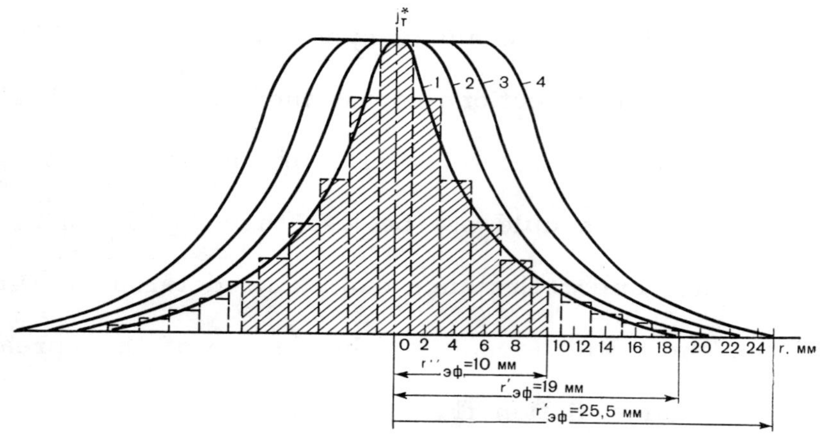

Figure 5.5 Distribution of the flux densities of radiation j_T^* inside a 10 mm-layer
of the soft part of bread as a function of the radius of the cross section recorded
by different spectrometers [39, 118]: 1,4 - IKS-12, IKS-14, IKS-21; 2,3 - SF-4A.

Photometers with an integrating sphere make it possible to measure the thermal radiational characteristics of strongly scattering substances with a detector
having a small registering area. But in this case the sample must have certain
minimal dimensions. It is important that the shielding of diffuse radiation reflected
or transmitted by the sample should be prevented. The minimal dimensions of
the sample used in measuring the thermal radiational characteristics are calculated
in the same way as those of the detector.

The above-mentioned method has been used to calculate the minimal dimensions
of samples of different foodstuffs (Table 5.2). Here the scattering and spreading
of the directional radiation flux in the samples are accounted for in determining 90%
of their transmittance. When determining their reflectance the minimal dimensions

of the samples and those of the receiving area of the radiation detector will be
somewhat smaller since the degree of spreading in this case is less than in the
case of transmittance. When we know the optimal dimensions of a sample we can
choose a method for determining its thermal radiational characteristics which will
reduce to a minimum the error due to the scattering and spreading of the cross
section of the directional radiation flux in the sample.

Table 5.2   Minimal dimensions of the sample (mm) [39, 118]

Sample	T_λ measured with the aid of:	
	SF-4A	IKS-12, IKS-14, IKS-22
Soft part of bread (ℓ=10 mm,,w=42%)	22x24	20x28
Bread crumbs (ℓ=3.0 mm, w=40.8%)	19x22	18x26
Brown bread crust (ℓ=3.0mm,w=36.2%)	17x20	16x24
Beets (ℓ=3.0 mm. w=80%)	21x23	19x27
Carrots (ℓ=3.0 mm, w=68%)	21.5x23.5	19.5x27.5
Apple (ℓ=3.0 mm, w=84%)	24x26	22x30
Potato (ℓ=5 mm, w=73%)	23.5x25.5	21.5x29.5

5.2   A comparison of different methods for measuring the spectral and integral thermal radiational characteristics of substances

Depending on the irradiation conditions and the methods used for measuring
the radiation scattered by the sample, there can be in principle nine different
methods for determining the transmittance and reflectance of foodstuffs (Table 5.1).
A complete review of these methods covering a period up to 1971 has been carried
out [39, 44, 54, 86]:

I - Method for measuring the bidirectional reflectance and transmittance of
light-scattering substances. Devices for determining the spatial distribution of radiation (the indicatrixes of scattering and transmission) are known as goniophoto-
meters or goniospectrophotometers (Fig. 5.6, a-d) [141, 215, 216, 222]. By mea-
suring $T_\lambda(\theta; \phi; \theta_T, \phi_T)$ and $R_\lambda(\theta, \phi; \theta_R, \phi_R)$ we can determine the hemispherical
transmittance and reflectance under any irradiation conditions and at any scattering
indicatrix of the medium. The magnitudes of $R_\lambda(2\pi)$ and $T_\lambda(2\pi)$ under different

irradiation conditions are obtained with the aid of the analytical relationship be-

tween the indicatrix and the reflectance or transmittance of a layer of the sample.

II - Method for measuring the directional-diffuse $R_\lambda(\theta,\phi: \omega_R)$ and $T_\lambda(\theta,\phi: \omega_T)$ of light-scattering substances. This method calls for the measurement of the radiation flux reflected or transmitted either by the sample or by a standard with-in a limited solid angle $\Delta\omega<2\pi$. The sample is irradiated with a directional flux (Fig. 5.6, e-j) [63, 125, 128, 205, 212, 217], with the reflection (transmission) indicatrix of the sample and that of the standard being identical. If this condi-tion is not fulfilled, there will be errors due to the difference between the indi-catrix of reflectance of the standard and that of the sample at all the wavelengths employed. A large error is inevitable with this method, and therefore it can be used only in the 0.25-2.50 μm spectral range.

III - Method for measuring the directional-hemispherical $R_\lambda(\theta,\phi; 2\pi)$ and $T_\lambda(\theta,\phi; 2\pi)$ (Fig. 5.7) [33, 39, 63, 95, 107, 116, 139]. According to the rule of reversibility the measured quantities are equal to the hemispherical coefficients of radiance $\rho_\lambda(2\pi; \theta_R,\phi_R)$ and $t_\lambda(2\pi; \theta_T,\phi_T)$. With the aid of the equations in Table 1.2 these coefficients can be used in determining the hemispherical $R_\lambda(2\pi)$ and $T_\lambda(2\pi)$ under any irradiation conditions. In order to measure the entire hemi-spherical radiation reflected and transmitted by the sample the following devices are used: integrating spheres, specular hemispheres, ellipsoids and paraboloids, as well as radiation receivers with a large detection area (Fig. 5.7, i). In litera-ture usually no account is taken of the large errors due to the spreading of the cross section of the directional radiation flux in the sample or of its indicatrix which differs from the ideal one.

IV - Method for measuring $R_\lambda(\omega; \theta_R,\phi_R)$ and $T_\lambda(\omega; \theta_T,\phi_T)$. This method has not been fully carried out in practice [149, 153, 165, 198, 205, 207, 210, 218].

V - Method for measuring $T_\lambda(\omega; \omega_T)$ and $R_\lambda(\omega; \omega_R)$. This method is based on the similarity between the reflectance indicatrix (or transmittance indicatrix) of the sample and that of the standard. Measurements are made under the condition of imperfectly diffuse irradiation (Fig. 5.8, a) [219, 220].

VI - Method for measuring $T_\lambda(\omega; 2\pi)$ and $R_\lambda(\omega; 2\pi)$ has not been applied in practice.

VII - Method for measuring $R_\lambda(2\pi; \theta_R, \phi_R)$ and $T_\lambda(2\pi; \theta_T, \phi_T)$. This method is based on experimental determinations of the reflectance and transmittance indicatrixes or the coefficient of radiance $\rho_\lambda(2\pi; \theta_R, \phi_R)$ [116, 140, 162, 177, 181, 209, 214, 221]. The surface of the sample is irradiated with hemispherical diffuse radiation. The method also makes use of the analytical interrelationship between the indicatrix and the reflectance (or transmittance) of a layer of the medium (Table 1.2). It is carried out with the aid of Eq. 1.57. The principal methods employed in practice for irradiating the surface of a sample with uniform diffuse radiation differ from one another, depending on the type of incident radiation used - total or monochromatic (Fig. 5.8, b-i).

VIII - Method for measuring $R_\lambda(2\pi; \omega_R)$ and $T_\lambda(2\pi; \omega_T)$. This method is of no practical interest.

IX - Method for measuring the bihemispherical reflectance $R_\lambda(2\pi; 2\pi)$ and transmittance $T_\lambda(2\pi; 2\pi)$ of light-scattering substances. This method involves the hemispherical irradiation of the sample and the measurement of the irradiation transmittance by the sample with the aid of two integrating spheres (Fig. 5.8, j) [150].

A simpler method for measuring bihemispherical transmittance in the 0.4–1.4μm region is carried out with the aid of a specular ellipsoid (Fig. 5.11, b). For measuring bihemispherical reflectance in the infrared spectral region 1–38 μm an attachment in the form of an integrating sphere can be used [191] (Fig. 5.8, k). The surface of the sphere and that of the standard, which are coated with gold, provide diffuse reflection. Measurements of $R_\lambda(2\pi; 2\pi)$ are made relative to the standard.

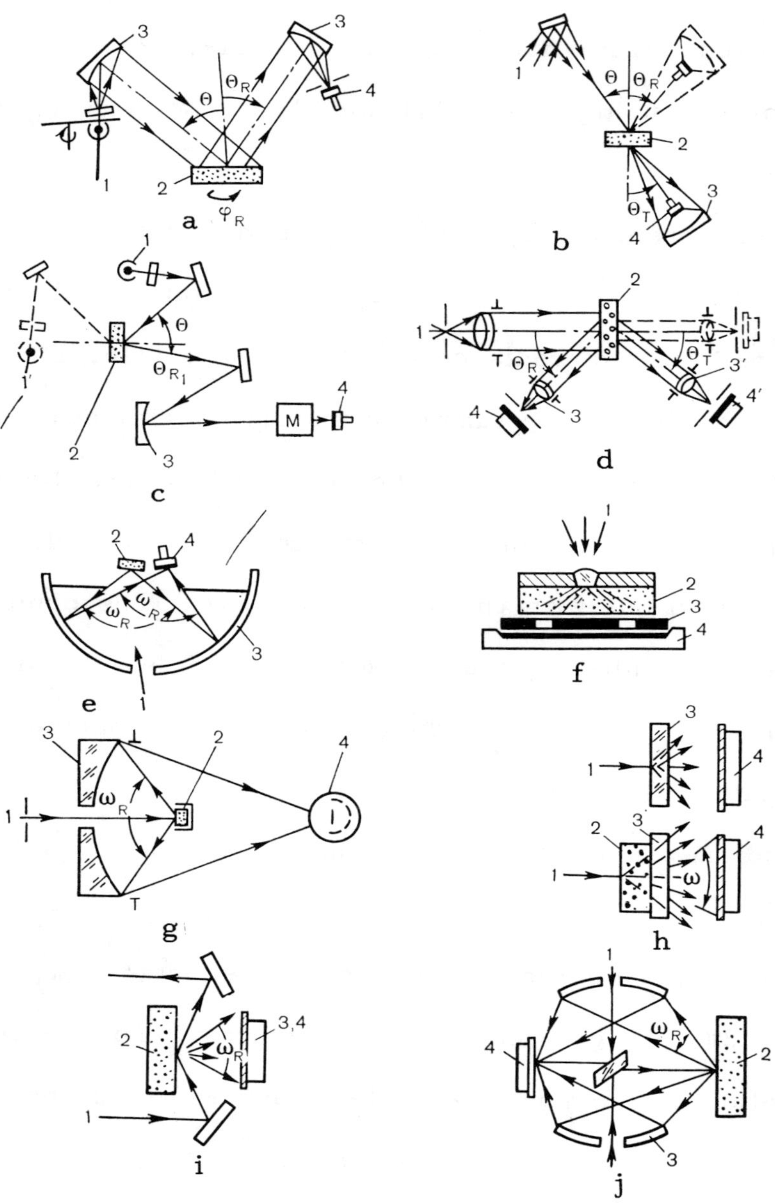

Figure 5.6   Schematic diagrams of devices for measuring the bidirectional- and directional-diffuse thermal radiational characteristics in the case of directional irradiation: a-d - goniophotometer for measuring the indicatrix of reflectance and transmittance; e,f - devices with an iris and a ring diaphragm, respectively, for determining the angular structure of reflected and transmitted radiation; g-j - devices for measuring directional-diffuse characteristics; 1 - source; 2 - sample; 3 - measuring device; 4 - radiation detector.

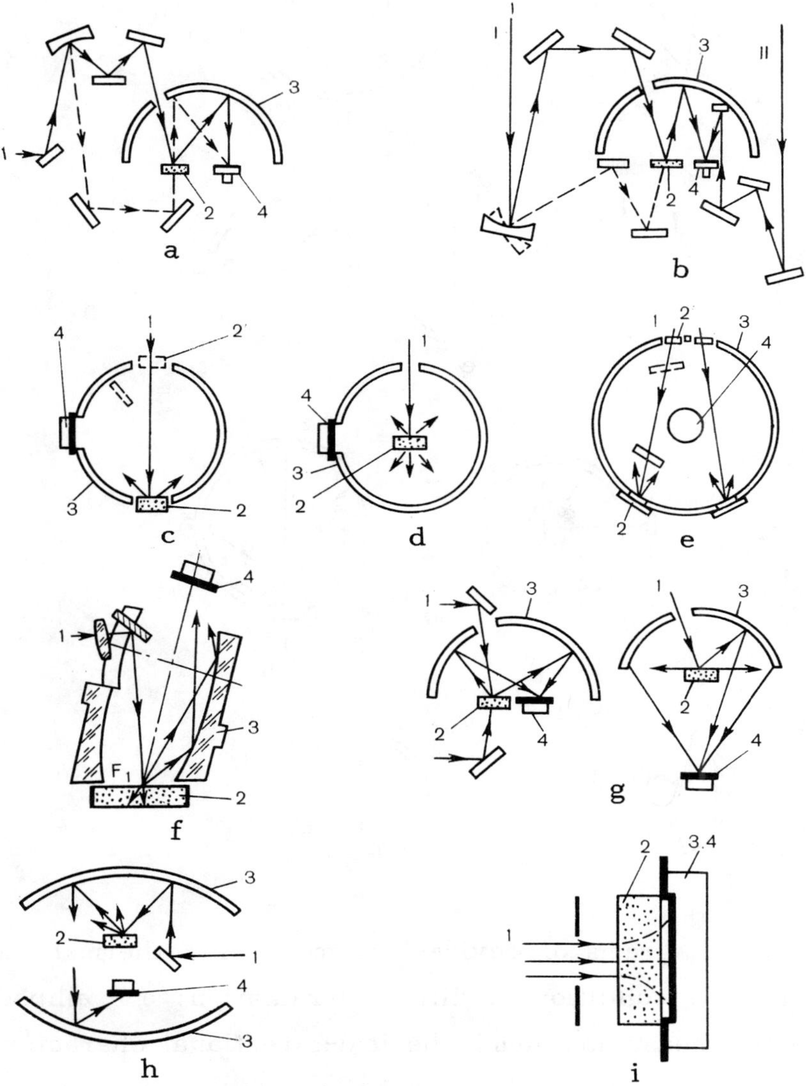

Figure 5.7 Schematic diagrams of devices for measuring the directional-hemispherical characteristics by different methods: a,b - specular hemisphere; c-e - integrating sphere: f,g - specular truncated ellipsoids; h - specular paraboloids; i - detector with a large radiation-receiving area; 1 - source; 2 - sample; 3 - measuring device; 4 - radiation detector.

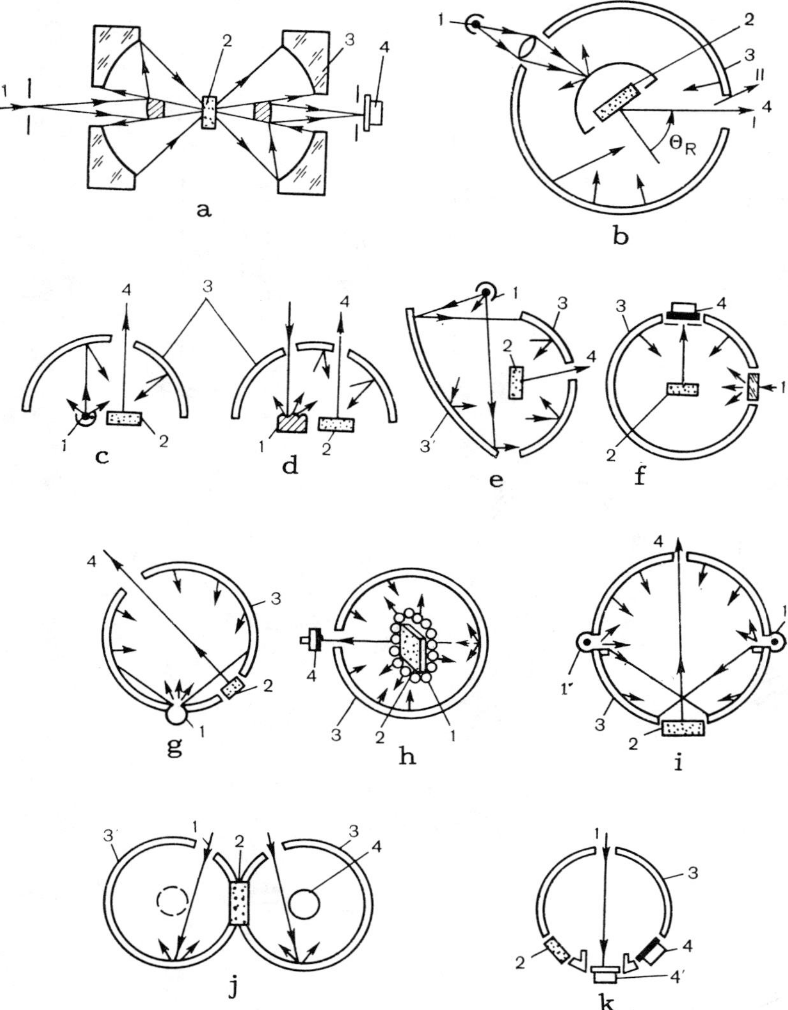

Figure 5.8 Schematic diagrams of devices for measuring thermal radiational characteristics under the condition of diffuse irradiation: a - ellipsoids; b,f, g-i - integrating sphere for measuring hemispherical-directional characteristics; c,d, - inverted specular hemisphere; e - two paraboloids (axial and extra-axial); j,k - integrating sphere for determining bihemispherical characteristics; 1 - source; 2 - sample; 3 - measuring device; 4 - radiation detector.

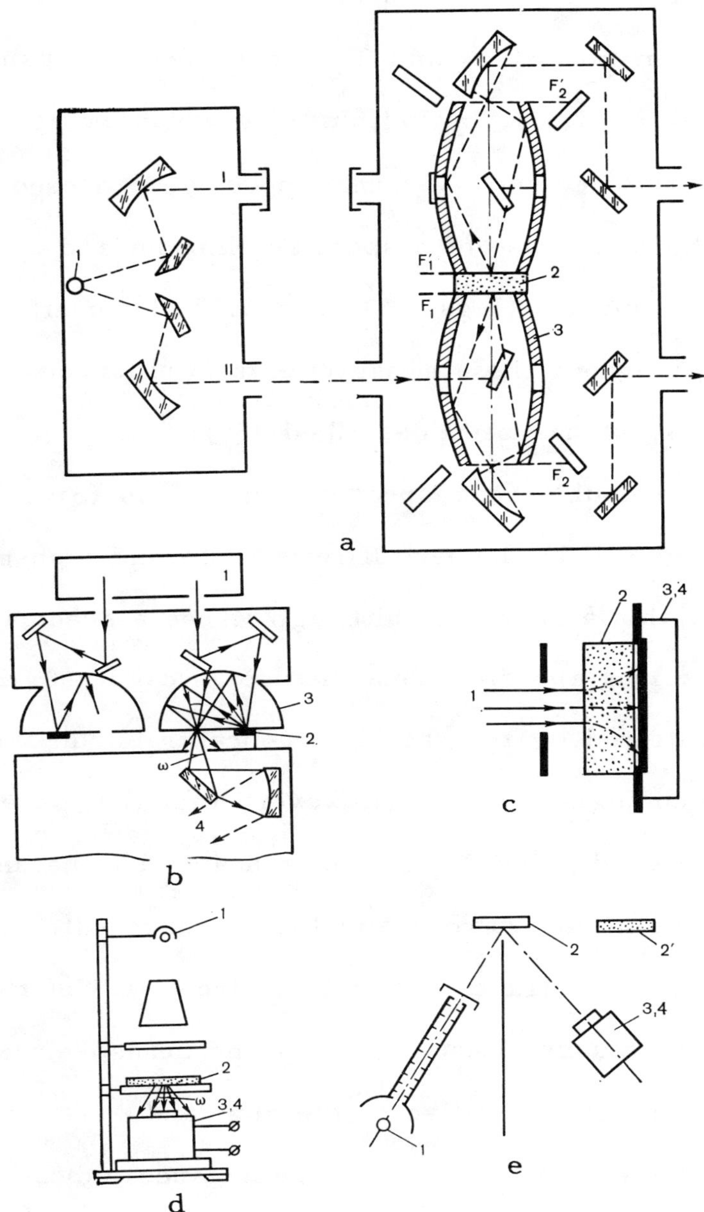

Figure 5.9 Schematic diagrams of devices for measuring the thermal radiational characteristics of foodstuffs subjected to total radiation flux: a,b,c – attachements for IKS-14A for measuring the ratio T_λ'/R_λ', R_λ' and T_λ', and $T_\lambda(\theta; 2\pi)$, respectively; d – device for measuring transmittance T; e – device for measuring bidirectional reflectance $R'(\theta; \theta_R)$: 1 – source; 2 – sample; 3 – measuring device; 4 – radiation detector.

When using the methods considered above for measuring the reflectance and transmittance of foodstuffs one can determine these characteristics only separately. The simultaneous determination of R_λ and T_λ can be done by using attachments (Figs. 5.11, 5.14, 5.24 and 5.25), but within a very narrow spectral range. Another method for measuring R_λ and T_λ simultaneously, proposed by the present authors [39, 44], combines the use of a specular ellipsoid and a detector with a large radiation-receiving surface (Figs. 5.11 and 5.14). An attachment with two specular ellipsoids carrying out these measurements simultaneously in the 0.75-15.0 μm range (Fig. 5.9, a) has been described [88].

An attachment for a double-beam spectrometer IKS-14 (Fig. 5.9, b) is used in determining the directional-hemispherical reflectance and transmittance of foodstuffs [97]. However, this attachment, although it has a specular hemisphere, cannot be used in measuring the directional-hemispherical $R_\lambda(\theta; 2\pi)$ and $T_\lambda(\theta; 2\pi)$. The radiation scattered by the sample is concentrated in the hemispherical flux and focused on a conjugated point. However, radiation penetrates the monochromator only within a solid angle $\Delta\omega$, which is limited by the angular dimensions of the toric mirror in the monochromator and the entrance slit. The rest of the radiation is screened by the entrance slit and the front wall of the IKS-14 monochromator. So, what is actually measured is not the hemispherical but the bidirectional $R(\theta;;\theta_R)$ and $T(\theta;\theta_T)$ relative to the standard. The readings of the instrument depend on the extent to which the reflectance indicatrix of the sample is similar to that of the standard.

The transmittance of foodstuffs is measured with the aid of a detector having a large radiation-receiving surface (Fig. 5.9, c) [75, 76, 83, 107].

Many researchers still measure the thermal radiational characteristics of such products as bread, dough, fruit-paste candy, etc. separately in order to determine their integral transmittance (Fig. 5.9, 9) [76, 83, 97]. A shortcoming of this approach is that in the case of thermolabile foodstuffs they become overheated when exposed to total radiation. Furthermore, the radiometer registers only a portion of the transmitted flux within a narrow solid angle, which introduces a large error in determining the values of T. The bidirectional reflection coefficients for total radiation are determined with the aid of a device shown in Fig. 5.9, e) [97, 225], which has the same shortcomings as the device mentioned earlier.

The most promising devices for determining the R_λ and T_λ of foodstuffs are those which permit their simultaneous measurements. However, the attachments for these devices (Figs. 5.9, 5.11) are also imperfect and require further improvement.

When determining the thermal radiational characteristics of foodstuffs by different methods, besides the errors due to the spreading of the directionl radiant flux in the sample, there are several factors which tend to lower the precision of the measurements. Many devices operate under the condition of zonal nonuniformity of the flux density of the radiation incident upon the sample. This is explained by the fact that the optical systems employed do not have lenses for maintaining uniform radiance of the slits of the monochromator with respect to their height. The zonal nonuniformity of the irradiation of the sample's surface can lead to considerable errors when measuring R_λ and T_λ of heterogeneous capillary-porous substances containing large particles and pores. In such cases it is necessary to increase the dimensions of the cross section of the incident radiation. Steps should be taken to ensure zonal uniformity of the image of the source on the surface of the sample.

The most frequently employed methods are those involving the use of a specular hemisphere, an integrating sphere, or a specular rotational ellipsoid.

When a specular hemisphere is used for measuring $R_\lambda(\theta: 2\pi)$ and $T_\lambda(\theta; 2\pi)$ the errors due solely to the spreading and scattering of the cross section of a narrow radiant flux in the sample can be as high as 96% (Table 5.7). Further research is obviously needed in order to improve the method. When an integrating sphere is used for measuring R_λ and T_λ the dimensions of the sample should be selected with account taken of the scattering and spreading of the cross section of the parallel radiant flux incident upon its surface. The specular rotational ellipsoid is usually employed together with detectors having a large radiation-receiving surface (32 mm in diameter in the case of FESS-U-10). Here, too, when deciding on the dimensions of the sample, the scattering and spreading of the radiant flux in the sample should be taken into consideration. The most precise measurements of the thermal radiational characteristics of foodstuffs are obtained with the use of an integrating sphere when the dimensions of the sample are chosen correctly.

5.3   Reflection standards for the infrared spectral region

Most of the spectrometers used today for light-scattering substances are reflectometers designed for hemispherical ($\omega=2\pi$) or directional ($\theta,\phi,d\omega$) irradiation of the sample being investigated.

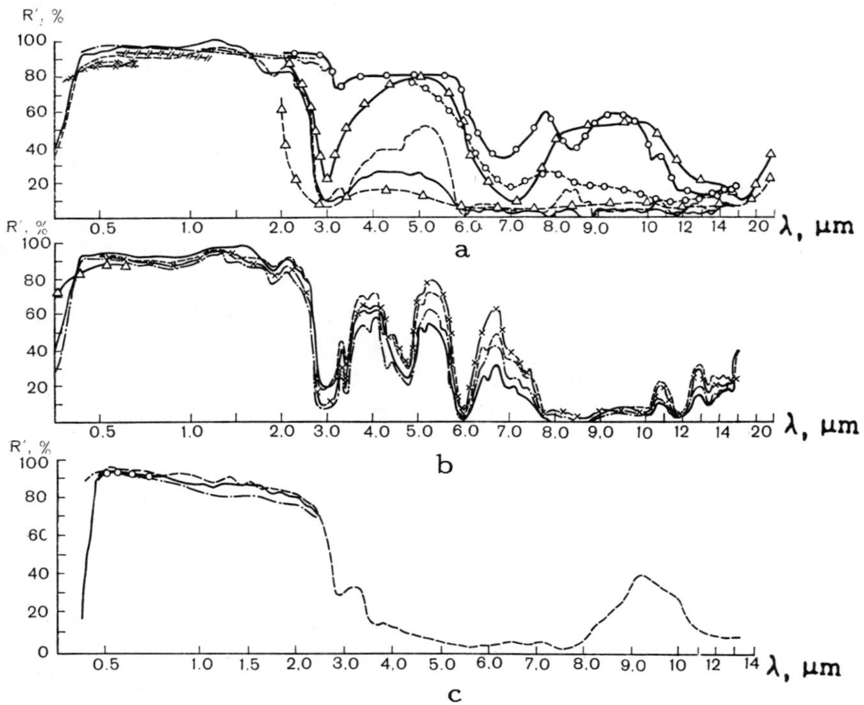

Figure 5.10   Spectra of reflection standards [121]: a - MgO; b - BaSO$_4$; c - opal glass MS-14.

They can be used for measuring either R_λ and T_λ or ρ_λ and t_λ (coefficients of radiance under different irradiation and collimation conditions). Having determined one of these coefficients, one can calculate the others (Table 1.2).

For the infrared spectrum at $\lambda > 3.0$ μm plane mirrors coated with aluminium, gold, or silver are used as the reflection standard; and for the near-infrared and visible spectral region up to 3.0 μm — either mirrors or various diffusely reflecting materials — opal glass MS-14, neutral glass ONS, and MgO, Al_2O_3, $BaSO_4$, polytetrafluoroethylene, and other coatings. Reflection standards enable us to determine the reflectance of the sample under certain conditions [163]. For a perfectly diffuse reflector $R_{\lambda ef}=1$ only in the case where the hemispherical method is used for measuring the reflected radiation under different irradiation conditions. For a plane perfect mirror $R_{\lambda ef}=1$ in the case where the hemispherical method of measurement is used for all kinds of irradiation. Under certain conditions it may also equal unity when other methods for measuring the reflected radiation are used; thus, under directional and diffuse irradiation conditions $R_{\lambda ef}=1$ when ω_R, θ_R and ϕ_R are equal to the "specular" images of ω, θ and ϕ, respectively. In all other cases $R_{\lambda ef}$ for a mirror equals zero, which indicates the unsuitability of mirrors as standards, and hence also the impossibility of measuring R_λ in such instances.

When $R_{\lambda ef}=1$ under all kinds of irradiation conditions the ratio of the radiant flux reflected by the sample to that reflected by the reference standard equals numerically the reflectance of the sample R_λ under the same irradiation conditions, which are: $R_\lambda(2\pi;\ 2\pi)$, $R_\lambda(\omega;\ 2\pi)$, and $R_\lambda(\theta,\phi;\ 2\pi)$, as well as $R_\lambda(\theta,\phi;\ \omega_R)$ and $R_\lambda(\theta,\phi;\ \theta_R,\phi_R)$. In all other cases the ratio of the radiant flux reflected within a certain solid angle in a direction determined by the sample to that reflected by the standard can be measured under the same irradiation and collimation conditions. The magnitude of this ratio depends not only on the optical properties of the sample, but also on the degree of similarity between the form of the scattering indicatrix and that of the reference standard. Numerically it corresponds to the reflection coefficient (factor) $R_{\lambda f}$, which differs from the reflectance of the sample R_λ.

In Table 5.3 and Fig. 5.10 are summarized data on the reflectance $R_\lambda(\theta; 2\pi)$ of some of the most widely used reflection standards with a diffuse indicatrix: opal glass MS-14, Al_2O_3, Ag-, Al-, Au-coated plane mirrors [2, 33, 74, 95, 116, 121, 181, 223, 224].

Table 5.3 Reflectance of different standards

λ (μm)	MS-14	Al_2O_3	Al	Ag	Au
0.40	0.900	0.780	0.917	0.956	0.380
0.45	0.920	0.863	0.915	0.970	0.375
0.50	0.940	0.889	0.911	0.978	0.477
0.55	0.940	0.908	0.907	0.983	0.845
0.60	0.940	0.930	0.902	0.986	0.922
0.65	0.930	0.923	0.898	0.988	0.960
0.70	0.920	0.918	0.889	0.989	0.969
0.75	0.920	0.927	0.876	0.990	0.973
0.80	0.920	0.920	0.860	0.991	0.979
0.85	0.900	0.920	0.860	0.992	0.982
0.90	0.890	0.914	0.889	0.993	0.984
0.95	0.882	0.937	0. 915	0.993	0.989
1.00	0.875	0.925	0.932	0.993	0.991
1.05	0.867	0.921	0.938	0.993	0.991
1.10	0.860	0.917	0.945	0.993	0.992
1.20	0.880	0.906	0.958	0.994	0.992
1.50	0.880	0.766	0.966	0.994	0.992
2.0	0.780	0.639	0.970	0.994	0.993
2.5	0.690	0.510	0.972	0.994	0.993
3.0	0.320	0.440	0.974	0.994	0.994
4.0	0.150	0.772	0.976	0.994	0.994
5.0	0.070	0.680	0.977	0.994	0.994
10.0	0.300	0.050	0.981	0.995	0.994
15.0	0.100	0.334	0.983	0.995	0.994

Neutral glass ONS (four different grades: ONS1, ONS2, ONS3, and ONS4) is used in comparative measurements of the color of the reflecting objects, R_λ and ρ_λ, as well in checking the operating stability of spectrometers and colorimetric photometers. The average values of the reflection coefficient R_λ for different grades of glass - ONS1, ONS2, ONS3, and ONS4 - at λ=0.55 μm are 0.75, 0.50,

0.30 and 0.17, respectively [226].

5.4   Instruments for the simultaneous measurement of the thermal radiational characteristics of foodstuffs subjected to directional irradiation

Several methods and devices have been developed by the present authors for the simultaneous measurement of the hemispherical characteristics $R_\lambda(\theta,\phi; 2\pi)$ and $T_\lambda(\theta,\phi; 2\pi)$ of light-scattering substances [44]. One method involves the use of two radiation detectors for measuring the reflectance and transmittance under identical irradiation conditions. A special attachment for spectrometers SF-4 and SF-4A makes it possible at the same time to measure the directional-hemispherical characteristics $R_\lambda(1; 2\pi)$ and $T_\lambda(1; 2\pi)$ (at $\theta=0°$, $\mu=1$) (Fig. 5.11) [39].

With this method one can obtain in a single experiment $R_\lambda(1; 2\pi)$ and $T_\lambda(1; 2\pi)$ and consequently also $A_\lambda(1; 2\pi)$ for the same sample under identical irradiation conditions. The method is especially helpful if changes in the optical properties of the sample cannot be reproduced in time (for example, the nonuniform distribution in the sample of moisture, density, the coloring substance, etc). The drawbacks of this method are the same as those of the method based on the use of an ellipsoid truncated by two focal planes and of the method involving the use of a detector with a large radiation-receiving area. The main disadvantage here lies in the different conditions of irradiating the surface of the photocell which registers the radiation transmitted through the sample (or the reference standard). This results in errors due to the dependence of the photoelectric current on the incidence angle of the radiation falling on the photocell's surface and to the fact that the photoconductive cell of the detector is not uniformly sensitive over the surface (in some instances it is much more sensitive near the area of contacts).

To overcome these shortcomings a method has been developed which calls for the use of a single radiation detector; the method is designed for carrying out complex measurements of $T_\lambda(2\pi; 2\pi)$ and $R_\lambda(1; 2\pi)$ in the spectral region 0.4-

1.4 μm [39]. In measuring $T_\lambda(2\pi; 2\pi)$ both the sample and the transmission standard are exposed to diffuse radiation flux. Owing to the fact that the detector is also irradiated all the time with a diffuse flux, the above-mentioned errors are eliminated [44].

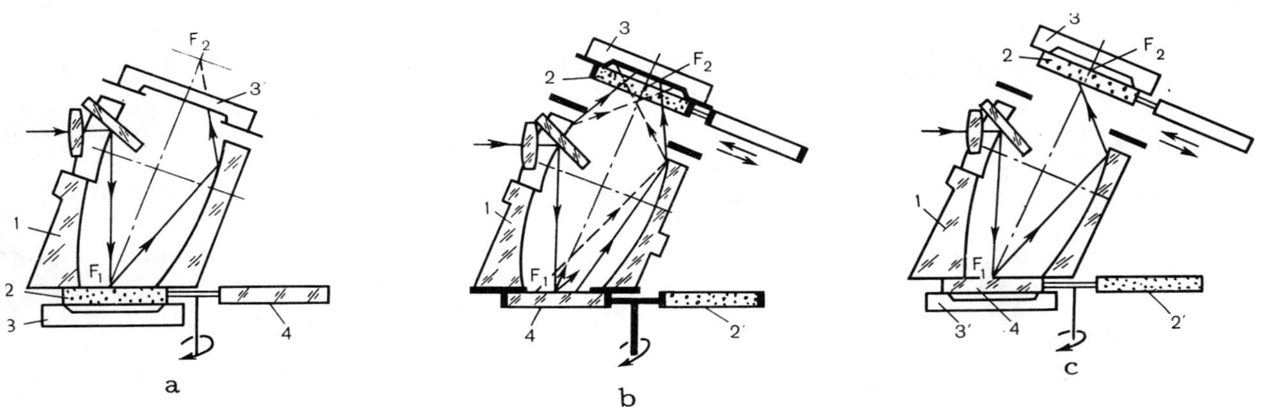

Figure 5.11 Schematic diagrams of optical attachments for an SF-4A spectrometer which permit the simultaneous (a) and complex (b, c) measurements of the hemispherical reflectance and transmittance of foodstuffs: 1 - specular ellipsoid; 2,2' - sample; 3,3' - radiation detector (photocell FESS-U-10); 4 - reflection standard (opal glass MS-14).

A device for the simultaneous measurement of the $R_\lambda(1; 2\pi)$ of a sample under directional irradiation conditions and its $T_\lambda(2\pi; 2\pi)$ under diffuse irradiation conditions is shown in Fig. 5.11, b. A reflection standard (4) is used for providing diffuse irradiation of the sample when its bihemispherical transmittance $T_\lambda(2\pi; 2\pi)$ is being measured. This method makes it possible, with the aid of one and the same radiation detector, at the same slit width to measure both the $R_\lambda(1; 2\pi)$ and $T_\lambda(2\pi; 2\pi)$ of different radiation-scattering substances. It takes twice less time to carry out the experiment, and the errors inherent in methods involving separate measurements are avoided.

The devices are so designed that they can be mounted together with the PDO-1 attachment (Fig. 5.11, c). Such an arrangement permits complex measurements of three quantities, $R_\lambda(2; 2\pi)$, $T_\lambda(1; 2\pi)$, and $T_\lambda(2\pi; 2\pi)$, by the methods described above at a required wavelength and with identical slit width of the monochromator.

A disadvantage common to these methods is that no measurements of the absolute values of R_λ and T_λ can be made. The extent of the error in measuring $R_\lambda(1; 2\pi)$ depends on the precision of determining the R_λ of the standard and and its stability in time. To overcome this difficulty a more precise method involving the use of an integrating sphere has been developed [44]. It is designed for measuring the absolute values of $R_\lambda(\theta; 2\pi)$, $T_\lambda(\theta; 2\pi)$, and $A_\lambda(\theta; 2\pi)$ (without the use of standards) in relation to the incidence angle θ of the radiation at an arbitrary scattering indicatrix of the sample in the 0.4-1.4 μm region.

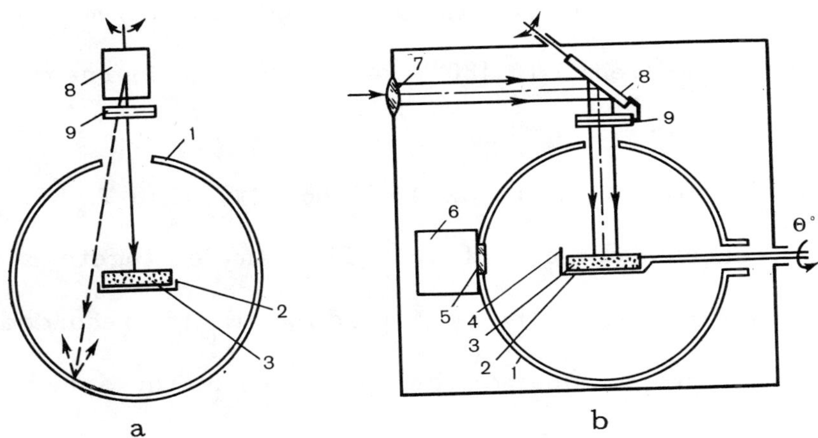

Figure 5.12 Schematic diagram of an attachment for an SF-4A spectrometer for investigating hemispherical T_λ' and R_λ' as a function of the incidence angle [44].

This method excludes the errors due to the difference between the reflection indicatrix of the standard and that of the sample and to the "aging" of the standard. The device also makes it possible to determine the thermal radiational characteristics of different foodstuffs exposed to polarized radiation.

Samples 40x40 mm^2 (Fig. 5.12, (3)) are placed inside an integrating sphere (1) 150 mm in diameter. The holder of the samples can be closed from below with a light-tight cap (2), which together with a screen (4) and the sphere is coated outside with diffusely reflecting Al_2O_3 (Table 5.3). The holder can be fastened in the center of the sphere or moved outside it. Under these circumstances the radiant flux focused by a lense (7) and a rotating mirror (8) will be incident upon the surface of the sample or directly upon the wall of the sphere. The flux reflected (transmitted) by the sample and reflected from the wall of the sphere is registered by a detector (6) (photoelectric multiplier FEU-62). To provide uniform irradiation of the detector's surface the detector is covered with opal glass (5) (MS-13). The rotation angle of the screen lying within the limits $\theta=0 \div 85°$ is read from the scale. The screen is so located that the radiation directly reflected or transmitted by the sample is not registered by the detector.

In investigating the thermal radiational characteristics of a sample exposed to polarized radiation a polaroid (9) is attached to the rotating mirror. The polaroid can rotate at an angle from 0 to 180° around the axis perpendicular to its surface.

The complex determination of $R_\lambda(\theta; 2\pi)$, $T_\lambda(\theta; 2\pi)$, and $A_\lambda(\theta; 2\pi)$ at any scattering indicatrix is carried out as follows. The sample, together with the cap, is moved outside the sphere. Then the first reading, N_1, is recorded of the photoelectric current, which is proportional to the irradiance of the sphere's surface E_1 due to the radiation reflected from the sphere. The holder with the sample and the cap are then moved to the center of the sphere. In this case the irradiance of the sphere's surface E_2 is due to the radiation reflected by the sample and is recorded as a second reading, N_2. The cap is then removed and a third reading, N_3, is recorded, which is proportional to the irradiance of E_3 due to the fractions of radiation reflected and transmitted by the sample.

The absolute value of the transmittance of the sample is defined as follows [44]

$$T_\lambda(\theta; 2\pi)=(N_3-N_2)/N_1. \qquad (5.6)$$

The reflectance of the sample with diffuse reflection at $\theta=0°$ is [44]:

$$R_\lambda(\theta; 2\pi)=(N_2 S)/N_1(S-4S_1); \qquad (5.7)$$

where, S_1 - area of the inlet of the sphere whose total surface area is S.

For samples with specular reflection the radiation losses through the sphere's inlet can be reduced by rotating the sample holder by an angle of 305°, and we have:

$$R_\lambda(\theta; 2\pi)=N_2/N_1. \qquad (5.8)$$

Since for the attachment employed S_1 is less than 0.1% of the total surface area of the integrating sphere, we can use an approximate relationship (Eq. 5.8) also for scattering substances.

The method described here can be used to determine R_λ and T_λ as a function of the incidence angle and the polarization of radiation. The size of the sample (40x40 mm) makes it possible to avoid errors in measuring R_λ and T_λ due to the spreading and scattering of the directional radiant flux in the sample (Figs.5.4 and 5.5). The reproducibility of R_λ and T_λ values for one and the same sample of different foodstuffs has been found to be ±1.5%.

5.5 Instruments for measuring the thermal radiational characteristics of coarse-grained friable foodstuffs

In analyzing the spectral thermal radiational characteristics of coarse-grained friable foodstuffs errors may arise due to the comparability of the dimensions of the cross section of incident radiation (for SF-4A and IKS-14 the area S is 6x10 and $(1\div2)$x15 mm^2, respectively) and the dimensions of the optical inhomogeneities of the substances being investigated. To eliminate such errors it is necessary either to use fluxes having a very large cross section or to move the sample so as to obtain characteristics of the sample averaged with respect to the surface. A method has been described [23] for measuring the spectral directional-hemispherical reflectance $R_\lambda(\theta; 2\pi)$ of any coarse-grained friable foodstuff (Fig. 5.13).

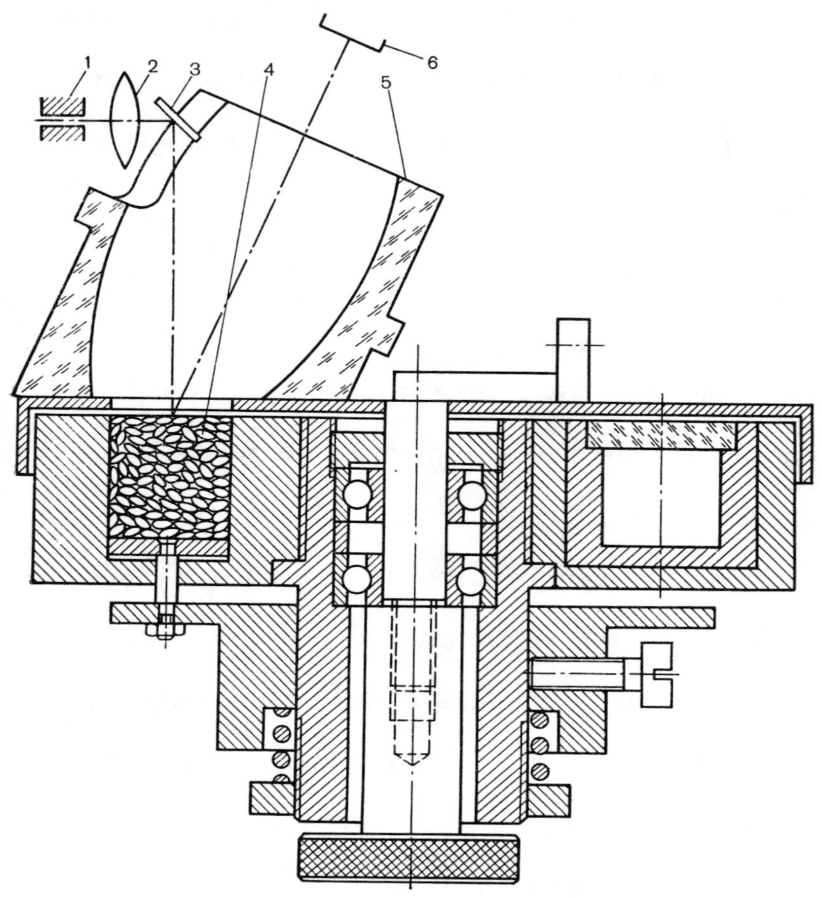

Figure 5.13 Design of an attachment for PDO-1 for measuring diffuse reflection from a layer of coarse-grained friable substance [23].

According to this method the sample moves relative to the stationary beam of light whose cross section is of limited dimensions, or the other way round. In these instances the magnitude of the reflected radiant flux varies, depending on the optical properties of a heterogeneous layer of the sample located in different ways relative to the incident radiation, and acquires fluctuating values as the sample is moved. Such fluctuating radiation reflected by the sample, upon entering the radiation detector, produces a fluctuating photoelectric current. The average magnitude of this fluctuating current is proportional to the average reflectance of the sample.

The monochromatic radiant flux from the exit slit (1) (Fig. 5.13) of SF-4A is focused by the lense (2) and the plane mirror (3) at a 90° angle to the surface of the sample (4). The diffuse radiation reflected by the sample with the aid of a specular ellipsoid (5) is focused on the surface of the photo cell FESS-U-10 (6). The photoelectric current is recorded by a reading device UF-206 which measures currents up to 1 μA.

Measurements of $R_\lambda(\theta; 2\pi)$ are made relative to the standard MS-14 (see Table 5.3). The N_R reading from UF-206, which is made by opening the monochromator's slit at 1000 division when radiation is reflected from MS-14, is taken to be 100%. Then, by rotating the supporting plate at a certain velocity so that the signal from UF-206 will remain approximately constant, recordings are made of N_R which correspond to the relative reflectance of the sample.

The absolute value of the reflectance is determined from the following relationship:

$$R_\lambda(\theta; 2\pi)=(N_R/N_{Rexp})R_{\lambda exp}(\theta; 2\pi). \qquad (5.9)$$

If it should be necessary to measure R_λ under the condition of diffuse irradiation, simultaneous and complex measurements are carried out with the aid of devices shown in Figs. 5.24 and 5.25 [48, 181].

An apparatus for PDO-1 for the simultaneous measurement of the reflectance and transmittance of individual grains is shown in Fig. 5.14 [23]. It consists of an attachment (1), a rotating plate (2) with the samples (3), two reference standards (4) and (5) (one for measuring reflection, and the other - transmission), and two radiation detectors (6) and (7) (photo cells FESS-U-10).

The grains are held by round plates which have been subjected to sandblasting and oxidized to a black color in order to reduce their reflectance to a minimum. Each plate has six holes 2.8 mm in diameter; they are located close to one another in the area of the photometric flux in the focal plane of the ellipsoid. The holes serve as the holders for individual grains. The samples are cut from the middle of a grain so that their diameter is somewhat greater than the opening in the plate. The samples are thus tightly held in the plate and do not fall out during the experiment.

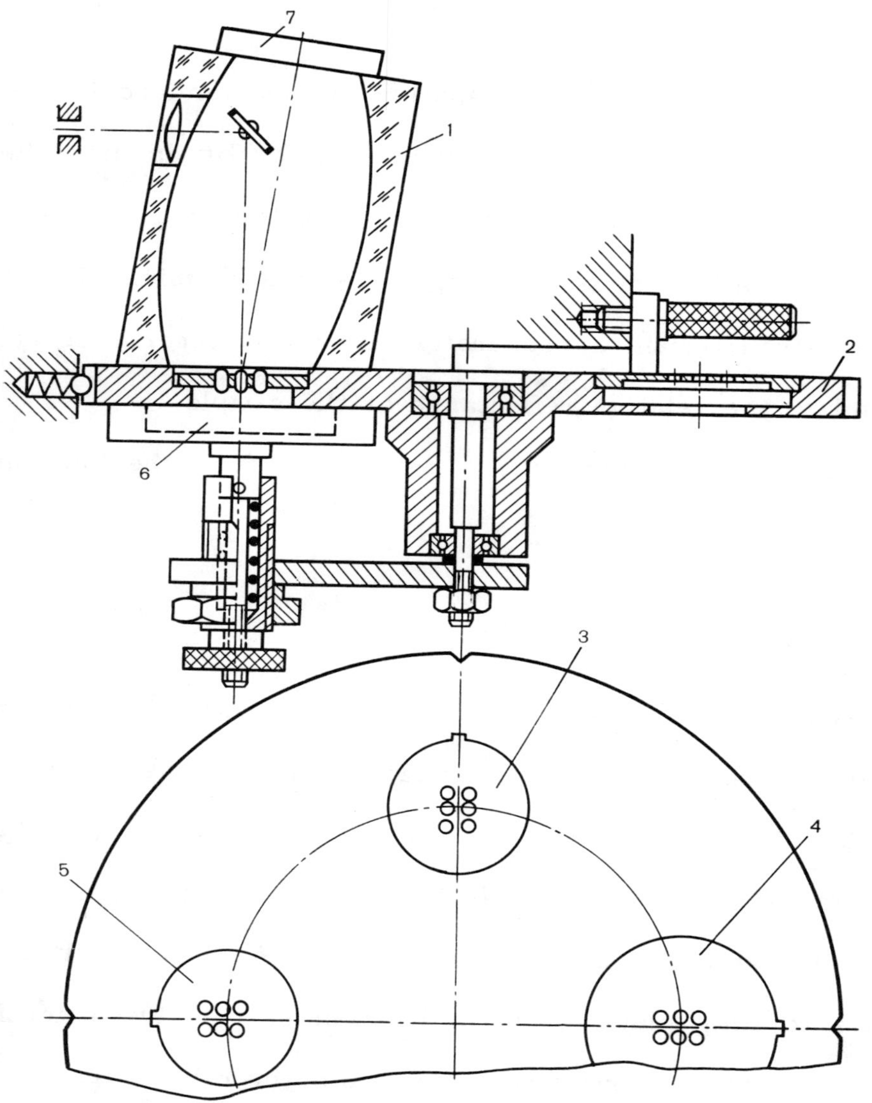

Figure 5.14 Design of an attachment for PDO-1 for the simultaneous measurement of $R_\lambda(\theta: 2\pi)$ and $T_\lambda(\theta; 2\pi)$ of individual grains.

5.6 Double-beam instruments for measuring the hemispherical thermal radiational characteristics of foodstuffs at different incidence angle of the radiant flux

Methods involving the use of a specular hemisphere and an integrating sphere are the most widely used methods for determining the thermal radiational characteristics of foodstuffs under conditions of directional irradiation at a certain incidence angle θ. The first of these methods has certain drawbacks and requires further improvement.

When designing double-beam infrared spectrometers an integrating sphere is more suitable for purposes of measurement. The method based on an integrating sphere gives the most precise values for R_λ and T_λ for diffusely scattering substances in the 0.2-2.6 µm spectral range. An integrating sphere is a hollow sphere whose inner wall is coated with diffusely scattering substances: $BaSO_4$, Al_2O_3, MgO, SiO_2, different enamels, etc [33, 54]. Because of the selective nature of the reflection of these coatings the use of an integrating sphere is limited to the 0.2-2.6 µm spectral range.

There are patented methods [166, 186] for making such coatings as an MgO layer, a mat gold layer, an elementary sulfur layer, or a combination of these layers, a sulfur layer dispersed from ethanol, and a thick sulfur layer (~3.0 mm) obtained from a benzene solution. These coatings provide a selective non-diffuse reflection indicatrix with a narrow range of intense reflection from 0.2 to 2.6 µm, 2-5 µm, 5-10 µm and 10-15 µm. Therefore, such coatings are insufficiently effective and are not a solution to the problem.

Several methods are known for preparing nonselective diffusely reflecting coatings for integrating spheres which are effective in the 1.0-1000.0 µm infrared spectral range. Such coatings include the following:

- a wavy surface with a certain pitch. It is made by pressing a surface with a wavy glass sphere having 7 mm steps [176]; by gluing with epoxide resin small spherical beads (60-80 mesh) to the surface [191]; or by gluing glass beads ~400µm in diameter to the surface [133];

- a rough surface; it is made by sandblasting the surface, or by gluing abrasive paper with grains ~83 μm in diameter to the surface [191];

- a specially made wavy surface [39].

Wavy or rough surfaces obtained by these methods, after being sprayed with gold, silver, or aluminium, are nonselective with respect to diffuse reflection in the near, middle and far infrared spectral regions, and are therefore effective coatings for integrating spheres.

The method for obtaining a wavy glass surface has some disadvantages. It is extremely labor-consuming to press out numerous undulations by local heating of the glass to the softening point. The step (7 mm) is too large in comparison with the 1-5 μm for the wavelength in the near infrared region. This results in a weak diffuse reflection indicatrix. The deviation of the reflection indicatrix from a perfectly diffuse one in the short-wave infrared region amounts to 20-60% at large observation angles.

Rough surfaces in integrating spheres obtained by sandblasting or gluing abrasive paper to the surface are less effective (Table 5.4) in comparison with wavy surfaces owing to the former's strong absorption. And the reflectance of a rough surface is less strong, while the reflection indicatrix is of a highly selective nature and has a strong specular component in the far infrared region.

Data in Table 5.4 and 5.5 indicate that aluminium coatings are highly effective on a wavy surface. The characteristics of the indicatrix are determined with the aid of the IPO-12 attachment, and the reflection coefficients - by the methods involving the use of an integrating sphere, a specular ellipsoid, and a hemisphere [39].

The efficiency of an integrating sphere E^* is determined by the ratio of the measured radiant power F_1 to that entering the sphere F_2:

$$E^* = F_1/F_2 = S_2 R_{\lambda 1}/S_1(S_1(1-R_{\lambda 1}): \qquad (5.10)$$

where, S_2 - area of radiation detector (usually, $S_2/S_1 = 0.01$).

Table 5.4 Spectral characteristics of integrating spheres with different types of surface coatings at different wavelengths (μm)

Surface	0.5	1.0	5.0	10.0	15.0	25.0	Ref.
Al-coated	0.864	0.916	0.969	0.973	0.974	0.975	39
Al-coated sand-blasred glass (particle size 45 μm)	-	0.90	0.95	0.94	0.82	0.59	178
Au-coated abrasive paper (particle size 83 μm)	0.447	0.713	0.790	0.784	0.737	-	213
Ditto	-	-	-	0.953	-	-	191
Ditto (particle size 37 μm)	0.443	0.806	0.808	0.854	0.847	-	213
Ditto	-	-	-	0.918	-	-	191
Al plane mirror	0.905	0.932	0.977	0.981	0.983	0.986	223
Au plane mirror	0.477	0.991	0.994	0.994	0.994	0.994	224

Table 5.5 Angular reflection characteristics of an Al-coated wavy surface at different reflection angles [39]

Wavelength (μm)	10°	20°	30°	40°	50°	60°	70°	80°
1.0	0.992	0.965	0.666	0.450	0.360	0.298	0.280	0.212
2.0	0.993	0.962	0.642	0.450	0.360	0.308	0.282	0.218
15.0	0.991	0.961	0.587	0.361	0.287	0.239	0.209	0.122

Data in Table 5.4 can be used for comparing the efficiency of an integrating sphere with different coatings. According to Eq. 5.10 the efficiency of the sphere with aluminium coating at λ=10.0 μm is six times greater than that of a rough gold surface (E_λ^*=0.03 at R_λ=0.854; E^*=0.18 at R_λ=0.973, respectively).

The method involving the use of a specular hemisphere [145, 180] is the most widely employed for measuring directional-hemispherical $R_\lambda(\theta; 2\pi)$ and $T_\lambda(\theta; 2\pi)$ in the infrared spectral region when irradiation is carried out with a narrow flux at a small incidence angle (θ=5-10°). However, when using attachments in conjunction with a single- or double-beam spectrometer [33, 54], no account is taken of all the factors which determine the precision of the measurements. Some of these factors, such as radiation losses at the entrance slit and screening by the holder of the detector, are usually compensated for by introducing correction factors, k_R and k_T. Optical aberrations are reduced to a minimum by observing the condition $\delta \geq 0.1\rho$.

At the same time errors due to imperfect irradiation conditions are not taken into consideration: irradiation with a nonparallel diverging flux with zonal irregularities in the irradiance of the sample; zonal and angular dependence of the absorption coefficient of the coating of the detector; and the reflection and absorption of radiation by the entrance widow of the detector. Also not considered are the errors due to the spreading of the cross section of a narrow radiation flux leaving the sample and to the nonuniformity of its reflection and transmittance indicatrixes. It is assumed that the magnitudes of the correction factors k_R and k_T, depending on the characteristics of the attachments and the radiation detector, lie in the range of 1.06 to 1.8 [39, 54, 63, 107, 139, 166] and are constant at all the wavelengths in the infrared region. This introduces uncontrollable sources of errors into the measurements. It is more difficult to take into account the angular dependence of the absorption coefficient of the highly sensitive photo cell at different wavelengths.

All radiation-registering devices are also characterized by the nonuniform distribution of zonal sensitivity in the radiation-receiving area [81, 185]. The sensitivity of the thermal registering devices is as a rule somewhat greater (by 5-20%) near the area of contacts, and then drops sharply to zero. The surface of the photo cells is not uniformly sensitive, the sensitivity being greater near the area of contacts.

The effect of the nonspherical nature of the hemisphere and of the zonal nonuniformity of its coating in the case of samples with a diffuse indicatrix can be determined experimentally with the aid of reference standards (MS-14 glass and mirors). But for samples with an imperfect diffuse indicatrix a regular negligible error is introduced. The optical aberrations cause a considerable distortion of the shape and dimensions of the image, and lead to the nonuniform zonal distribution of the radiant flux in the image; i.e., on the surface of the radiation detector.

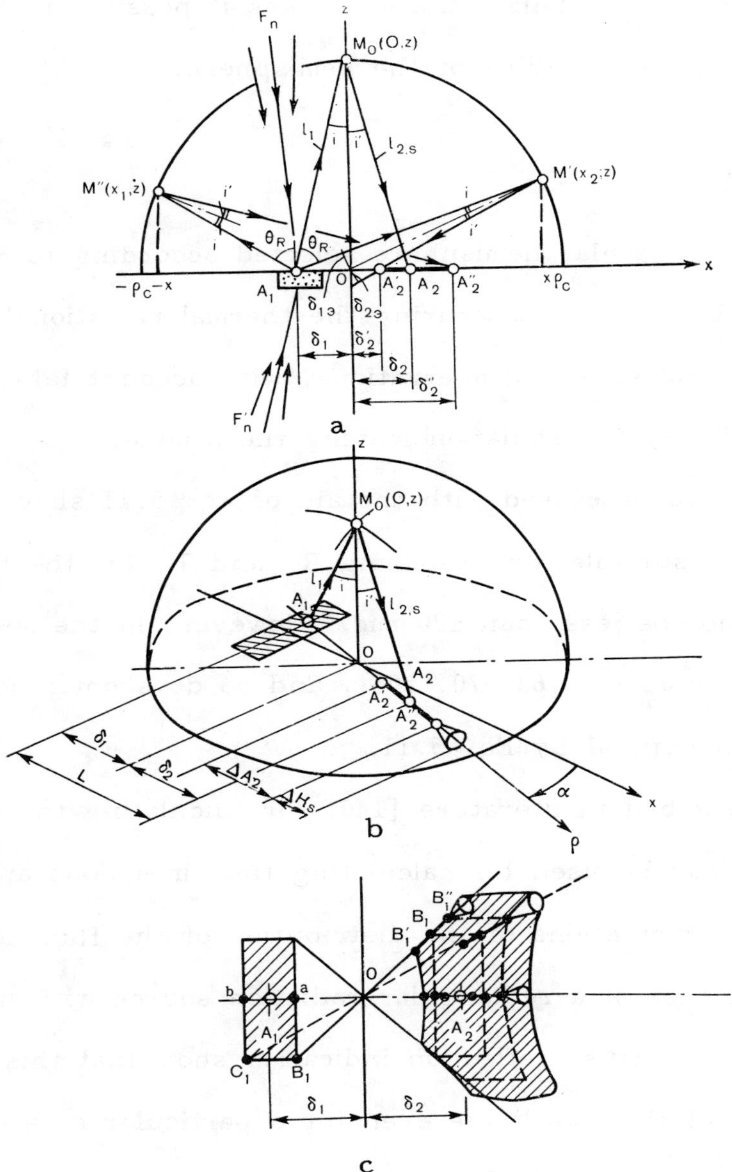

Figure 5.15 Determining the image of a rectangular source with the aid of a specular hemisphere [39, 65, 71, 137, 140]: a - sagittal astigmatism for rays coming from point A at different angles θ_R (or θ_T); b - image of a point source distorted by sagittal astigmatism ΔA_2 and sagittal aberration ΔK_s; c - schematic diagram showing how an image for a rectangular source is constructed.

Errors due to optical aberrations can be reduced to a minimum under the condition $\delta \leq 0.1\rho$. However, it is necessary in this case to compensate for the spreading of the cross section of the outgoing radiation flux (Figs. 2.15 and 5.2). The shift in δ (Fig. 5.15) from the center of the sample to the center of the specular hemisphere should not be less than r_{ef} of the radiant flux reflected or transmitted by the sample: $\delta > r_{ef}$. This limitation makes it possible to arrive at a new condition defining the minimal radius of the hemisphere:

$$\rho_{min} > 10 r_{ef}. \qquad (5.11)$$

The radius of the specular hemisphere selected according to Eq. 5.11 is the minimal one that can be used for measuring the thermal radiational characteristics of foodstuffs with minimal spherical aberrations, with account taken of the spreading of the cross section of the radiation leaving the sample.

Data from Table 5.2 processed with the aid of Eq. 5.11 show that the minimal radius of a hemisphere suitable for measuring R_λ and T_λ for the foodstuffs under investigation should not be less than 120 mm. However, in the setups used in practice ρ equals 55 or 75 mm [63, 70. 106], and so does not meet the requirements of the condition defined by Eq. 5.11.

The methods descibed in literature [140] for calculating the aberrations in a perfect hemisphere can be used for calculating the dimensions and the shape of the distorted image. Calculations of the distribution of the flux density within the image [65] in the case of a rectangular radiation source with uniform radiance $E(\rho,\alpha)=1$ for a perfectly diffuse reflection indicatrix show that this distribution is a complex function of the coordinate even for a particular case.

To elucidate the main causes of optical aberrations (astigmatism and coma), we shall analyze a ray diagram of the fluxes scattered by the sample (reflected or transmitted) into the hemisphere (Fig. 5.15. a and b). Since the symmetry of the hemisphere is of a circular nature we shall, for the sake of simplicity, consider the cross section which is perpendicular to the equatorial plane of the hemisphere. The sample and the radiation detector are located at conjugated points A_1 and A_2, respectively, inside the hemisphere. When irradiated from above (with flux F) the sample reflects, and when irradiated from below (with flux F') it transmits the flux into the hemisphere. These scattered radiant fluxes F_R and F_T are focused by the specular surface of the hemisphere at conjugated points A_2 and are characterized by the complex spatial distribution of their intensity. Depending on the indicatrix of reflection and of transmission, the rays leaving at θ_R and θ_T angles from one point A_1 (located at distance δ_1 from the center of the hemisphere O) are focused at several conjugated points – A_2', A_2, A_2'' etc at distances δ_2', δ_2, respectively. The equality $\delta_2=\delta_1$ of the displacements is possible only for a ray scattered by the sample in the direction of the hemisphere apex $M_0(O,z)$ when $\ell_1=A_1M_0$.

The reflection from the specular surface of the hemisphere of an arbitrary ray scattered by the sample fulfills the law of reflection: the incidence angle i equals the reflection angle i' ($i=i'$). Therefore, the rectangular triangles A_1M_0O and A_2M_0O with a common side M_0O and the same angles at the apex $i=i'$ are equal to one another, and the displacements of A_1 and A_2 are also equal, $\delta_2=\delta_1$, since $A_1O=A_2O$. For all other directions of θ_R and θ_T, with the rays reflected at points $M'(x>0)$, the shift of A_2' points from the center O is smaller, $\delta_2'<\delta_2=\delta_1$; with the rays reflected at points $M''(x<0)$, the shift of A_2 points is greater, $\delta_2''<\delta_2=\delta_1$. These inequalities are obtained by geometrical construction under the conditional transfer of the apex M_0 until it is superimposed with M' and M'' points, and by the effective displacements $\delta_{1ef}=\delta_{2ef}$ (Fig. 5.15, a).

The main ray ℓ_1 is incident upon the specular surface at point $M_0(O,z)$ at an angle i to the normal (radius), and the distance to the luminous point A_1 equals ℓ_1. Here, the astigamtic tangential and sagittal focal distances parallel to the main ray are $\ell_2\tau$ and $\ell_2 s$, respectively, and are related to the spatial radius ρ as follows:

$$1/\ell_1 + 1/\ell_{2,\tau} = -2/\rho\cos i; \qquad\qquad (5.12)$$

$$1/\ell_1 + 1/\ell_{2,s} = -2\cos i/\rho. \qquad\qquad (5.13)$$

From Fig. 5.15,a it can be seen that when ray ℓ_1 is incident upon point $M_0(O,z)$ the relation $\ell_1=\rho/\cos i$ is valid, and it follows from Eq. 5.13 that $\ell_{2,s}=\ell_1$; i.e., the A_1 source and the sagittal astigmatic image (ΔA_2 line, Fig. 5.15,b) lie in the same plane which is perpendicular to the axis. The length of the line ΔA_2 can approximately be taken as $(\ell'_{2,s} - \ell_{2,\tau})$. On the basis of this we can determine the maximum permissible value of ΔA_2 and the maximum distance $L=\delta_1+\delta_2$ between the A_1 source and the A_2 image at given values of the astigmatism and of the specular radius ρ of the hemisphere.

However, the presence of a coma limits the possible value of L regardless of the radius ρ of the hemisphere.

For a spherically symmetrical system under 1:1 magnification the linear dimension of the sagittal coma, regardless of ρ, is:

$$\Delta K_s=(3/16)L. \qquad\qquad (5.14)$$

The total expansion of the image Δab of the source having ab width, with account taken of the coma and the difference between the astigmatism of the outer and the inner parts of the image (Fig. 5.15,c), is:

$$\Delta ab=\Delta K_s + \Delta a/2 + \Delta b/2. \qquad\qquad (5.15)$$

Thus, owing to optical aberrations the image of the arbitrary point A_1 is in the form (Fig. 5.15,b) of a segment of a line ΔA_2 (astigmatism) with a divergent cone (coma) ΔK_s, the length and the radiance of which depend on the scattering indicatrix of the sample and on the shift of the image δ_1 from the center of the sphere. The outer points of the image A'_2 and A''_2, when radiation is scattered into the hemisphere ($\omega_R=2\pi$; $\omega_T=2\pi$), can be determined from Eqs. 5.12 and 5.13 which are for rays scattered in the directions θ_R and $\theta_T=\pm\pi/2$ when the incidence angle i=0, cosi=1, and $\ell_1=\rho\pm\delta_1$.

By determining ΔA_2 for each A_1 point of the E_1 source, we can construct the image of E_2 when the irradiation point E_1 is of an arbitrary shape. In Fig. 5.12,c is shown an approximate diagram of the image E_2 for a rectangular E_1 source. This source represents a cross section of the radiant flux incident upon the sample (the image of the slit of the monochromator on the surface of the sample). Under real conditions it is necessary to take into account the shape and the dimensions of the cross section of the radiant flux leaving the sample. As shown above (Fig. 5.4), the effect of the spreading of the cross section of the radiant fluxes leaving the sample is of a complex nature. It differs considerably from that shown in Fig. 5.15, c.

The shift δ_2 of the image from the center of the hemisphere is defined as follows [140]:

$$\delta_2=\delta_1\rho^2/(\rho^2+2x\delta_1)=\delta_1/[1+2\delta_1(x/\rho^2)]; \qquad (5.16)$$

where, x - coordinate of the point M(x,z) on the surface of the hemisphere into which the radiant flux scattered by the sample in the θ_R and θ_T directions ($x_1\le x\le x_2$; here, $0>x_1\ge-\rho$, $0<x_2\le\rho$) enters.

If the reflection indicatrix of the sample $r(\theta,\theta_R)$ or the transmission indicatrix $t(\theta,\theta_T)$ is known, the coordinate x can be expressed through the shift δ_1, ℓ_1 and $\sin\theta_R$ or $\sin\theta_T$:

$$x=\ell_1\sin\theta_R+\delta_1. \qquad (5.17)$$

It follows from Eq. 5.16 that at $0>x>-\rho$ the magnitude of the shift is: $\delta_2'>\delta_1$, and at $0<x-\rho_c$ - $\delta_2'<\delta_1$. For a particular case at x=0, for a flux reflected at point M_0, the shift is : $\delta_2'=\delta_1$. The position of the outer points A_2' and A_2'' is determined from Eq. 5.16 when x equals ρ and $-\rho$, respectively:

when $x=\rho$, $\delta_2'=\delta_1/[1+2(\delta_1/\rho)]$;   (5.18)

when $x=-\rho$, $\delta_2''=\delta_1/[1-2(\delta_1/\rho)]$.   (5.19)

It follows from Eqs. 5.18 and 5.19 that with an increase in the shift δ_1 of the point A_1 from the center of the hemisphere the magnitude of the segment of its image ΔA_2 increases since δ_2' decreases and δ_2'' increases. Consequently, the aberrations for a point source can be calculated with the aid of a known relationship [140]:

$$|\sigma_{x=-\rho}| + |\sigma_{x=+\rho}|=4\rho f^2/(\rho^2-4f^2);$$ (5.20)

where, f - the distance between the center and the source.

Next we shall determine the distribution of the radiant flux $E_2(\rho_2,\alpha)$ incident upon the surface of the infrared detector located in the center of the image. According to the law of radiance:

$$dE_2(\rho_2,\alpha,\theta_2)=dF_2(\rho_2,\alpha)\cos\theta_2/dS_2:$$ (5.21)

where, $dF_2(\rho_2,\alpha)$ - the radiation flux element incident upon point A_2, θ_2 - incidence angle at point A_2, dS_2 - area element according to the polar system of coordinates, which is equal to $\rho_2 d\rho_2 d\alpha$.

When determining the magnitude of the incident flux dF_2 account should be taken of all the optical properties of the sample: R_λ and T_λ; the scattering indicatrixes $r(\theta,\theta_R)$ and $t(\theta,\theta_T)$; the spreading of the flux's cross section, determined by the functions of the distribution of the reflected and transmitted flux densities $j_R^*(\rho_1,\alpha)$ and $j_T^*(\rho_1,\alpha)$.

The radiation flux reflected from the sample at point A_1 under the condition of directional irradiation, $\Delta F(\theta)$, is:

$$E_1(\rho_1,\alpha m\theta_R)=\Delta F(\theta)R_\lambda(\theta,2\pi)r_\lambda(\theta,\theta_R)j_R^*(\rho_1\alpha)/S_1;$$ (5.22)

where, S - the cross section area of the incident radiation flux (the cross section area of the image of the monochromator's slit on the surface of the sample). Then, the radiation flux element reflected in the θ_R direction, with account taken of Eq. 5.22, is defined as:

$$dF_1(\rho_1,\alpha,\theta_R)=(1/\pi)E_1(\rho_1,\alpha,\theta_R)\cos\theta_R dS_1 d\omega; \qquad (5.23)$$

where, $dS_1=\rho_1 d\rho_1 d\alpha$.

The incident radiant flux dF_2, with account taken of the reflection of the dF_1 flux from the surface of the specular hemisphere, is defined as:

$$dF_2(\rho_2,\alpha)=R_{\lambda c}(i)dF_1(\rho_1,\alpha,\theta_R). \qquad (5.24)$$

By substituting the magnitude of the radiation flux from Eq. 5.24 into Eq. 5.21 and by taking into account Eqs. 5.22 and 5.23, Eq. 5.21 for the irradiance at point A_2 can be written in the following form:

$$dE_2(\rho_2,\alpha,\theta_2)=(1/\pi)R_{\lambda c}(i)E_1(\rho_1,\alpha,\theta_R)\cos\theta_2\cos\theta_R(\rho_1 d\rho_1/\rho_2 d\rho_2)\varphi\omega; \qquad (5.25)$$

where, $R_{\lambda c}$ - reflection coefficient of the hemisphere's coating.

By intergrating Eq. 5.25 with respect to the hemisphere and by expressing the trigonometric functions through ρ_1 and ρ_2, the radiant flux at point A_2 of the image can be described by a relationship similar to that derived for a particular case of a perfectly diffuse radiation-emitting source, E_1, at $R_{\lambda,c}=1$:

$$E_2(\rho_2,\alpha)=(1/4\pi)\int_{\gamma(\rho_2,\alpha)}^{\beta(\rho_2,\alpha)}E_1(\rho_1,\alpha,\theta_R)f(\rho_1,\rho_2)d\rho_1. \qquad (5.26)$$

The $f(\rho_1,\rho_2)$ function for the diffuse source E_1 at $R_{\lambda c}=1$ [65] is of a sufficiently simple type:

$$f(\rho_1,\rho_2)=\{\rho^5(\rho_2-\rho_1)[4\rho_1\rho_2^2-\rho^2(\rho_1+\rho_2)^2]^{\frac{1}{2}}\}/(\rho_2-\rho_1)^2\rho_1^2\rho_2^3. \qquad (5.27)$$

The limits for integrating Eq. 5.26 are determined by the geometric dimensions of the radiation-emitting source in the ρ and α system of coordinates.

The distribution of $E_2(\rho_2,\alpha)$ has been determined [65] by calculating (with the aid of a computer) the radiation flux E_2 of the image under the condition of rectangular irradiation $E_1=1=$const for a series of δ_1/ρ values. As an example, these results are shown in Fig. 5.17.

However, for foodstuffs the said assumptions are not valid. This is due to the fact that the scattering indicatrixes of foodstuffs (Fig. 1.8) are not perfectly diffuse ones, and the $E_1(\rho_1,\alpha,\theta_R)$ function cannot be considered to be equal to unity since j_R^* and j_T^* are complex in nature (Fig. 2.15). Furthermore, the distribution of the flux density on the surface of the monochromator's exit slit is determined by the spread function of the spectrometer and can be approximately expressed by the Gauss distribution function [33].

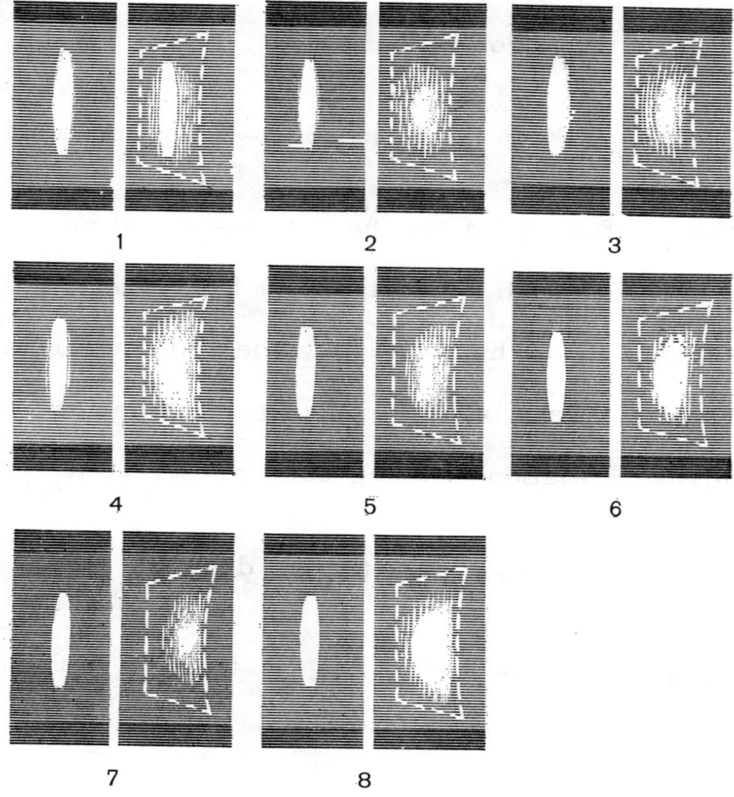

Figure 5.16   Distribution of radiation flux at conjugated points of a diametric plane of a hemisphere for the image of a source and for the image created during reflection by different substances [39, 50, 118]: 1 - mirror; 2 - foam polystyrene ($\ell=$ 3 mm); 3 - enamel VL-548 ($\ell=0.25$ mm); 4 - polytetrafluoroethylene ($\ell=2.2$ mm); 5 - paper ($\ell=0.09$ mm, 75 g/m^2); 6 - apple pulp ($\ell=5$ mm, w=85.4%); 7 - soft part of bread (grade 1 flour, $\ell=5$ mm, w=40.7%); 8 - potato starch ($\ell=2$ mm, w=12.3%).

Therefore, we have developed an experimental method for determining the radiation flux $E_2(\rho_2, a)$ inside the image area [39, 50, 118]. The photographic method has been used for investigating the combined effect of all the above-mentioned factors on the shape and dimensions of the image as well as on the distribution of the radiant flux inside the image. The samples and the radiation detector located at conjugated points A_1 and A_2 inside the hemisphere (Fig. 5.15) were covered with a special film (kinoinfra MRTU-43 No. 134162), which is sensitive to infrared radiation. The obtained pictures were of the type shown in Fig. 5.16. For purposes of comparison the images calculated according to Eqs. 5.18 and 5.19, with corrections made for aberrations, are also shown in the figure.

As can be seen from Fig. 5.16, the distribution of the radiant flux inside the image, E_1, and that inside the radiation detector, E_2, is nonuniform and depends on the optical properties of the sample. Observance of the condition $\delta_1 \leqq 0.1\rho$ makes it possible to reduce the optical observations to a minimum. The image thus obtained of the radiation "source" reflected by the reference standard (Fig. 5.16, frame 1) has clearly defined boundaries, but its shape is somewhat distorted. At the same time, the spreading of the narrow cross section of the beam emitted by the sample and intensified by optical aberrations of the system largely accounts for the nonuniform distribution of the radiation flux E_2 (Fig. 5.16, frames 2-8; Fig. 5.17).

To determine the function of the source E_1 and quantitatively estimate the spreading of the cross section of the narrow beam emitted by the sample, the flux densities for different cross sections of the radiation reflected and transmitted by the sample were measured with the aid of a special device (Fig. 5.1).

In the case of directional point-source irradiation, with R_λ and T_λ being measured by the specular hemisphere method, the dimensionless functions of E_1^* are determined, in accordance with Eq. 5.22, by the dependence on r of the relative flux density of the radiation which is reflected by the sample, $j_R^*(r)$, and that which is transmitted by the sample, $j_T^*(r)$, and by the scattering indicatrixes

$r(\theta;\theta_R)$ and $t(\theta;\theta_t)$, respectively.

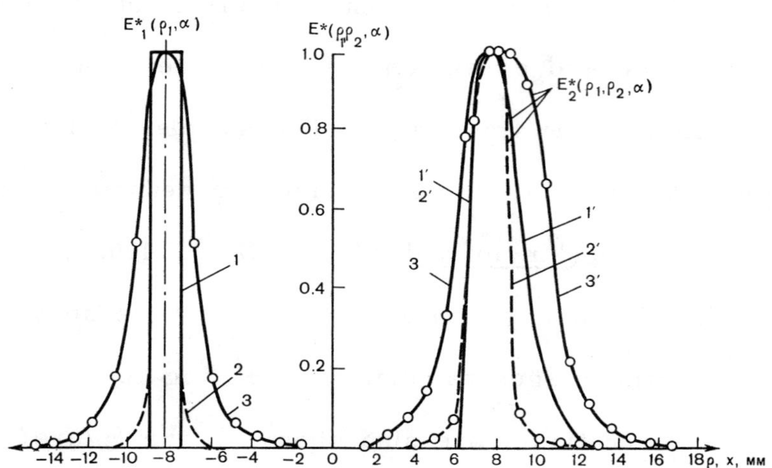

Figure 5.17 Distribution of the dimensionless values of the radiation flux $E^*(\rho_1, \rho_2, \alpha)$ in the image of the source $E_1^*(\rho_1, \alpha)$ (curves 1-3) and in the image of $E_2^*(\rho_1, \rho_2, \alpha)$ curves 1'-3') obtained by analytical and experimental methods [39, 50, 118]: 1,1' - Eq. 5.26 for a rectangular source $E_1^*=1=$const.; 2,2' - experimental data for the reflection standard: 3,3' - experimental data for potato starch.

The relative distribution of the flux density of the reflected radiation, $j_R^*(r)$, and of that of the transmitted radiation, $j_T^*(r)$, depending on the radius r of the cross section of the emitted flux (Fig. 2.15), indicates a considerable spreading of the cross section of a narrow beam in the investigated substances. This is due to multiple scattering of radiation on the optical inhomogeneities in a layer of the sample. The spreading of the cross section of a narrow beam and its scattering, intensified by the original angular divergence of the incident flux, leads to a notable spreading of the radiation flux emitted by the sample. The effective cross section of the radiation flux reflected by the sample in this case is also much larger than that of the incident radiation flux.

For a point-source radiation flux the radius of the cross section r_{ef} of the flux emitted by the sample, depending on the optical properties and the thickness

of the sample, reaches 8-10 mm and can be >14 mm (Figs. 2.15 and 5.2). The
cross section of the radiation flux emitted by the sample increases with an increase
in the thickness of the sample (Fig. 5.17, curves 1-3), and exceeds the dimensions
of the receiving area of the radiation detectors (12 mm in diameter [26]; 3×10 mm^2
[39]) used in the specular hemisphere method. Such radiation losses which depend
on the layer's thickness, are the source of considerable unaccounted-for errors.

The shape and dimensions of the image of the source E_1^*, at different thicknesses of a layer of the soft part of bread, are illustrated in Fig. 5.4 (the dimensions of the image of the slit width on the sample are 2×15 mm^2). As can be seen,
the dimensions of the image of the source, E_1^*, (24×42 mm^2) considerably exceed
those of the slit width on the sample. This can result in the violation of the condition under which the sample's shift from the center of the sphere should be: $\delta_1 \leq$
$0.1\ \rho$, since $r_{ef} > \delta_1$ [39, 62, 63].

Thus, we have obtained the values for the radius of the effective cross section
r_{ef} of the radiation fluxes emitted by the sample and of the dimensiones of the image of the source E_1^*, which are corrected for beam spreading. These values enable us to formulate a new condition (Eq. 5.11) for the minimal radius of a specular
hemisphere which can be used for measuring R_λ and T_λ at minimal aberrations:

$$\rho_{min} > 10 r_{ef}.$$

Eq. 5.11 also defines the condition under which the specular hemisphere method
can be used for measuring the R_λ and T_λ of light-scattering substances. In the
case where the radius of the effective cross section r_{ef} of the radiation emitted by
the sample equals or exceeds 0.1ρ, the errors involved will be large, and so the
specular hemisphere method should not be used here for determining the values of
R_λ and T_λ.

The obtained results explain why the values for the correction factor are too
high: $k_T = 2.0 \div 3.0$. This factor was introduced when the values for T_λ obtained
by the specular hemisphere method were compared with those obtained by other

methods (methods based on an integrating sphere or a photocell with a large detection area). This point is that the $r_{ef}(T)$ of the radiation flux transmitted by the sample for most foodstuffs is >10 mm (Fig. 2.15), while the shift of the slit image from the center of the hemisphere δ_1 is 7.5 mm or greater at $\rho=75$ mm. Therefore, the condition $\delta_1 < r_{ef}(T)$ is violated since $r_e hf(T) > \delta_1$, as is the condition $\delta_1 \leqq 0.1\rho$ for the minimum of aberration. At the same time the $r_{ef}(R)$ of the reflected radiation for a number of substances is smaller than $w_1=7.5$ mm $[r_{ef}(R) < \delta_1]$. Here the conditions for the measurements are fulfilled, and the magnitude of the coefficient k_R can be taken to be approximately 1.5 for the attachment used with the IKS-12 spectrometer.

If we have experimentally obtained functions $j_R(r)$ and $j_T(r)$, $r(\theta, \theta_R)$ and $t(\theta, \theta_t)$, we can use analytical methods for calculating the distribution of the radiation flux E_2^* at conjugated points on the image when measuring R_λ and T_λ.

As an example, in Fig. 5.17 a comparison is made of the functions E_1^* and E_2^* obtained by analytical methods (curves 1,1') and by experimental methods (curves 2,2' and 3,3'). The experimental data for the dimensionless functions of the source $E_1^*(\rho_1', \alpha)$ in the case of reflection from a mirror (curve 2) and from potato starch (curve 3) are obtained by the method described at the beginning of this chapter. The functions $E_2^*(\rho_1', \rho', \alpha)$ for the image (curves 2' and 3') are obtained by photometric measurements of the relative degree of emissivity produced in negatives by a laser beam 1.0 mm in diameter with the aid of an integrating sphere and the FEU-62 photometer.

A comparison of curves 1' and 3' shows that the spreading of the cross section of the reflected flux emitted by potato starch affects considerably the distribution of the radiation flux E_2^* and increases the size of the image. This distribution depends not only on the geometrical characteristics of the hemisphere and on the scattering indicatrix (or transmission indicatrix) of the sample. It also depends largely on the distribution of the radiant flux E_1^* emitted by the sample.

When we know the distribution of E_2^* and the shape and dimensions of the image (Figs. 5.16 and 5.17) obtained for a number of absorbing-scattering substances, we can choose correctly the shape and dimensions of the radiation detector and estimate the errors of measurements.

To eliminate the above-mentioned imprecision of the specular hemisphere method an improved procedure has been developed [39, 118]. It involves the use of an attachment for the IKS-12 spectrometer and a double-beam spectrometer based on the IKS-14 spectrometer. Unlike other attachments [63, 70, 145, 180] this one has, instead of a radiation detector in the focal plane of the specular hemisphere, a special integrating sphere operating in the infrared spectral region 1-1000 μm.

A schematic optical diagram of the attachment for IKS-12 for measuring the thermal radiational characteristics of foodstuffs irradiated at $\theta = 5 \div 10°$ is shown in Fig. 5.18. In determining $R_\lambda(\theta; 2\pi)$ and $T_\lambda(\theta; 2\pi)$ the sample (3) is located near the center of the specular hemisphere (1). At the conjugated point on the other side of the center is the integrating sphere (2) (50 mm in diameter). Its inner wavy surface provides diffuse reflection: the reflectance of the surface with a diffuse reflection indicatrix is 0.91-0.97 in the near- and middle infrared region.

The integrating sphere has two openings. A radiation detector 3×10 mm^2 is placed in the first opening (10 mm in diameter) [62, 63]. The dimensions of the second opening (20×12 mm^2) (4) have been calculated with the aid of Eqs. 5.18 and 5.19, with account taken of spherical aberrations, and have been made more precise with the aid of experimental photographic measurements of the image (Fig. 5.16).

As can be seen from Figs. 5.16 and 5.18, the cross section of the radiation flux reflected or transmitted by the sample (3) is smaller than the opening (4) (in Fig. 5.18 the opening is shown with dashed lines). Therefore, the measured flux wholly enters the integrating sphere where, being repeatedly reflected from the walls, it uniformly illuminates the inner surface of the sphere, and the illumina-

tion is recorded by the radiation detector. In this way the effect of optical aberrations of flux spreading in the sample, of the reflection and transmission indicatrixes, and of the nonuniform angular and zonal sensitivity of the radiation detector on the radius of the detector is excluded. The experimental method is the
same as the method that has been described by the present authors [39].

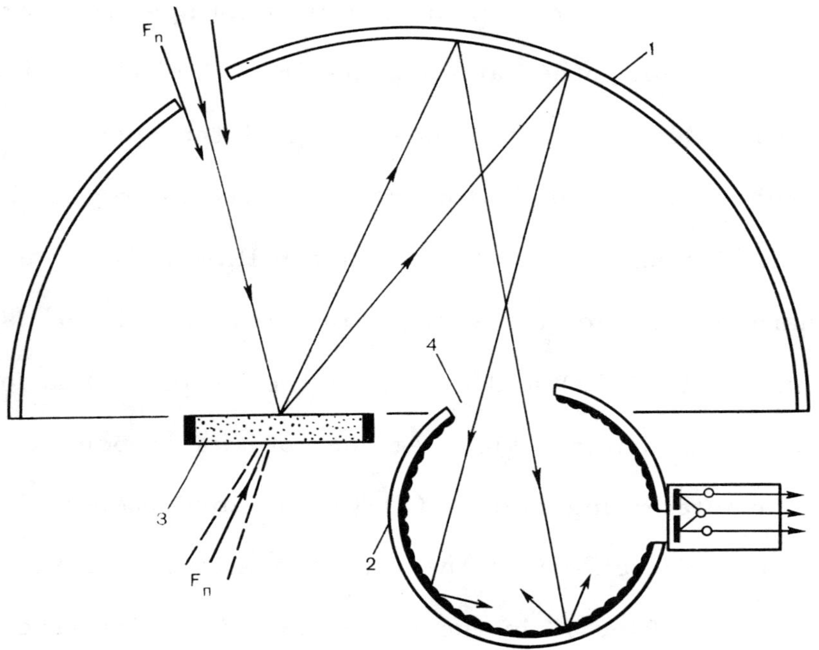

Figure 5.18 Schematic optical diagram of an improved attachment used in the single-beam specular hemisphere method [39, 118].

The IKS-14A spectrometer with a double-beam attachment is used for measuring the thermal radiational characteristics of foodstuffs according to an improved
specular hemisphere method (Fig. 5.19). The method of measuring $R_\lambda(1:2\pi)$ and
$T_\lambda(1:2\pi)$ is based on the known specular hemisphere method [145, 180]. With the
improved double-beam method the specular hemisphere (6) (150 mm in diameter) is
used in conjunction with a special integrating sphere (1) (50 mm in diameter) operating in the infrared spectral region. The inner surface of the integrating sphere
is coated with a nonselective diffuse reflecting substance whose reflectance in the

1–1000 µm range is 0.91–0.97.

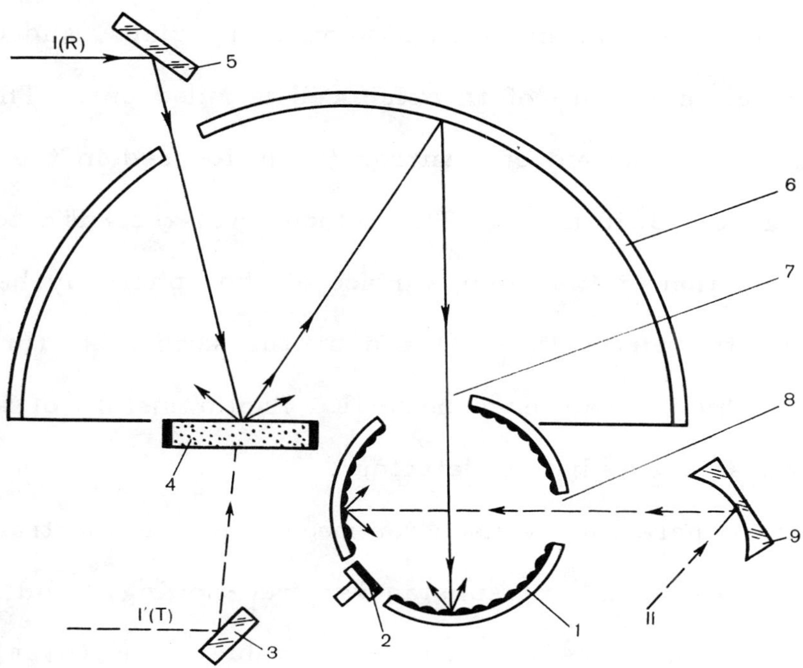

Figure 5.19 Schematic optical diagram of an improved double-beam attachment for the IKS-14A spectrometer used in conjunction with a specular hemisphere and an integrating sphere [39].

An optical diagram of the double-beam specular hemisphere and the procedure for measuring R_λ and T_λ with its aid have been shown [39]. Here, instead of a radiation detector, a special integrating sphere with three openings is used. The dimensions of the entrance opening (7) have been calculated with the aid of Eqs. 5.18 and 5.19, with account taken of spherical aberrations, and have been made more precise with the aid of photographic measurements of the image (Fig. 5.16). The dimensions of the entrance opening (7) in the integrating sphere are 20x12 mm^2.

As can be seen from Fig. 5.16, the cross section of the radiation reflected or transmitted by the sample (4) in the image is smaller than the dimensions of the opening (7). Therefore, the measured flux wholly enters the integrating sphere where it is repeatedly reflected from the walls. The resultant uniform illumination is recorded by the radiation detector (2). Thus, the readings are not affected by the nonideal irradiation conditions: the effect of optical aberrations of beam spreading in the sample, of the reflection and transmission indicatrixes, and of the non-uniform angular and zonal sensitivity of the photocell is ruled out. The radiation of the reference beam II with the aid of a mirror (9) is focused in the integrating sphere through opening (8) (3×10 mm^2). The detector (2) records alternately at 8.8 cycles/sec the illumination of the inner surface of the sphere by beam I after it has been reflected by the mirror (5 or 3) and by the sample (4) (or transmitted by the sample), and by the reference beam II. Photoresistors of the type 6AN and 40AN are used as the radiation detectors.

The device described here can be used for measuring the spectral reflectance and transmittance of different kinds of substances: free-flowing, solid, or liquid. For purposes of comparison we have summarized the values of $R_\lambda(\theta;2\pi)$ and $T_\lambda \cdot (\theta;2\pi)$ for different substances obtained by the double-beam specular hemisphere method at $\lambda=2.5$ μm, with no corrections being made for factors K_R and K_T (Table 5.6). It can be seen that the improved method helps considerably to reduce error and obtain $R_\lambda(\theta;2\pi)$ and $T_\lambda(\theta;2\pi)$ values which are close to the real ones. Here the radiation losses are due solely to the escape of scattered radiation through the entrance opening.

Because of the presence of an entrance opening in the hemisphere the correction factor should be introduced when calculating R_λ and T_λ. The losses of reflected radiation through the entrance opening (20 mm in diameter), when the reflection is of a diffuse nature, amount to about 3% of the total reflected energy. Losses due to the specular component can be large since most of this component can leave the hemisphere through the entrance opening. Such losses can be prevented by proper location of the entrance opening in the hemisphere.

The signal of the detector from radiation reflected by the sample is proportional to the illumination inside the sphere caused by the reflected flux that is being registered:

$$N_1 \equiv F_\lambda E_\lambda^* R_\lambda(\theta;2\pi)R_2(1-R_1+R_\lambda \mid R_2^2 R_1 - R_\lambda R_2^2 R_1^2 + \ldots). \qquad (5.28)$$

The signal from the reference beam is:

$$N_{II} = F_\lambda E_\lambda^* R_3(1-R_1+R_\lambda R_2^2 R_1 - R_\lambda R_2^2 R_1^2 + \ldots); \qquad (5.29)$$

where, R_3, R_2, and R_1 - reflectance of the mirror (9), the hemisphere (6), and the integrating sphere (1), respectively; E_λ^* - efficiency of the sphere defined by Eq. 5.10.

Such location of the sample, the sphere, and the hemisphere excludes the effect of multiple reflection between the sample and the sphere on $R_\lambda(\theta;2\pi)$. This is due to the fact that N_I/N_{II} ratio equals $R_\lambda(\theta;2\pi)R_2/R_3$, which equals the reflectance of the sample $R_\lambda(\theta;2\pi)$ when $R_2=R_3$. The quantity $R_\lambda(\theta;2\pi)$ determined in this way is smaller than the real one only owing to radiation losses through the entrance opening.

When measuring T_λ', the sample is irradiated with the aid of a mirror (3), and the detector's signals will be proportional to the radiant power of flux I' and II:

$$N_I \equiv F_\lambda E_\lambda^* T_\lambda(\theta;2\pi)R_2(1-R_1+R_\lambda R_2^2 R_1 - R_\lambda R_2^2 R_1^2 + \ldots); \qquad (5.30)$$

$$N_{II} \equiv F_\lambda E_\lambda^* R_3(1-R_1+R_\lambda R_2^2 R_1 - R_\lambda R_2^2 R_1^2 + \ldots). \qquad (5.31)$$

When $R_2=R_3$, the ratio $N_I/N_{II}=T_\lambda(\theta;2\pi)$; the magnitude of the transmittance is smaller than the real one only owing to radiation losses through the entrance opening in the hemisphere.

Therefore, the proposed double-beam method makes it possible to determine qualitatively the spectrum of diffuse transmission and reflection. The absolute quantities $R_\lambda(\theta;2\pi)$ and $T_\lambda(\theta;2\pi)$ are determined by multiplying the ordinates of the spectrum by the corresponding correction factors K_R and K_T, which have been found to be $K_R=K_T=1.05$.

Table 5.6　Reflectance $R_\lambda(\theta;2\pi)$ and transmittance $T_\lambda(\theta;2\pi)$ for some substances determined by the specular hemisphere method [39] (I) and by an improved version of the method (II) [118]

Substance	I		II	
	R_λ	T_λ	R_λ	T_λ
Polytetrafluoroethylene, ℓ=2.2 mm	0.020	0.848	0.034	0.908
Paper, ℓ=0.09 mm (75g/m^2)	0.208	0.128	0.268	0 144

Table 5.7　Transmittance and reflectance of some substances at different cross sections of the radiation flux emitted by the sample [118]

Substance	R_λ'	$R_{\lambda,3}'$	$\varepsilon(\%)$	T_λ'	$T_{\lambda,3}'$	$\varepsilon(\%)$
Polytetrafluoroethylene, ℓ=9.0 mm	0.899	0.325	63.8	0.49	0.002	95.9
Apple, ℓ=8.25 mm (w=83.4%)	0.898	0.282	68.6	-	-	-
Soft part of bread, ℓ=14 mm (w=11.3%)	0.780	0.118	84.9	-	-	-

To estimate the errors due solely to the spreading of the cross section of a narrow beam unaffected by spherical aberrations in measurements carried out with a RTE detector with a radiation receiving area of 3x10 mm, the $R_\lambda(\theta;2\pi)$ and $T_\lambda(\theta;2\pi)$ of substances were measured at different cross sections of the radiation flux emitted by the sample (Table 5.7). The measurements were made with the aid of a special device consisting of an integrating sphere, an iris diaphragm, and a LG-56 laser which generates a narrow parallel beam 2 mm in diameter at λ=0.63 μm. The $R_{\lambda,3}'$ and $T_{\lambda,3}'$ were measured with a diaphragm 3 mm in diameter, which corresponds to the radiation-receiving area of the RTE detector (3x10 mm^2) usually employed in the specular hemisphere method. The R_λ' and T_λ' were measured with the diaphragm (30 mm in diameter) being completely opened.

As can be seen from Table 5.7, the errors due solely to the spreading of the cross section of the radiation flux emitted by the sample are as high as 96%. Therefore, measurements of $R_\lambda(\theta;2\pi)$ and $T_\lambda(\theta;2\pi)$ made according to the improved method are more precise than those carried out according to the specular hemisphere method.

A double beam spectrometer has been developed for improving the precision of determining the directional-hemispherical thermal radiational characteristics of foodstuffs in relation to the incidence angle of the radiation flux in the infrared region. The spectrometer is designed as an attachment for the IKS-14A spectrometer and is used in conjunction with photoresistors operating in the range of 0.7-2.50 μm [39, 118]. The double-beam method makes it possible quickly and effectively to obtain at once the reflection and absorption spectrum of a sample. Like the SF-4A attachment, it can be employed for determining the absolute quantities of $R_\lambda(\theta;2\pi)$, $T_\lambda(\theta;2\pi)$, and $A_\lambda(\theta;2\pi)$ as a function of the incidence angle θ of the radiation at an arbitrary scattering indicatrix, without resorting to the use of reference standards.

The attachment consists of a special integrating sphere (1) (120 mm in diameter) whose inner surface is characterized by nonselective diffuse reflection in the short- and long-wave infrared region from 1 to 1000 μm (Fig. 5.20). The sphere has three openings. Through one of them, with the aid of a holder (2), the sample (3) is placed in the center of the sphere. In a second opening is a radiation detector (7), and a third opening serves as an inlet for the radiation incident upon the sample. The bottom of the sample can be shielded with a cap that is blackened inside (4). A screen (5) protects the radiation detector from direct exposure to radiation scattered by the sample. To obtain uniform illumination of the detector in the spectral region up to 2.0 μm an opal glass (6) is placed in front of it. When experiments are carried out at $\lambda > 2.0$ μm the glass plate (6) is removed. The sample holder, the screen, and the cap are made of polished aluminum.

With this attachment one can measure samples 30 mm long and 20 mm wide and from 1 to 10 mm thick. The optical diagram of an IKS-14A spectrometer has been modified for work with the attachment [39, 118]. Instead of a specular hemisphere (6) with a sphere (1) (Fig. 5.19) an integrating sphere (1) is used (Fig. 5.20). This integrating sphere is so placed that flux I and flux II (Fig. 5.19) are focused on the sample (3) (Fig. 5.20) with the aid of mirrors (8, 9, 10, 11). This provides monochromatic irradiation of the sample, thereby preventing the heating up of the sample during measurements. This is very important when measuring the thermal radiational characteristics of foodstuffs.

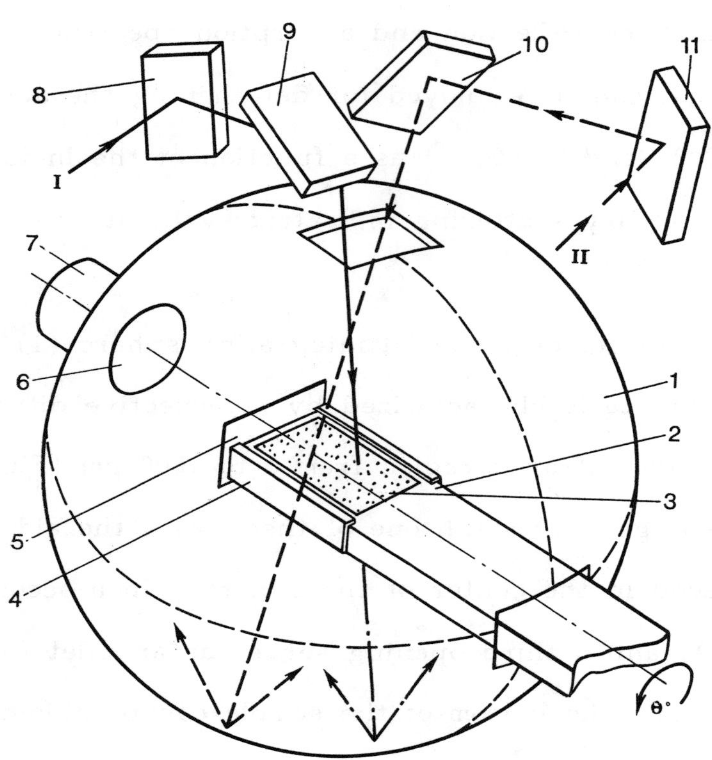

Figure 5.20 Optical schematic diagram of a double-beam attachment for the IKS-14A spectrometer for measuring $T_\lambda(\theta;2\pi)$ and $R_\lambda(\theta;2\pi)$ as a function of the incidence angle θ in the infrared region [39, 118].

To measure $R_\lambda(\theta;2\pi)$, $A_\lambda(\theta;2\pi)$, and $T_\lambda(\theta;2\pi)$ radiation flux I is focused (x1 magnification) on the outer surface of the sample through the entrance opening into the integrating sphere with the aid of plane mirrors (8 and 9). Radiation flux II is focused (x1 magnification) on the inner surface of the integrating sphere through the entrance opening, bypassing the sample, with the aid of plane mirrors (11 and 10). The determination of $R_\lambda(\theta;2\pi)$ and $A_\lambda(\theta;2\pi)$ of the sample at any scattering indicatrix is carried out as follows. The sample holder together with the sample and the cap is moved outside the sphere. Radiation fluxes I and II in this case fall on the inner surface of the sphere and are recorded as 100% by the detector. When I is blocked with a shutter the detector's reading is taken as 0%. The holder with the samples and the closed cap are then placed in the center of the sphere. In this case the irradiance of the sphere's surface is due to radiation reflected by the sample, and the detector records the reflectance of the sample $R_\lambda(\theta;2\pi)$.

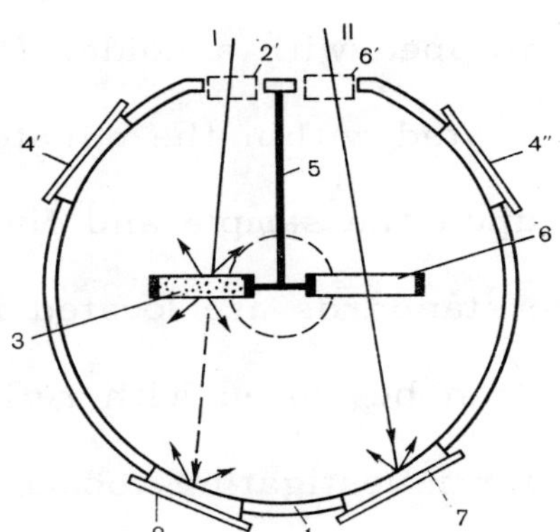

Figure 5.21   Schematic diagram of a device used with the SF-10 integrating sphere for measuring the absorptance $A_\lambda(\theta:2\pi)$ of foodstuffs [17, 39].

When the sample is placed in the center of the sphere, the cap (4) is removed, and the sphere is illuminated with the radiation flux reflected and transmitted by the sample. The registered spectrum corresponds to the sum $R_\lambda(\theta;2\pi)+T_\lambda(\theta;2\pi)$ of the sample. The difference between the 100% line and the recorded spectrum equals the absorptance since $A'_\lambda=1-(R'_\lambda+T'_\lambda)$. From the known $R_\lambda(\theta:2\pi)$ and $A'_\lambda(\theta;2\pi)$ we can readily calculate the transmittance of the sample $T_\lambda(\theta:2\pi)$: $T'_\lambda=1-(A'_\lambda+R'_\lambda)$.

The proposed device makes it possible to obtain directly a qualitative reflection or absorption spectrum of the sample. As compared with the single-beam attachment for SF-4A, the device considerably broadens the spectral region accessible to investigation, and there is no need for manual manipulation of the mirror (8), (Fig. 5.12). The integrating sphere and the holder for large samples are used to measure the R_λ', T_λ', and A_λ' of foodstuffs, with corrections made for the scattering and spreading of the directional radiant flux in the samples.

Next we shall describe an improved double-beam spectrometer SF-10 (SF-14) for measuring the absorptance $A_\lambda(\theta;2\pi)$ of foodsruffs. In carrying out such measurements the samples (2 and $2'$)(Fig. 5.21) are placed on the surface of the sphere (1). The spectra corresponding to the sum $R_\lambda(\theta;2\pi)+T_\lambda(\theta;2\pi)$ are directly recorded in the visible spectral region 0.40-0.75 μm. The difference between the recorded spectrum and the 100% line is the absorptance of the sample $A_\lambda(\theta;2\pi)$. The improved device is equipped with a holder (5) for sample (3) and absorption standards (6), which are located within the equatorial plane of the sphere (1) so that flux I is incident upon the sample and flux II – upon the standard. The samples and the reference standards are located in the holder (5) through peepholes ($4'$ and $4''$), which can be closed with reflection standards (7).

This device has been used [17] for investigating tobacco of different degrees of ripeness. It has been established that a relationship exists between $A_\lambda(\theta;2\pi)$ and the quality of tobacco in terms of its ripeness.

5.7 Instruments for measuring the hemispherical thermal radiational characteristics of foodstuffs subjected to diffuse irradiation

The conditions for infrared irradiation of different foodstuffs have been analyzed in Chapter 2. In industrial thermal radiation processing units, ovens and closed chambers the foodstuffs are subjected to diffuse or mixed (diffuse and directional) irradiation. It is therefore necessary to measure the bi-hemispherical $R_\lambda(2\pi;2\pi)$ and $T_\lambda(2\pi;2\pi)$ of substances irradiated with diffuse radiation flux. But most of the experimental results reported in literature have to do with directional irradiation at small angles ($\theta=5 \div 10°$) [19, 20, 84]. These results for directional-hemispherical $R_\lambda(\theta;2\pi)$ and $T_\lambda(\theta;2\pi)$ can be used in calculating radiant heat transfer corresponding only to these conditions.

The errors in calculating radiant heat transfer on the basis of data for $R_\lambda(\theta;2\pi)$ and $T_\lambda(\theta;2\pi)$ in the case of directional irradiation at $\theta \cong 5 \div 10°$ (and not on the basis of bi-hemispherical $R_\lambda(2\pi;2\pi)$ and $T_\lambda(2\pi;2\pi)$) can exceed 30-40% and depend on the scattering indicatrix of the substances investigated. In Table 5.8 are summarized the results of a comparison of the thermal radiational characteristics calculated with the aid of Eqs. 3.1 and 3.2 and those calculated with the aid of Eqs. 4.16 and 4.17 for different irradiation conditions for a layer of finite thickness $k_\lambda \ell = 1.0$ in the case of isotropic scattering $\chi\lambda(\gamma)=1$.

Table 5.8 Reflectance and transmittance under diffuse irradiation conditions [39]

Survival probability of photon Λ	$\theta=0°$ $R_\lambda(\theta;2\pi)$	$\omega=2\pi$ $R_\lambda(2\pi;2\pi)$	Relative deviation (%)	$\theta=0°$ $T_\lambda(\theta;2\pi)$	$\omega=2\pi$ $T_\lambda(2\pi;2\pi)$	Relative deviation (%)
0.1	0.017	0.025	32.0	0.380	0.150	153.3
0.5	0.107	0.162	34.0	0.451	0.236	91.0
0.9	0.274	0.403	32.1	0.601	0.422	42.5

An increase in $R_\lambda(2\pi;2\pi)$ and a decrease in $T_\lambda(2\pi;2\pi)$ under diffuse irradiation conditions (Table 5.8) are explained by the fact that with an increase in the incidence angle θ $R_\lambda(\theta;2\pi)$ increases, while $T_\lambda(\theta;2\pi)$ decreases. This phenomenon has been confirmed experimentally [48] by investigating spectral dependences $R_\lambda(\theta;2\pi)$ and $T_\lambda(\theta;2\pi)$ as a function of the incidence angle θ for a number of foodstuffs (Figs. 6.22 and 6.23).

An attachment for the IKS-21 spectrometer has been developed [39, 118] for measuring the spectral and angular characteristics of foodstuffs subjected to hemispherical irradiation in the infrared region from 1 to 1000 μm (Fig. 5.22). The spectral and angular characteristics of the nonselective, diffusely reflecting coating of the sphere (6) and the screen (11) are shown in Tables 5.4 and 5.5. To ensure identical conditions of irradiation at any observation angle, and to prevent errors caused by the rotation of the sample holder (such errors occur with the use of a known device [140]), a stationary, diffusely scattering screen (11) is installed. This eliminates errors due to the possible nonuniform spraying of the coating of the cylindrical, diffusely reflecting surface of the sample, to the nonuniform optico-geometrical properties of the coating and its selectivity, and to the heating up of the sample.

When the integral radiation flux enters the sphere (6), it is twice reflected from the same areas of the stationary screen (11) and once from the sphere's coating, providing uniform irradiance inside the sphere. At any observation angle a stable diffuse radiation flux is incident upon the surface of the sample (8), which is thrice reflected from the scattering surface (11) and from the sphere (6). The rotational sample holder is fastened to a hollow tube 8 mm in diameter through which water is circulated to keep the sample at a constant temperature. This makes it possible to measure the radiance coefficient as a function of temperature. The holder can be rotated within the polar angle θ from 0 to 90° in 5°-intervals. For carrying out measurements in relation to the azimuthal angle ϕ the sample can be rotated around the axis perpendicular to its surface.

The integral radiation flux from the entrance (1) is focused by mirrors (3, 4, 5) on the stationary screen (11). To exclude the effect of radiation reflected from the sample the readings are modulated with the aid of a circuit breaker (2). The radiation reflected from the sample passes through an exit opening (14x28 mm^2) in the sphere and is focused by a flat mirror (9) and a spherical mirror (12) on the entrance slit (10) of an ISK-21 monochromator. The surface of the integrating sphere serves as the reference standard.

With a slight shift in the position of the flat mirror (9) to (9') (shown with dashed lines in Fig. 5.22) the radiation flux which is repeatedly reflected from the surface of the integrating sphere can be directed into entrance slit (10). The radiation flux from the sample and reference flux both propagate within the limits of the same solid angle formed by the mirror (12) of the IKS-21 illuminator. After that the modulated signals received by the detector are recorded. The temperature of the sample is recorded with the aid of a Cr-Cu thermocouple and a PP-63 potentiometer.

The values for the irradiance coefficient at different observation angles are used to determine the hemispherical $R_\lambda(\theta,\phi; 2\pi)$ of the sample under different irradiation conditions.

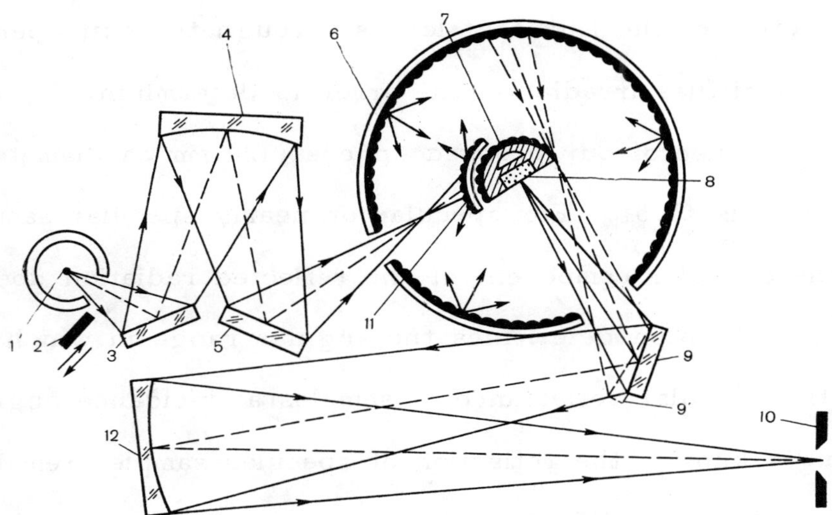

Figure 5.22 Schematic diagram of an attachment for the IKS-21 spectrometer for measuring the irradiance coefficient under diffuse irradiation conditions [39, 118].

According to the Helmholtz principle of the reversibility of radiant flux in isotropic media, at the same incidence angles θ and ϕ and reflection angles θ_R and ϕ_R, we have:

$$\rho_\lambda(2\pi; \theta_R, \phi_R) = R_\lambda(\theta, \phi; 2\pi). \qquad (5.32)$$

This equation serves as a theoretical basis for the method of measuring the coefficients of nonuniform diffuse reflection $R_\lambda(\theta, \phi; 2\pi)$ during directional irradiation with a narrow flux determined by the angles θ and ϕ within the limits of the solid

angle $d\omega$.

Thus, the attachment described above can be used to measure the intensity of

the radiation reflected by the sample in the direction of the angles θ_R under hemi-

spherical irradiation conditions, and to compare it with the intensity of the radia-

tion incident upon the surface of the sample. The ratio of the detector's signals is

proportional to the hemispherical radiance coefficient $\rho_\lambda(2\pi; \theta_R, \phi_R)$ and to the

hemispherical reflectance $R_\lambda(\theta, \phi; 2\pi)$.

The presence of an exit opening also leads to errors in determining R_λ. The

reason is that the irradiation of the sample is less intense than hemispherical ir-

radiation by a fraction of the flux that leaves through the exit opening. For a

sample subjected to diffuse irradiation the error in determining R_λ due to the loss

of radiation, with the use of our propsed sphere (120 mm in diameter, with an exit

opening 14x28 mm^2), is <1.5%. For specular or nearly specular samples the error

is negligible if the specular component of the reflected radiation does not get into

the opening. This condition determines the angular range in which measurements

can be made of the specular reflectance of samples at incidence angles <13.5°. At

the same time, as is known, the reflection of specular samples remains practically

the same at incidence angles <15-20°.

The reproducibility of $\rho_\lambda(2\pi; \theta_R)$ values for one and the same sample lies

within ±1.0%.

The bihemispherical $R_\lambda(2\pi : 2\pi)$ are determined by integrating the values for

$\rho_\lambda(2\pi; \theta_R, \phi_R)$ according to Eq. 1.46 and Table 1.2 with respect to $\omega_2 = 2\pi$.

It is difficult, however, to apply the experimental-analytical method in practice.

The reason lies in the complex nature of the integration functions $\rho_\lambda(2\pi; \theta_R, \phi_R)$

and $R_\lambda(\theta, \phi; 2\pi)$ with respect to the solid angle $\omega_R = 2\pi$ even when the reflection

indicatrix is symmetrical in form.

The attachment for IKS-21 makes it possible to carry out not only absolute

measurements, but also relative measurements of the spectral coefficient $R_\lambda(\theta, \phi;$

2π) and the integral coefficient $R_\lambda(\theta,\phi;\ 2\pi)$. For this purpose a plane mirror is placed behind the exit slit of the IKS-21 monochromator to focus the integral flux on the exit slit. The quantities for integral R' are measured in a similar way as those for R'_λ.

The spectral reflection coefficient of the sample in the case of relative measurements at a given observation angle is calculated according to the following relationship:

$$R_\lambda(\theta.\phi;\ 2\pi)=(N_o/N_{ref})R_{\lambda ref}(\theta,\phi;\ 2\pi);\qquad\qquad (5.33)$$

where, N_o and N_{ref} - readings corresponding to the reflection of the sample and that of the reference standard, respectively; $R_{\lambda ref}(\theta,\phi;\ 2\pi)$ - reflection coefficient of the standard.

In Figs. 5.23 and Table 5.9 are shown data obtained by different methods. The characteristics for $1'$, $2'$, and $3'$ were obtained by the specular hemisphere method (with the use of an IKS-12 spectrometer and an LiF prism in the 2.0-50 μm range and a NaCl prism in the 5.0-10 μm range). The correction factor for the entire spectral range of 2-10 um was taken to be 1.5. The characteristics for 1, 2, and 3 were obtained with the aid of an attachment for IKS-21 with the use of of a NaCl prism only and without introducing the correction factor.

As can be seen, the experimental results obtained by different methods are in good agreement with one another. Some differences arise owing to the fact that the correction factor, determined on the basis of a comparison of the values for R'_λ at $\lambda=1.2\div1.4$ μm and considered to be 1.5 during the measurements made with an IKS-12 spectrometer, does not remain constant as assumed when there is a change in the wavelength. It changes with a change in the wavelength of the reflection indicatrix and with a change in the cross section of the reflected radiation flux emitted by the sample as a result of its scattering and spreading inside the layer. Therefore, losses in the reflected radiation. which are not recorded by the detector, are dependent on the wavelength of the radiation.

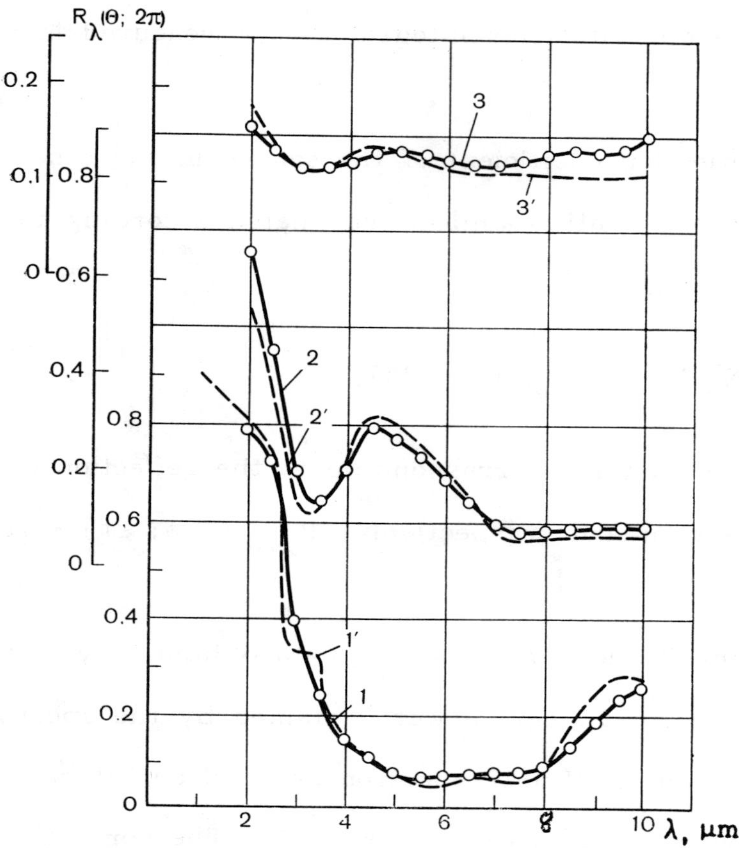

Figure 5.23   The dependences of the hemispherical radiance and reflection coefficients $\rho_\lambda(2\pi;\ \theta_R)=R_\lambda(\theta;\ 2\pi)$ on the wavelength, measured by the integrating sphere method (with the use of an IKS-21 spectrometer) (solid lines 1, 2, 3) and by the specular hemisphere method (dashed lines 1', 2', 3') for different substances [39, 118]: 1,1' - opal glass MS-14; 2,2' - VL-548 enamel; 3,3' - bread crust.

A device for the simultaneous measurement of the bihemispherical thermal radiational characteristics of foodstuffs subjected to diffuse irradiation is shown in Fig. 5.24 [48]. It consists of an integrating sphere and two radiation detectors with a large radiation-receiving area. The device is worked out on the assumption that the radiation fluxes passing through the sample and the diffuse flux incident upon the sample undergo diffusion to the same degree.

For the simultaneous determination of the absolute bihemispherical transmittance

$T_\lambda(2\pi; 2\pi)$ and reflectance $R_\lambda(2\pi; 2\pi)$ in the spectral range of 0.4–1.4 μm the device used consists of an integrating sphere (4) (100 mm in diameter) with a screen (1), three openings (Fig. 5.24,a) and two FESS-U10 photocells (2 and 8) (32 mm in diameter). The inner surface of the sphere is coated with a diffusely scattering substance (BaSO$_4$). The monochromatic radiation flux emerging from the slit in the SF-4A monochromator falls upon the inner surface of the sphere through the entrance opening (12 mm in diameter), and following multiple reflection it uniformly illuminates the sample (7) (20 mm in diameter). The illumination is recorded by the photocell (2). The opening (18 mm in diameter) in front of the stationary photocell (2) is closed with a diffusely scattering opal glass (3) (MS-13). The movable photocell (8) with a large energy-receiving area (32 mm in diameter) measures the diffuse radiation flux after it is transmitted by the sample (7). The photocell can be installed directly under the sample or under the 20-mm opening in the sphere (4). The magnitude of the signal from the photocell is measured with the aid of a UF-206 recorder, which can register currents up to 1μA.

Table 5.9 Experimental hemispherical radiance coefficients of an opal glass (MS-14) measured with the aid of an attachment for IKS-21 [39. 118] (left column) and by the specular hemisphere method [116] (right column)

Wavelength $\lambda(\mu m)$	$\rho_\lambda(2\pi; \theta_R)$ at $\theta_R=10°$	
2.0	0.77	0.78
2.1	0.76	0.75
2.2	0.75	0.73
2.3	0.73	0.71
2.4	0.70	0.69

The dimensions of the sphere and of the opening for holding the sample are so chosen that there is an additional illumination $\Delta E = 3.0 R_{\lambda 0} E$ [33. 39] inside the sphere. It is due to the radiation flux reflected by the sample , and is 2.5 times greater than in Duntley's device ($\Delta E = 1.33 R_{\lambda 0} E$) [150]. This can increase detector's signal with a better spectral resolution and improve the precision of the measurements. A shortcoming of the device is that there may be some difference between the conditions for the illumination of the photocell (8) (Fig. 5.24,a) which registers the radiation transmitted by the sample and by the reference standard. And this may result in errors.

To eliminate this shortcoming a modified Duntley method (a method based on two integrating spheres) has been used to measure $R_\lambda(2\pi : 2\pi)$ and $T_\lambda(2\pi; 2\pi)$ without an additional optical system.

Like Duntley's device, the attachment for SF-4A used for carrying out measurements (Fig. 5.24,b) consists of two optically identical integrating spheres (4 and 4y'). Two photocells (2 and 8) (FEU-62) are used for measuring the illumination of the sphere. The lower sphere (4') can be replaced with a blackbody (a sphere with a blackbody inner surface). The sample (7) and the reflection and transmission reference standards (6) (opal glass MS-14) are fastened to a rotating holder (5).

The integrating sphere (4') and the semitransparent, diffusely scattering glass (3) provide identical conditions for illuminating the photocell (8) by the radiation flux transmitted by the sample and the standard. In this way the above-mentioned errors inherent in the single-sphere method are excluded. The stationary screens (1) scatter the rays that are reflected or transmitted by the sample in the direction of the photocell.

The method for the simultaneous measurement of $R_\lambda(2\pi; 2\pi)$ and $T_\lambda(2\pi; 2\pi)$ with the aid of a single integrating sphere with two photocells (Fig. 5.24,a) involve the following operations.

1. The lower photocell (8) is placed directly in the opening (4) of the sphere in the absence of the sample (7). The signals N_{fR} and N_{fT} from the upper (2) an lower (8) photocell respectively are proportional to the irradiance of the sphere $E(R_f)$ and depend on the reflectance of the photocell R_f:

$$E(R_f)=(F/S)[R_s/(1-\phi'R_s)-R_fR_s(S_o/S)],\qquad(5.34)$$

$$N_{fR}=\alpha E(R_s),\quad N_{fT}=\beta E(R_f);\qquad(5.35)$$

where, ϕ' - the coefficient which determines the actual portion of the total surface area of the sphere S that participates in multiple reflection, and which takes into account the absorption of radiation by the openings and other i elements and is constant for a given sphere and independent of the R_o of the sample:

$$\phi'=1-(1/S)\sum_{i=1}^{n}(1-R_i)S_i-(S_o/S).\qquad(5.36)$$

2. The sample (7) is placed in the opening (4) of the sphere; the signals N_R and N_T from the upper (2) and the lower (8) photocell respectively are proportional to the irradiance of the sphere:

$$E(R)=(F/S)[R_s/(1-\phi'R_s)-RR_s(S_o/S)],\qquad(5.37)$$

$$N_R=\alpha E(R),\quad N_T=\beta E(R)T.\qquad(5.38)$$

In most cases air is used as the transmission standard ($N_{fT}\equiv N_{refT}$); the simultaneous solution of Eqs. 5.34-5.38 yields the following relationship for the absolute bihemispherical transmittance of the sample:

$$T_\lambda(2\pi;\ 2\pi)=(N_T/N_{fT})(N_{fR}/N_R).\qquad(5.39)$$

The absolute bihemispherical reflectance of the sample is determined on the basis of Eqs. 5.34 and 5.37, with account taken of Eqs. 5.35 and 5.38, respectively:

$$R_\lambda(2\pi;\ 2\pi)=(1-N_f/N_R)[(1-\phi'R_s)/R_sS_o/S+(R_fN_f/N_R)-\Delta R_\lambda];\qquad(5.40)$$

where, ΔR_λ - correction which takes into account the reflection from the lower photocell of the radiation flux transmitted by the sample. Approximately this correction is:

$$\Delta R_\lambda = T_\lambda^2 (2\pi; \ 2\pi) R_{\lambda f}.$$ (5.41)

For relative measurements of the sample's reflectance recording is made of the signal from the upper photocell (2) in the case where the reflection standard (6) (opal glass MS-14) is installed in the opening of the sphere:

$$N_{stR} = \alpha E(R_{st}) = \alpha(F/S)[R_s/(1-\phi'R_s) - R_{st}R_s(S_o/S)].$$ (5.42)

The transmittance of the sample $R_\lambda(2\pi; \ 2\pi)$ is determined on the basis of the following relationship, with account taken of Eqs. 5.34–5.38:

$$R_\lambda(2\pi; \ 2\pi) = [(N_R - N_{fR})/(N_{stR} - N_{fR})](N_{stR}/N_R)R_{\lambda st}(2\pi: \ 2\pi) + \Delta R_\lambda.$$ (5.43)

The correction ΔR_λ in this case is:

$$\Delta R_\lambda = R_f[1 - T_\lambda^2(2\pi; \ 2\pi) - (N_R - N_{fR})N_{stR}/(N_{stR} - N_{fR})N_R].$$ (5.44)

It follows from Eq. 5.44 that at large values of T_λ and R_λ (greater than 0.8) the correction ΔR_λ is insignificant owing to the low reflectance of the photocell ($R_f \leq 0.05$); and since the difference in the square brackets becomes less than 0.2, $\Delta R_\lambda \leq 0.001$.

The method of measuring $T_\lambda(2\pi: \ 2\pi)$ and $R_\lambda(2\pi; \ 2\pi)$ with the aid of two integrating spheres (Fig. 5.24,b) involves the following operations.

1. The openings (20 mm in diameter) in the spheres (4 and 4') are superimposed on each other in the absence of the sample (7). The signal N_{sR} from the upper photocell (2) and the signal N_{stT} from the lower photocell (8) are proportional to the irradiance:

for the upper sphere, $N_{sR} = \alpha E_1(R'_s);$ (5.45)

for the lower sphere, $N_{stT} = \beta E_2(R'_s) = \beta E_1^2(R'_s)S_o/F;$ (5.46)

where, R'_s – effective coefficient characterizing the fraction of the radiation reflecte into the sphere (4) from the sphere (4') through the opening whose area is S_o.

2.   The sample (7) is placed in the superimposed openings of the sphers (4 and 4') (Fig. 5.24,b); recording is made of the signals N_r and N_T from the lower and the upper photocell, respectively:

$$N_R = \alpha E_1(R'); \tag{5.47}$$

$$N_T = \beta E_2(T, R') = \beta T E_1^2(R')S_o/F. \tag{5.48}$$

The absolute bihemispherical transmittance is determined on the basis of the following relationship, with account taken of Eqs. 5.45-5.48:

$$T_\lambda(2\pi; \ 2\pi) = (N_T/N_{stT})(N_{sR}^2/N_R^2). \tag{5.49}$$

To determine the reflectance $R_\lambda(2\pi; \ 2\pi)$ of the sample the following measurements must be carried out.

3.   The sphere (4') under the sample (7) is replaced by an absorbing sphere. In this case the irradiance depends on the reflectance of the sample, and the signal from the photocell is:

$$N_R = \alpha E_1(R). \tag{5.50}$$

4.   The sample (7) is removed, and measurements are made of the irradiance R_A of the absorbing blackbody.   In this case the signal from the photocell is:

$$N_{Ra} = \alpha E_1(R_A). \tag{5.51}$$

The bihemispherical reflectance of the sample, with account taken of Eqs. 5.45-5.51, is determined on the basis of the following equation:

$$R_\lambda(2\pi; \ 2\pi) = (1 - N_{RA}/N_R)/[(1-\phi'R_s)/(R_sS_o/S)]. \tag{5.52}$$

For relative measurements the absolute reflectance $R_\lambda(2\pi; \ 2\pi)$ is calculated according to the following relationship:

$$R_\lambda(2\pi;\ 2\pi)=(N_R-N_{RA})/(N_{Rst}-N_{rA})[(N_{Rst}/N_R)R_{\lambda st}(2\pi;\ 2\pi);\qquad\qquad(5.53)$$

where, $N_{R\ st}$ - the signal from the photocell (2) when the reflection standard is located in the opening of the sphere (4).

The methods described here have been used in direct measurements of the hemispherical $R_\lambda(2\pi;\ 2\pi)$ and $T_\lambda(2\pi;\ 2\pi)$ of selectively absorbing-scattering substances. These quantities obtained with the aid of a single sphere (Fig. 5.24,a) differ from those obtained with the aid of two spheres (Fig. 5.24,b) by ±0.005 on the average. This indicates that the single-sphere method is sufficiently precise.

An advantage of the single-sphere method as compared with the two-sphere method and Duntley's method (Fig. 5.8, j) is that the transmission and reflection signals are of the same order of magnitude since they come from the same sphere. With the two-sphere device the illumination in the second sphere is by two orders of magnitude less intense than in the first. As a result, when measuring $T_\lambda(2\pi;\ 2\pi)$, the photocell reading is about 100 times lower. One must make sure, therefore, that the photometric measurements are sufficiently precise.

Another device has been developed [39, 118] for the complex measurements of the thermal radiational characteristics of foodstuffs subjected to directional and diffuse irradiation (Fig. 5.25). It is based on two methods: the integrating sphere method for the simultaneous determination of $R_\lambda(2\pi;\ 2\pi)$ and $T_\lambda(2\pi;\ 2\pi)$ exposed to diffuse radiation: and the method for the separate determination of $R_\lambda(\theta;\ 2\pi)$ and $T_\lambda(2\pi;\ 2\pi)$ by means of comparison under directional conditions.

The attachment shown in Fig. 5.25 consists of two integrating spheres (10 and 2) 100 mm in diameter. For operation in the 0.2-2.5 µm spectral region the inner surface of the spheres is coated with $BaSO_4$ to provide diffuse reflection. For operation in the infrared region of 1-1000 µm spheres providing diffuse nonselective reflection are used. An FEU-62 photomultiplier for the 0.2-1.4 µm spectral region and special thermoelectric cells [62] for the infrared region $\lambda>1.0$ µm serve as the radiation detectors. The inner surface of the sphere (8), also 100 mm in diammeter, is coated with black enamel which provides a mat absorbing surface for measuring $R_\lambda(\theta; 2\pi)$. One of the spheres (sphere 2) can be replaced with a blackened sphere when measuring $T_\lambda(\theta; 2\pi)$. The diameters of the openings in the spheres (10 and 2) take into account the spreading of the cross section of the radiation emitted by the sample, and are 30 and 50 mm, respectively. The transmission standard (7) is in the form of a truncated segment of a spherical surface with two openings (30 and 50 mm in diameter).

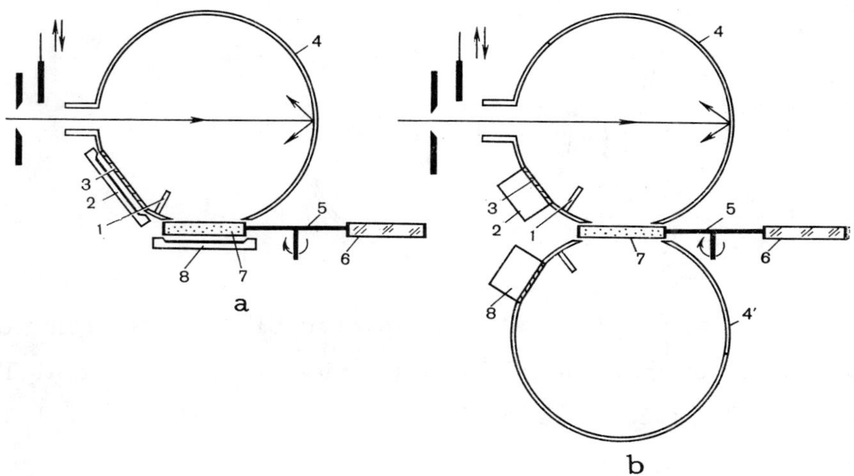

Figure 5.24 Schematic diagram of an attachment for a SF-4A spectrometer for the simultaneous (a) and complex (b) measurements of the bihemispherical reflectance and transmittance of radiation-scattering substances [48].

The holder (6) can support three samples of different thicknesses (from 0.5 to 10 mm) and a transmission or reflection standard (50 mm in diameter). By moving the holder in the direction shown with arrows measurements can be made of all three samples. The upper integrating sphere (2) is stationary, while the lower sphere (10) is rigidly connected to a rotating device (9) to which an absorbing sphere (8) is fastened. The lower sphere can be replaced with the absorbing sphere (8) with the aid of the rotating device (9). The radiation detectors (5 and 11) are placed in openings 14 mm in diameter in the integrating spheres.

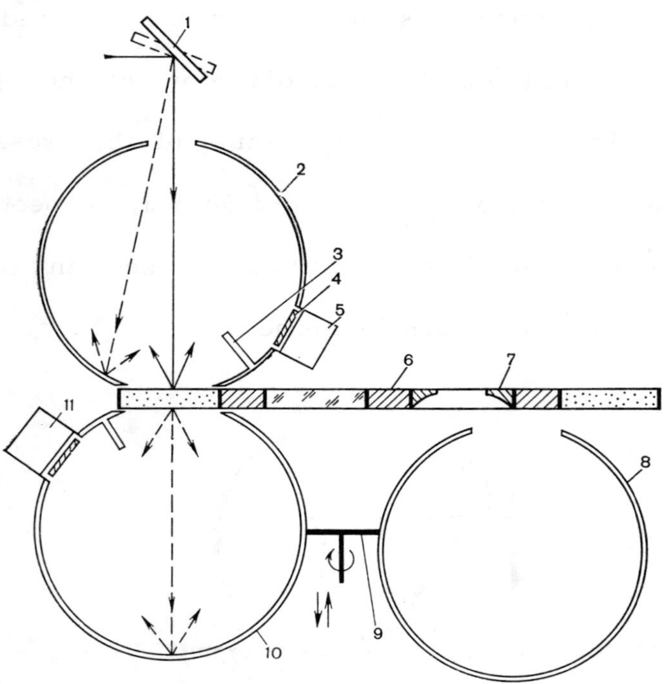

Figure 5.25 Schematic diagram of an attachment for SF-4A spectrometer for the complex measurements of directional-hemispherical $R_\lambda(\theta; 2\pi)$ and $T_\lambda(\theta; 2\pi)$ and bihemispherical $R_\lambda(2\pi; 2\pi)$ and $T_\lambda(2\pi; 2\pi)$ characteristics [39, 118].

MS-13 opal glass plates (4) are installed to provide uniform irradiation of the radiation detectors. The parallel radiation flux can be focused on the walls of the integrating sphere (2) with the aid of a rotating mirror (1). This results in diffuse irradiation of the sample when its bihemispherical characteristics are measured. When measuring the directional-hemispherical characteristics, the parallel flux is focused on the center of the sphere. The screens (3) prevent the detectors from being exposed to radiation directly reflected or transmitted by the sample. The surface of the screens is coated with the same substance as the inner walls of the integrating spheres.

The method involving the use of this device for measuring the bihemispherical reflectance and transmittance of foodstuffs is similar to the method for their simultaneous measurement with the aid of two spheres.

1.   The openings (10 and 2) in the integrating sphere are superimposed on each other in the absence of the sample. Then the signals N_{sR} and N_{stT} from the upper (5) and the lower (11) radiation detector respectively are measured. In order to provide diffuse irradiation the monochromatic radiation flux is focused with a rotating mirror (1) on the inner surface of the integrating sphere (2) (dashed line in Fig. 5.25).

2.   The samples are placed in the openings in the spheres (10 and 2) with the aid of a holder (6). Following this, the signals N_R and N_T from the upper (5) and the lower (11) radiation detector respectively are recorded.

The absolute bihemispherical transmittance $T_\lambda(2\pi; 2\pi)$ is determined from Eq. 5.49. In order to measure the bihemispherical reflectance the sphere (10) under the sample is replaced by the absorbing sphere (8) with the aid of the rotaing device (9). And the signal N_{Rs} from the upper detector (5) is recorded. The sample is then removed, and the irradiance of the upper sphere (2) is measured, which depends on the reflection R_A of the absorbing sphere (8). The signal from the detector is equal to N_{RA}.

The bihemispherical reflectance in this case is determined from Eq. 5.52. When making relative determinations the absolute values of $R_\lambda(2\pi; 2\pi)$ are calculated with the aid of Eq. 5.53.

Measurements of the directional-hemispherical $R_\lambda(\theta; 2\pi)$ are carried out as follows:

1.   The opening on the integrating sphere (2) and the opening in the absorbing sphere (8) are superimposed on each other, and a reflection standard (MS-14 opal glass) is placed between them. The monochromatic radiation is focused by means of the mirror (1) on the center of the reference standard. The illumination inside the integrating sphere (2), which is due to the radiation flux reflected by the standard, is proportional to $R_{\lambda st}(\theta; 2\pi)$, and the readings of the detector (5) are equal to N'_{Rst}.

2. The reflection standard is replaced by the sample. The readings of the detector (5), N'_{Rst}, are proportional to the illumination of the inner surface of the sphere (2), which is due to the radiation reflected by the sample. The radiation transmitted by the sample in this case is absorbed by the blackened sphere (8). The directional-hemispherical reflection coefficient of the sample is calculated from the relationship:

$$R_\lambda(\theta;\ 2\pi) = N'_{Rst} R_{\lambda st}(\theta;\ 2\pi)/N'_{Rst}. \qquad (5.54)$$

To determine the directional-hemispherical transmittance of the sample the integrating sphere (10) is placed under it. The upper integrating sphere (2) is replaced by an analogous absorbing sphere. This is done to avoid errors in determining $T_\lambda(\theta;\ 2\pi)$ due to additional illumination of the sphere (10) following the reflection of radiation by the sample.

The transmission standard (7) is placed between the upper absorbing sphere and the lower integrating one (10), and the signal N'_{stT} from the radiation detector (11) is recorded. The sample is then placed between the two spheres, and the signal N'_T from the detector (11) is recorded, which is proportional to the irradiance of the sphere (10). The transmittance of the sample in this case is determined by the following relationship:

$$T_\lambda(\theta;\ 2\pi) = N'_T/N'_{stT}. \qquad (5.55)$$

The device has been used to measure the directional-hemispherical R'_λ and T'_λ and the bihemispherical R_λ and T_λ for some foodstuffs subjected to irradiation at 0.4-1.4 μm. Measurements have also been carried out in the infrared region of 1-1000 μm following adjustments of the device as described above.

The measurement of $R_\lambda(\theta;2\pi)$ and $T_\lambda(\theta;2\pi)$ and of $R_\lambda(2\pi;2\pi)$ and $T_\lambda(2\pi;2\pi)$ in a single experiment makes it possible to determine the parameters which interrelate the thermal radiational characteristics obtained under different conditions (Duntley's parameters C_1 and C_2). These parameters are determined when data are available on the hemispherical R_λ and T_λ or $R_{\lambda\infty}$ and T_λ and on the directional-hemispherical R_λ' and T_λ' or $R_{\lambda\infty}'$ and T_λ'. The parameter C_2 is then defined by Eqs. 5.49 and 5.55, with account taken of Eq. 4.77, as follows:

$$C_2 = T_\lambda'/T_\lambda = (N_T'/N_{stT}')(N_{stT}N_R^2/N_t N_{sR}^2. \qquad (5.56)$$

The parameter C_1 is determined from Eq. 3.77, with account taken of Eqs. 5.53-5.55. Duntley's parameters define the optical properties of the medium and the spatial distribution of intensity of the radiation fluxes and at the same time take into account the irradiation conditions of a layer of sample.

The complex measurement of the R_λ and T_λ of foodstuffs makes it possible to obtain the directional-hemispherical and bihemispherical reflectance and transmittance of the sample in a single experiment. In this way errors of measurement due to changes which occur in foodstuffs with the passage of time are avoided.

5.8 Instruments for measuring the integral thermal radiational characteristics of foodstuffs

A method based on a hemispherical radiation source and a double-beam measuring device has been developed for determining the thermal radiational characteristics of foodstuffs subjected to infrared irradiation (Fig. 5.26) [10]. With this method, during measurements of the thermal radiational characteristics of foodstuffs, diffuse radiation flux E is generated whose spectral composition is equivalent to the radiation in closed industrial infrared installations.

The measuring unit consists of "linear" emitters (1), a semicylindrical reflector (2), a movable hearth (3) for supporting a layer of sample (4), and two reflection or transmission reference standards. The semicylindrical reflector and the hearth are made of aluminum. The location of the emitters and the distance between them and the sample are the same as found in industrial units. As is known, each filament of a linear infrared radiator can be considered to be an independent source of diffuse radiation. The multiple reflection of incident radiation from all the objects involved in the heat transfer (the reflector, the sample, and the radiation sources) results in the formation of a diffuse radiation flux whose spectral composition is of a complex nature [39].

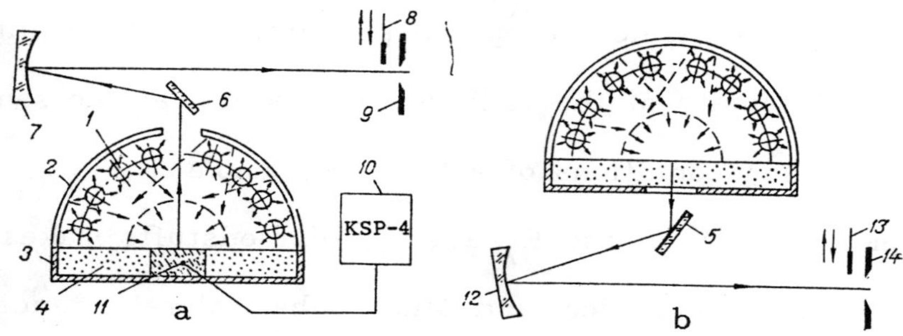

Figure 5.26 Schematic diagram of a hemispherical radiation source and a measuring unit [10]: 1 - radiation emitter; 2 - semicylindrical reflector; 3 - movable hearth; 4 - sample; 5,6 - plane mirrors; 7,12 - spherical mirrors; 8,13 - shutters; 9,14 - entrance slit of the monochromator in an IKS-14A spectrometer; 10 - KSP-4 potentiometer; 11 - thermocouple.

In our experiments the temperature of the sample was registered with thermocouples and a multiple-point automatic KSP-4 potentiometer. The thermocouples were located at various distances from the surface of the sample.

The double-beam device for measuring the thermal radiational characteristics of foodstuffs consists of two plane mirrors (6) and two spherical mirrors (7); two reflection or transmission standards (5), and the monochromator (1) of a double-beam spectrometer (Fig. 5.27). The device can be used for measuring the spectral and integral hemispherical radiance coefficients $\rho(2\pi; \theta_R, \phi_R)$ and $t(2\pi; \theta_T, \phi_T)$.

With the double-beam method one can determine the integral or spectral directional-hemispherical reflectance or transmittance in relation to the incident flux, which differs from the radiant flux of the infrared source in terms of radiant power and spectral composition. Here either "luminous" or "dark" infrared generators, such as KG lamps, panel electric heaters in the form of metallic or ceramic semicylindrical plates, can be used.

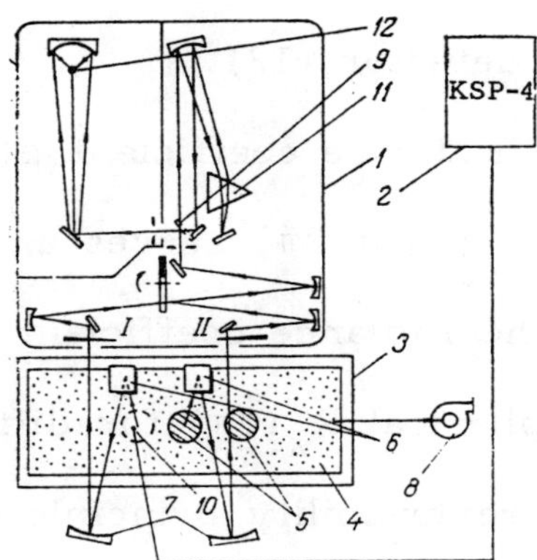

Figure 5.27 Schematic diagram of a device for measuring the thermal radiational characteristics of foodstuffs subjected to infrared diffuse irradiation [10]: 1 - monochromator of an IKS-14A spectrometer; 2 - KSP-4 potentiometer; 3 - movable hearth; 4 - sample; 5 - reflectance or transmittance standards; 6 - plane mirrors; 7 - spherical mirrors; 8 - fan; 9 - removable plane mirror; 10 - thermocouple; 11 - prism; 12 - radiation detector.

The method involved a comparison of two radiation fluxes, one of which is reflected (or transmitted) by the standard, and the other - reflected (or transmitted) by the sample. The surface of the standard is in the same plane as the surface of the sample. As the reflection reference standards various reflecting substances can be used: MS-14 opal glass; ONS neutral glass; MgO, Al_2O_3, or $BaSO_4$ coatings [54]. The wavy surface of the standard is a double structured aluminum coating which provides nonselective reflection of infrared irradiation [39].

The radiation fluxes from the sample and the standard are focused through windows on the hemispherical reflector (2) (Fig. 5.26), and are directed with the aid of mirrors (6 and 7) into the corresponding slits of the double-beam IKS-14A spectrometer. The fan (8) (Fig. 5.27) which is connected to a flexible hose, removes the vapors evolved from the heated sample, thus preventing the mirrors (6 and 7) from being becoming clouded.

When measuring the integral $\rho(2\pi;\ \theta,\phi_R)$ and $t(2\pi;\ \theta_T,\phi_T)$ a plane mirror (9) (Fig. 5.27) is installed behind the exit slit of the monochromator. With the aid of this mirror the radiation flux, bypassing the prism, is focused directly on the entrance slit and the radiation detector (12).

The ratio of the hemispherical radiance coefficient $\rho(2\pi;\ \theta_R,\phi_R)$ to the directional-hemispherical reflectance $R_\lambda(\theta;\ \phi,2\pi)$ serves as the theoretical basis of the method [156]. By measuring the radiance coefficient at different observation angles we can determine the hemispherical $R_\lambda(2\pi)$ for different irradiation conditions. According to the Helmholtz reversibility principle for isotropic media [156], when the incidence angles θ,ϕ and the reflection angles θ_R,ϕ_R are equal, we have:

$$\rho_\lambda(2\pi;\ \theta_R;\ \phi_R) = R_\lambda(\theta;\ \phi;\ 2\pi). \tag{5.57}$$

And for the transmittance we have:

$$t_\lambda(2\pi;\ \theta_T;\ \phi_T) = T_\lambda(\theta;\ \phi;\ 2\pi). \tag{5.58}$$

Eqs. 5.57 and 5.58 form the basis for measuring the coefficients of nonuniform diffuse reflection $R_\lambda(\theta;\ \phi;\ 2\pi)$ and transmission $T_\lambda(\theta;\ \phi;\ 2\pi)$ in the case of directional irradiation at angles $\theta,\ \phi$ with a narrow flux within the solid angle $d\omega$.

The intensity of monochromatic radiation incident upon a unit area of the sample dS from the area dS_i of the hemispherical radiation source in the vicinity of a certain point, which is visible from the center of the sample in the θ', ϕ' direction, is defined as follows:

$$dI_{\lambda,i} = B_{\lambda,i}(\theta';\ \phi')\cos\theta' d\omega; \tag{5.59}$$

where, $B_{\lambda,i}(\theta'; \phi')$ - radiance of the surface of the hemispherical radiation source at point $(r; \theta'; \phi')$ in the direction of the normal to the area element dS_i; $d\omega = dS_i/r^2$ - solid angle. The intensity of the monochromatic radiation which is reflected by the sample and that which is reflected by the reference standard in the direction of the exit opening in the hemisphere radiation source are defined as follows, respectively:

$$dI_{\lambda,R} = B_\lambda(\theta; \phi)d\omega, \qquad\qquad (5.60)$$

$$dI_{\lambda,st} = B'_{\lambda,st}(\theta'_{st}; \phi_{st}; \alpha)d\omega; \qquad\qquad (5.61)$$

where, $B_\lambda(\theta; \phi)$ - radiance of the sample surface in the θ, ϕ direction; $B_{\lambda,st}(\theta'_{st}; \phi'_{st}; \alpha)$ - radiance of the surface of the hemispherical radiation source at point $(r; \theta'_{st}; \phi'_{st})$ in the direction of the α reference flux towards the exit opening; α - angle between the reference flux and the normal to the surface of the hemispherical radiation source at point $(r; \theta'_{st}; \phi'_{st})$.

If the equality of the solid angles $d\omega = d\omega'$ is maintained, the reflectance of the sample equals the radiance ratio [39, 54]:

$$dI_{\lambda R}/dI_{\lambda,st} = B_\lambda(\theta, \phi)/B'_{\lambda,st}(\theta'_{st}; \phi'_{st}; \alpha) =$$
$$= \rho_\lambda(2\pi; \theta_R; \phi_R) = R_\lambda(\theta; \phi; 2\pi). \qquad\qquad (5.62)$$

Thus, the proposed method can be used for measuring the intensity of the radiation reflected by the sample in the direction of the angle θ_R in the case of hemispherical irradiation and for comparing it with the intensity of the radiation incident upon the sample. The ratio of the signals from the radiation detector is proportional to the hemispherical radiance coefficient $\rho_\lambda(2\pi; \theta_R; \phi_R)$ and thus to the hemispherical reflectance $R_\lambda(\theta; \phi; 2\pi)$.

Eq. 5.62 can be used for determining the real reflectance of the sample at any angles θ, ϕ only when the radiance of the surface of the hemispherical radiation source $B'_{\lambda st}$ is the same at all points facing the sample and only when the radiation incident upon the sample is uniformly diffused. Any deviation from these conditions leads to errors in determining R_λ. The magnitude of the errors depend on the scattering properties of the sample, on the degree of polarization of radiation by the sample and the monochromator, and on whether there are exit windows in the hemispherical radiation source [39].

On the basis of analogous considerations the following relationship is obtained:

$$t_\lambda(2\pi;\ \theta_T;\ \phi_T) = T_\lambda(\theta;\ \phi;\ \pi). \qquad (5.63)$$

Measurements of the spectral characteristics $R_\lambda(\theta;\ \phi;\ 2\pi)$ of foodstuffs are carried out as follows. Two identical reflection standards (5) (Fig. 5.27) are placed on the movable hearth. The integral radiation coming from the infrared generators, being reflected by the reference standards with the aid of mirrors (6 and 7), are focused on channels I and II of the monochromator. At a certain wavelength λ the N_{st} is recorded. The hearth is then shifted so that one of the standards moves from channel I to channel II, and the sample is now in channel I (Fig. 5.27). The radiation which is reflected by the sample and that which is reflected by the reflection standard enter channel I and channel II, respectively. The reading N_s in this case stands for the reflectance of the sample relative to the standard. The absolute reflectance of the sample at a given wavelength is calculated from the relationship:

$$R_\lambda = N_s R_{st}/N_{st}. \qquad (5.64)$$

When measuring the spectral transmittance of the sample, the hemispherical radiation source and the movable hearth are so located that the radiant fluxes transmitted by the sample and by the transmission standard are focused with the aid of mirrors (5 and 12) (Fig. 5.26, b) on channels I and II, respectively, of the monochromator. For this purpose three openings are cut in the movable hearth, two of which serve as transmission standards. The radiation flux transmitted by the sample passes through the third opening. The absolute spectral transmittance is calculated as follows:

$$T_\lambda = N'_s T_{st}/N'_{st};$$

(5.65)

where, N'_s and N'_{st} - readings corresponding to the transmission of the sample and of the standard, respectively; T_{st} - transmittance of the standard.

With the use of the mirror (9) (Fig. 5.27) the integral radiation fluxes from channels I and II, reflected and transmitted by the sample and by the standard, respectively, are focused directly on the radiation detector (12) (the prism (11) is bypassed).

The setup has been used in determining the temperature and the integral reflectance of carrots, potato, semolina, the pulp of apples, and wheat flour dough irradiated with a KG-1000 halogen lamp. Furthermore, the spectral reflectance of some foodstuffs has been investigated as a function of temperature under conditions of conductive heating (Fig. 5.28). It has been found that during conductive heating from 30 to 100°C the spectral reflectance of some foodstuffs at $\lambda=2$ μm remains practically unchanged (Fig. 5.28).

The changes in the temperature of some foodstuffs exposed to infrared radiation have been investigated (Fig. 5.29, a). The most rapid change in temperature has been observed in the case of semolina (curve 3). The almost linear increase in temperature is due to the low moisture content of semolina (7.2%); most of the radiant energy absorbed is used up in heating the sample, and very little in drying it. In the case of foodstuffs with a relatively high moisture content (the pulp of apples, carrots, and wheat flour dough) their temperatures rise gradually until they reach a certain level (about 100°C) at which they remain practically constant (curves 1,2,4,5). This corresponds to their constant desiccation rate. An increase in the temperature of potato after 4 minutes of heating corresponds to its frying stage. At this point there is a sharp rise in the temperature, and the sample increases in size, and its surface becomes deformed.

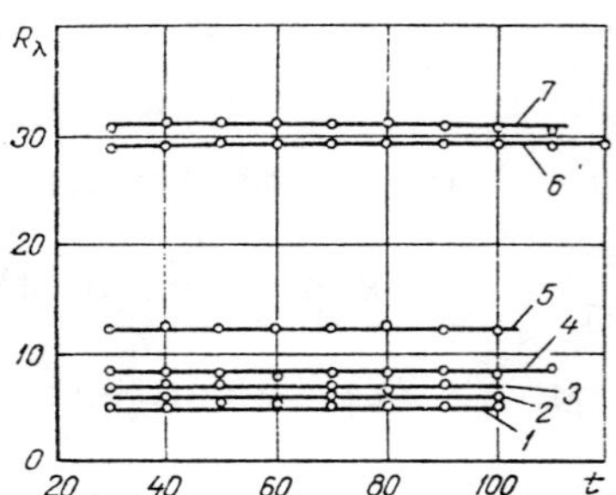

Figure 5.28 Hemispherical reflection coefficients (λ=2 µm) as a function of temperature under conditions of conductive heating of a 10 mm-thick layer of sample (τ in min; R in %) [10]: 1 - potato (W=78.8%); 2 - beet (W=80.2%); 3 - carrots (W=76.3%); 5 - bread crust (W=31.3%); 6 - oats (W=11.8%); 7 - wheat flour (W= 8.4%).

The integral reflectance of foodstuffs subjected to irradiation, unlike the spectral reflectance of foodstuffs subjected to conductive heating, does not remain constant (Fig. 5.29, b). For example, in the case of carrots its integral reflectance increases by 33% (curve 1); and in the case of the pulp of apples and wheat flour dough - by 22% each (curves 4,5). The changes in the reflectance are of a nonlinear nature. This is due to the complex physico-chemical processes taking place in foodstuffs exposed to infrared radiation. Thus, the low reflectance of water (1.6-4.8%) in the 1.0-15.0 µm spectral region decreases the reflectance of the water-containing foodstuffs, especially on their surface. Because of a decrease in the concentration of water the reflectance of carrots, potato, and wheat flour dough (curves 1,2,5) increases rapidly during the initial period of infrared irradiation.

The maxima on curves 2 and 5 at the end of the infrared irradiation corresponds to the moments when the surfaces begin to char. The difference between the integral reflectance which is measured and that which is calculated is only ±2%. Thus, for example, the integral reflectance of wheat flour dough measured in the early stages of baking is 36% (Fig. 5.29, curve 5) as compared to 37% calculated by a known method [39].

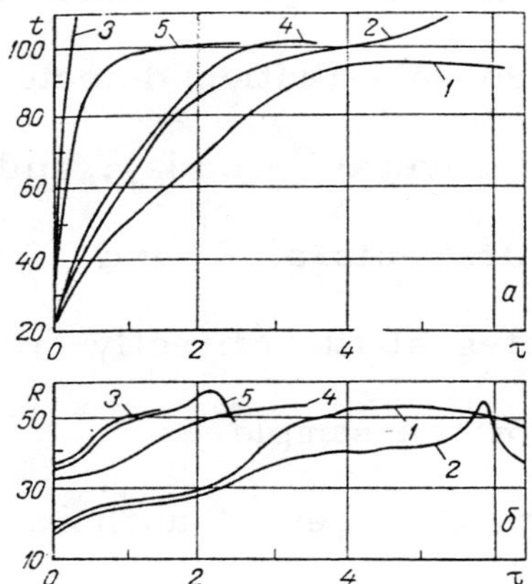

Figure 5.29 Changes in temperature (°C) and in the integral reflectance of foodstuffs (ℓ=15 mm) subjected to irradiation with a KG-1000 radiation source (τ in min; R in %) [10]: 1 - carrots (W=74.8%); 2 - potato (78.2%); 3 - semolina (W= 7.2%); 4 - pulp of apples (W=76.4%); 5 - wheat flour dough (W=42.2%).

The method described here can be used for investigating the spectral and integral thermal radiational characteristics of foodstuffs subjected to infrared irradiation. The data obtained are necessary for designing industrial infrared units.

We need to know the integral thermal radiational characteristics R, T, and A of different substances under different infrared irradiation conditions in order to calculate the heat transfer in processes of drying and thermal treatment by infrared radiation. These data can be obtained analytically, by calculating the spectral thermal radiational characteristics according to the method described in Chapter 3. A simpler method for determining the integral characteristics R' and T' is the experimental method.

Up to now most researchers resort to certain old methods (Fig. 5.9, d,e) in determining these integral characteristics. However, these methods are inadequate if we want to obtain the real values of R' and T' since the radiation detector registers only a part of the radiation reflected (or transmitted) by the sample. A large part of the radiant energy scattered by the sample does not enter the detector. Furthermore, in such measurements of R' and T' radiometers and thermopiles are usually employed as radiation detectors. Their delayed action is prolonged (there is a time lag of several seconds), and they have low sensitivity. To carry out precise measurements of R' and T' it is necessary to irradiate the sample with a powerful integral flux directly from the radiation source, which leads to a rapid heating up of the sample.

To eliminate these shortcomings the present authors have developed the methods of separate and simultaneous determination of spectral R'_λ and T'_λ (Figs. 5.19, 5.20, 5.22, 5.24, 5.25).

In measuring the integral thermal radiational characteristics under the conditions of directional irradiation at $\theta = 5 \div 10°$ (Fig. 5.19) (depending on the incidence angle (Fig. 5.20)) by the double-beam method, the spectrometers IKS-14 and IKS-14A are adjusted in the same way as in the determination of $R_\lambda(\theta; 2\pi)$ or $T_\lambda(\theta; 2\pi)$. The integral flux from the radiation source is focused alternately on channels I and II at a frequency of 8.8 cycles/s.

In measuring the integral $R(\theta; 2\pi)$ or $T(\theta; 2\pi)$ a reflection or a transmission standard is installed in the device. After the sample is mounted its R' or T' are recorded in percents in relation to the corresponding standard. The absolute $R(\theta; 2\pi)$ and $T(\theta; 2\pi)$ are calculated with the aid of the following equations:

$$R(\theta; 2\pi) = KR'R'_{sr}(\theta; 2\pi), \qquad (5.66)$$

$$T(\theta; 2\pi) = KT'; \qquad (5.67)$$

where, $K = 1.05$.

The methods described here, as compared with the known methods [19, 75, 84] for determining R' and T', have the following advantages. First, the over-heating of the sample is prevented because of low-powered irradiation (10^{-8} - 10^{-10} W), which can be achieved by the use of quick-response radiation detectors. Secondly, the time required for carrying out the experiment and processing the results is reduced. Thirdly, nearly all the hemispherical radiation reflected or transmitted by the sample is measured. And fourthly, the effect of the radiation emitted by the sample is completely excluded because the sample is irradiated by modulated radiation and also because the sensitivity of the detectors is low outside the operating spectral range.

Table 5.10   Reflectance and transmittance of some substances measured by the double-beam method (R'_m , T'_m ; %) and calculated with the aid of Eq. 4.117 (R'_c, T'_c ; %) [39].

Sample	R'_m	T'_m	R'_c	T'_c
Cornmeal (W=14.3%, ℓ=10 mm)	40.2	0.4	41.1	0.47
Foam plastic PS-4 (ℓ=8 mm)	78.5	-	80.5	-
Pinewood (ℓ=1.5 mm)	70.5	5.3	72.8	5.7

A comparison of the values of R'_λ and T'_λ of some substances obtained experimentally and those that are calculated is given in Table 5.10. As can be seen, there is little difference between these values. This shows that the double-beam method is sufficiently precise.

The attachment for IKS-21 (Fig. 5.22) can be used to measure the integral radiance coefficients $\rho(2\pi; \theta, \phi) = R(\theta, \phi; 2\pi)$ under the conditions of hemispherical irradiation with an integral flux generated by a given source. For this purpose the source is put in the place of the globule (1) and a plane mirror is installed outside the entrance slit (10). The mirror focuses the integral flux directly on the entrance slit; the flux is measured by the bolometers of the IKS-21 spectrometer. The procedure of measuring R' is analogous to that used for R'_λ.

In the case of diffuse irradiation with an integral flux $R(2\pi;2\pi)$ and $T(2\pi;2\pi)$ can be measured with the devices shown in Figs. 5.24 and 5.25. The spheres are coated with a material which makes them suitable for experiments in the 1-1000 μm infrared region. They serve as radiation detectors, each with a KRS-5 window (3×10 mm^2). With the aid of such a sphere the integral $R(2\pi;2\pi)$ of different substances are measured. The efficiency of this method can be seen from the data in Table 5.11

Table 5.11 Integral bihemispherical reflectance of some substances obtained experimentally (R_e) and calculated (R_c).

Sample	$R_e(2\pi;2\pi)$ T_{irrad}=1400 K	$R_c(\theta;2\pi)$	Reference
Black enamel 1519	0.085	0.074-0.055	2
Silver	0.990	0.978-0.950	74
Pinewood (ℓ=3.05 mm)	0.307	0.312	Eq. 4.117

5.9 Choice of radiation detectors in measuring the spectral and integral thermal radiational characteristics of foodstuffs

In a spectral analysis of light-scattering substances, in contrast to transparent nonscattering media, we need detectors with a large radiation-registering area ($S\geq2\times10$ mm^2). They should have quick-response time ($\tau\leq50$ μs) so that the modulated radiation can be recorded without delay, and have a high threshold sensitivity, D_λ^*. Of late photoelectric detectors are being more often used. Their threshold sensitivity lies in the range of 10^{-9}- 10^{-14} W$\cdot$Hz$^{-\frac{1}{2}}$; they are of the quick-response type (τ=0.1÷0.000011 μs), and have a large radiation-receiving area ($S\geq20$ mm^2). Their operating spectral range is practically unlimited. There are now industrially manufactured detectors which operate in the 0.25 to 1000.00 μm range. Detectors with emitted photoconductivity can operate only in the 0.4-

8.0 μm range. For measurements in the 10-100 μm range and higher detectors

should be of the mixed photoconducting type [81].

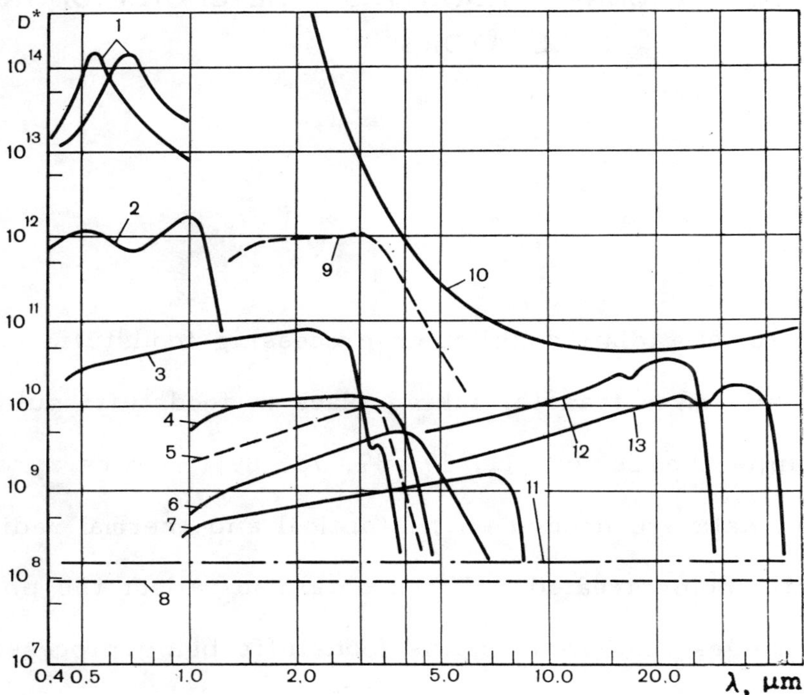

Figure 5.30 Specific spectral threshold sensitivity D_λ^* ($cm \cdot Hz^{\frac{1}{2}} \cdot W^{-1}$) of photoelectric and other types of radiation detectors as a function of the radiation wavelength [81, 152, 185]. Detectors that are not cooled: 1 - CdS; 2 - Tl_2S; 3 - PbS; 4 - PbTe; 5 - InAs; 6 - PbSe; 7 - InSb; 8 - thermal detectors, bolometers, and optical detectors OAP; 9 - 6AN (PbS); 10 - uniform narrow-band detector IU (λ, 2π, 290 K); 11 - pyroelectric detectors. Cooled detectors: 12 - Ge:Cu (15 K); 13 - Ge:Zn (4 k).

The dependence of D_λ^* on the radiation wavelength at room temperature and

at low temperatures is illustrated in Fig. 5.30. The results show that for the

spectral determination of R_λ and T_λ in the 0.4-8.0 μm range photoresistors of

the open type are suitable; and in the 8.0-25.0 μm range - bolometers or pyro-

electric radiation-detectors. More precise values of the integral characteristics

R and T can be obtained only by using radiation detectors with uniform spectral

sensitivity in the entire infrared spectral range.

EVALUATING THE QUALITY OF FOODSTUFFS ON THE BASIS OF THEIR OPTICAL AND THERMAL RADIATIONAL CHARACTERISTICS, THE CHOICE OF INFRARED GENERATORS

The design of infrared radiation units for processing foodstuffs is based on the principles of heat and mass transfer taking place in foodstuffs subjected to irradiation and in the units themselves [11, 19, 39, 75, 89]. To carry out the calculations involved here data are needed on the optical and thermal radiational properties of the foodstuffs being treated. These data also reflect the physico-chemical and biochemical changes that occur in the foodstuffs being processed and can therefore be regarded as indicators of their quality.

Simultaneous measurements have been made by the present authors and their colleagues of the R_λ, A_λ, and T_λ of different foodstuffs (wheat dough, bread, grains, groats, several kinds of starch, potato, carrots, beet, tea leaves, pears, apples, cocoa beans, milk, yeast powder, squash paste, different kinds of jam, etc) in the spectral range of 0.4-1.4 μm. In the 1.0-5.0 μm region the measurements were carried out by the double-beam method, and in the 2.0-15.0 μm region - by the single-beam method with the aid of specially designed attachments for the SF-4, SF-4A, IKS-12, IKS-14, IKS-14A, and IKS-21 spectrometers [10, 23, 39, 118]. The optical properties, $\overline{k}_\lambda$, s_λ, L_λ, and Λ_{st} were determined experimentally by analytical and graphic methods described in Chapter 3.

6.1 Spectral characteristics

<u>Optical properties of starch.</u> In designing infrared driers and infrared dex-

trinization equipment it is necessary to know the optical properties of starch. Starch grains are a main component in most foodstuffs of vegetable origin. The grains consist of layered crystalline carbohydrates and are optically anisotropic. Being a capillary-porous thermolabile substance, starch readily undergoes physico-chemical changes when it is heated. Therefore, measurements of its reflectance and transmittance must be carried out only under conditions of monochromatic irradiation.

In our studies [39, 61] of the thermal radiational characteristics and optical properties of starch we used starch derived from potato, corn, and rice. The nature of the dependence of $R_\lambda(\theta; 2\pi)$, $T_\lambda(\theta; 2\pi)$, $\bar{K}_\lambda$, s_λ, and L_λ on the radiation wavelength in all three cases and also in some other materials (Fig. 6.1, 6.2 and 6.25) is the same. The notable difference in the magnitude of these optical and thermal radiational characteristics of the three kinds of starch in the 0.4–1.5 μm region is due to the difference in the size of the grains, their form, and their moisture content. As can be seen from Fig. 6.2, the backward scattering coefficient s_λ and the transmittance T_λ for a layer of a sample of a certain thickness are the qualitative characteristics of the given kind of starch, which distinguish it from other kinds of starch. The transmittance of potato starch is the largest, and that of rice starch is the smallest. Rice starch has the largest backward scattering coefficient, and potato starch, the smallest. This is due to an increase in the optical density of the sample consisting of smaller grains. In the case of smaller particles a stronger scattering of radiation has been observed in the 0.4–1.5 μm spectral range.

To disinfect wheat grains by means of infrared irradiation, and to develop optical methods of quality control of wheat grains, it is necessary to know their optical properties and thermal radiational characteristics. Our investigations [23] show that the difference in the optical properties of different kinds of wheat grain is due to the different optical properties of their endosperm (Fig. 6.3).

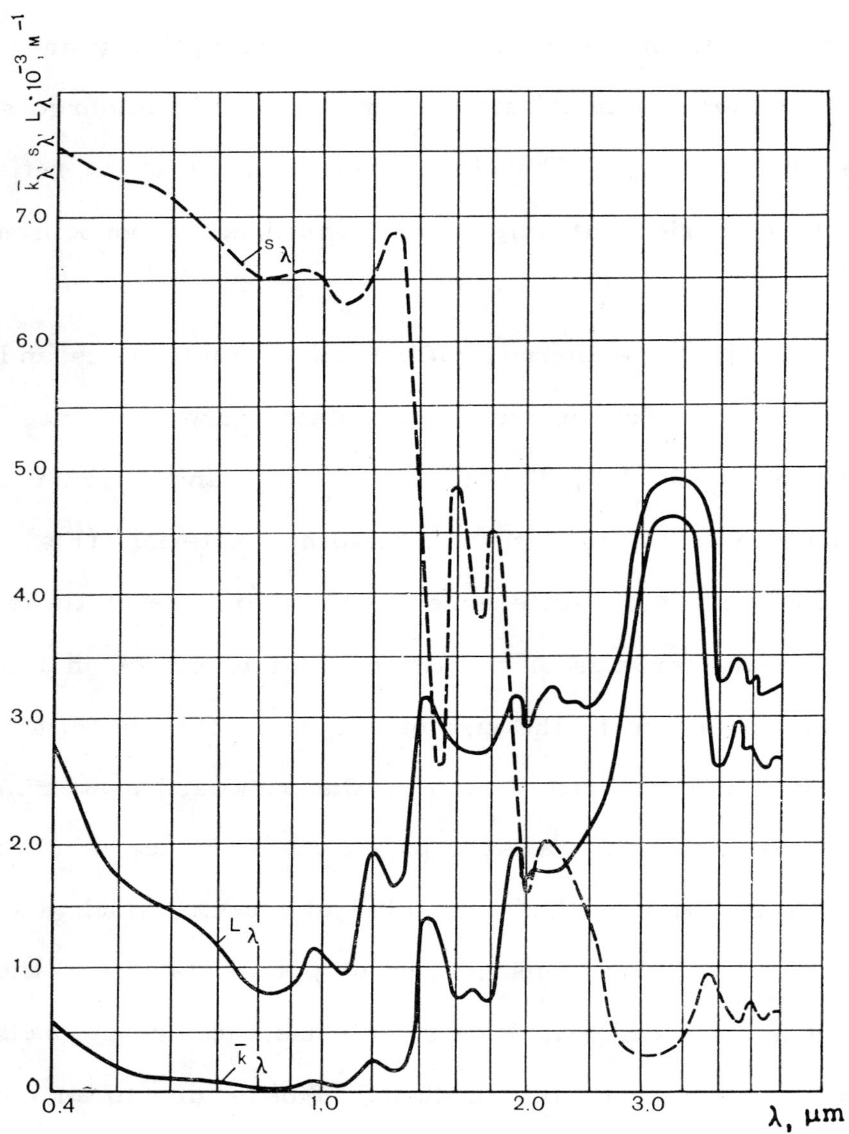

Figure 6.1 Optical properties of potato starch as a function of the radiation wavelength [39].

The nature of the dependence of $\bar{k}_\lambda$ and s_λ on λ is the same for the endosperm of different kinds of wheat grain. However, this dependence is more notable in the case of farinaceous endosperm with its large scattering coefficient, which increases with a decrease in λ. The Schuster number Λ for farinaceous endosperm in the 0.4–1.4 μm region reaches 0.97–0.99, while the specific absorption $\bar{k}_\lambda / s_\lambda =$

0.006÷0.033 (Tables 6.1 and 6.2). Therefore. farinaceous endosperm is condidered
to be a strongly scattering and practically nonabsorbing substance. The optical properties of farinaceous endosperm are qualitatively and quantitatively similar
to those of starch (Figs. 6.1, 6.5). This means that the starch grains play the
main role in the scattering of light. This scattering varies in a complex way
throughout the spectrum owing to the colloidal properties of the endosperm and
the inclusion in it of particles of different sizes.

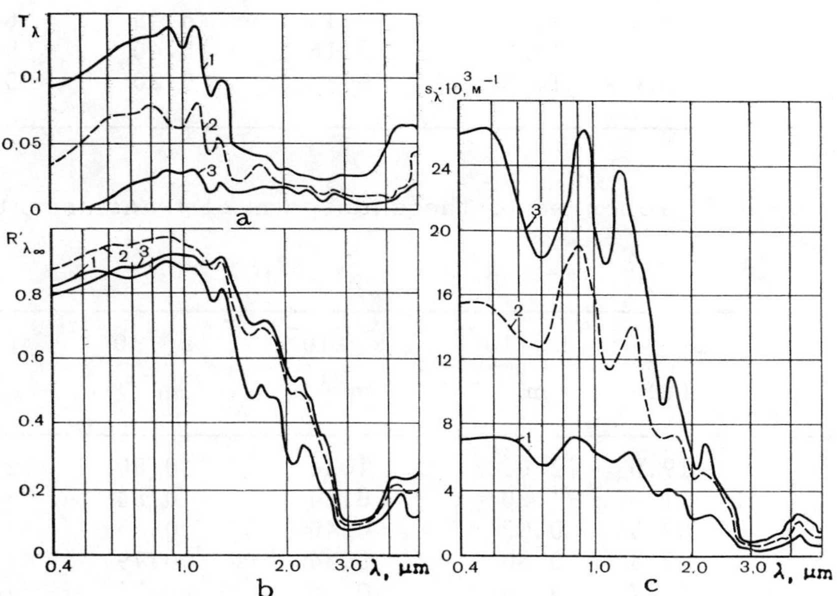

Figure 6.2 The dependence of the $R_{\lambda\infty}(\theta; 2\pi)$ (a) and $T_\lambda(\theta; 2\pi)$ of a layer of
sample (ℓ=1.0 mm) (b) and of the optical characteristic s_λ (c) on the radiation
wavelength λ (μm) in the case of [61]: 1 - potato; 2 - cornstarch; 3 - rice starch.

Owing to the dependence of the spectral reflectance and transmittance of the
husk on the radiation wavelength (Fig. 6.4), the maximum difference in the R_λ of
the endosperm lies in the 0.5-0.6 μm region, while that of the whole grain - in the
0.8-1.3 μm region. The low transmittance of the husk in the 0.5-0.6 μm range explains why the sharp difference in the reflectance of the endosperm with different
structures in the same range becomes unnoticeable in the case of the whole grain.

Table 6.1 Spectral optical properties of the endosperm of farinaceous wheat (W= 11.1%) [23]

Wavelength λ (μm)	$R_{\lambda\infty}$ (%)	$L_\lambda \cdot 10^{-3}$ m^{-1}	$\bar{k}_\lambda 10^{-3}$ m^{-1}	$s_\lambda \cdot 10^{-3}$ m^{-1}	$\varepsilon \cdot 10^{-3}$ m^{-1}	Λ
0.5	78.5	2.50	0.28	10.20	10.48	0.97
0.6	87.0	1.66	0.12	12.00	12.12	0.99
0.7	88.0	1.33	0.09	10.00	10.09	0.99
0.8	89.0	0.90	0.04	7.30	7.34	0.99
0.9	87.4	0.90	0.06	6.57	6.63	0.99
1.0	84.8	1.00	0.08	5.80	5.88	0.99
1.1	85.1	0.96	0.08	5.76	5.84	0.99
1.2	79.0	1.50	0.18	6.45	6.63	0.98
1.3	79.2	1.25	0.16	5.40	5.56	0.98
1.4	78.0	1.33	0.17	5.20	5.37	0.98

Table 6.2 Spectral optical properties of the endosperm of transluscent wheat (W= 11%) [23]

Wavelength λ (μm)	$R_{\lambda\infty}$ (%)	$L_\lambda \cdot 10^{-3}$ m^{-1}	$\bar{k}_\lambda \cdot 10^{-3}$ m^{-1}	$s_\lambda \cdot 10^{-3}$ m^{-1}	$\varepsilon \cdot 10^{-3}$ m^{-1}	Λ
0.4	19.0	2.00	1.34	0.80	2.14	0.37
0.5	31.0	1.20	0.64	0.84	1.48	0.57
0.6	39.5	0.92	0.40	0.85	1.25	0.68
0.7	41.3	0.80	0.34	0.79	1.13	0.70
0.8	42.7	0.76	0.30	0.76	1.06	0.71
0.9	41.0	0.72	0.30	0.70	1.00	0.70
1.0	39.0	0.73	0.32	0.73	1.05	0.70
1.1	40.0	0.69	0.30	0.66	0.96	0.68
1.2	35.0	0.80	0.38	0.63	1.01	0.62
1.3	36.0	0.76	0.36	0.63	0.99	0.63
1.4	34.0	0.84	0.40	0.65	1.05	0.62

The maximum difference in the R_λ of the whole grain with different structures appears in the same spectral region where the transmittance of the husk is high and where the differences in the R_λ of the endosperm are still large; i.e., in the 0.8–1.3 μm region. The larger T_λ and the smaller R_λ of the husk of durum wheat are due to the fact that the husk is less thick (0.12 μm) than the husk of soft wheat (0.24 μm). Therefore, the difference between the reflectance of translucent grain and that of farinaceous grain is greater in the case of durum wheat than in the case of soft wheat (Fig. 6.3).

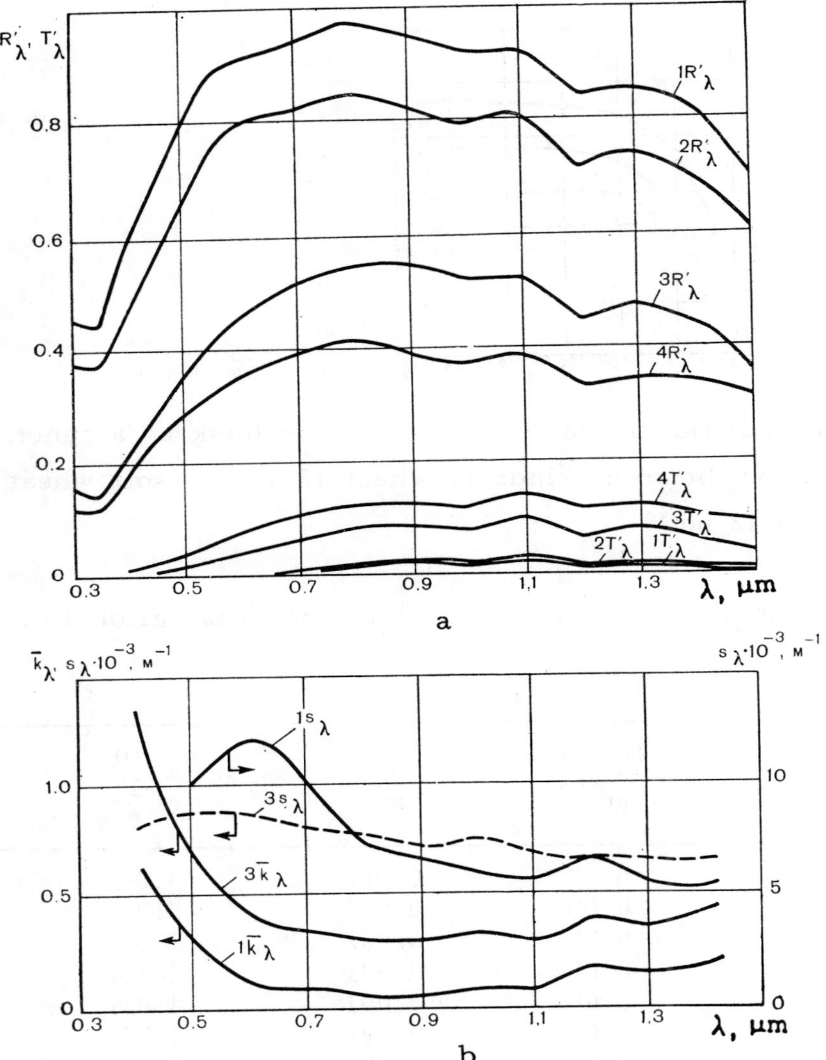

Figure 6.3 Thermal radiational characterisics (a) and optical properties (b) of the endosperm of translucent and farinaceous wheat as a function of the radiation wavelength λ (μm) [23]: 1,2 – farinaceous wheat (Mironovskaya 808); 3,4 – translucent wheat.

The optical properties of the husk of wheat grain (Table 6.3) [23] indicate that the husk is a practically nonabsorbing medium, that it is a scattering medium ($\Lambda > 0.99$). The coefficient of effective attenuation of the husk L_λ is small (7-10 times smaller than that of the endosperm; Tables 6.1-6.3). The sharp difference

in the optical properties of the husk and the endosperm is due to the different

composition of their tissues.

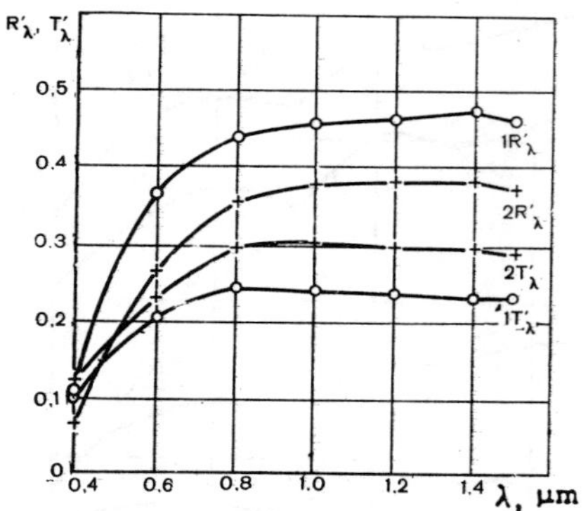

Figure 6.4   Thermal radiational characteristics of the husk as a function of the
radiation wavelength for different kinds of wheat [23]: 1 - soft wheat (ℓ=0.24 μm);
2 - durum wheat (ℓ=0.12 μm).

Table 6.3   Spectral optical properties of the husk of wheat grain (air-dried state)
[23]

Wavelength λ (μm)	$R_{\lambda\infty}$ (%)	$L_\lambda \cdot 10^{-3}$ m^{-1}	$\bar{k}_\lambda \cdot 10^{-3}$ m^{-1}	$s_\lambda \cdot 10^{-3}$ m^{-1}	$\varepsilon \cdot 10^{-3}$ m^{-1}	Λ
0.4	96.5	0.11	0.002	3.01	3.012	0.999
0.5	98.4	0.08	0.011	5.25	5.261	0.998
0.6	98.0	0.15	0.002	7.35	7.352	0.999
0.7	97.8	0.08	0.001	4.16	4.161	0.999
0.8	98.0	0.08	0.001	4.05	4.051	0.999
0.9	98.2	0.08	0.001	3.70	3.701	0.999
1.0	98.0	0.07	0.001	3.40	3.401	0.999
1.1	94.7	0.19	0.005	3.60	3.605	0.998
1.2	97.0	0.10	0.002	3.04	3.042	0.999
1.3	95.2	0.17	0.005	3.52	3.525	0.998
1.4	97.6	0.06	0.001	2.14	2.141	0.999

<u>Optical properties of buckwheat groats.</u>   In order to determine the kind of in-

frared source needed for breaking the fibers of the epidermis of buckwheat groats

in the production of concentrated buckwheat, not by boiling the groats, but by processing them in a fluidized bed with submerged radiators, we have measured

the $R_\lambda(\theta; 2\pi)$ of buckwheat groats that were boiled and dried (Fig. 6.5) [25].

Two samples were chosen: boiled and dried groats; and boiled and dried groats

that have been thermally treated once again.

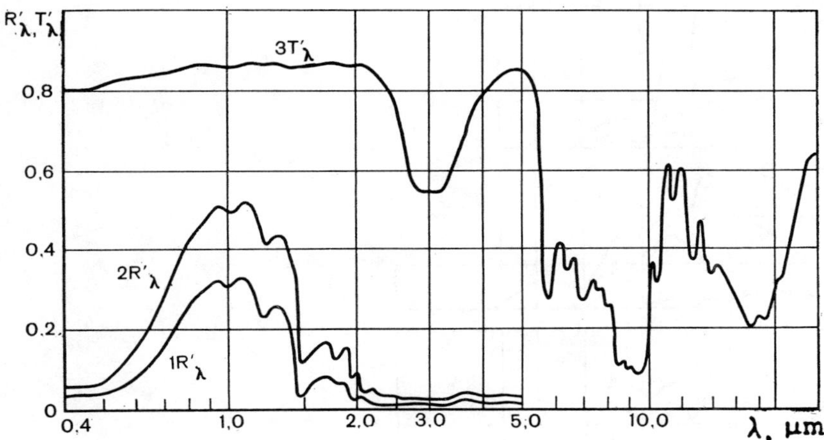

Figure 6.5   The reflectance and transmittance of a sample of buckwheat layer ($\ell=$ 8 mm and $\ell=0.043$ mm, respectively) as a function of radiation wavelength (μm) under different conditions of processing [25]: 1 - R_λ - boiled and dried groats that have been thermally treated once again; 2 - R_λ - boiled and dried groats; 3 - T_λ - sample of groats pressed together with KBr.

As can be seen from Fig. 6.5, following the second thermal treatment the re-

flectance of the groats decreased by 2-20% in the range from 0.4 to 5.0 μm.   The

qualitative spectrum of absorption can be seen from the spectrum of T_λ of a ground

sample of buckwheat obtained with the aid of an IKS-14 and a UR-20 spectrometer

(a 3.0 mg sample was pressed together with KBr into tablets, and a suspension of

ground groats in mineral oil was prepared).   The main absorption bands are in the

spectral region $\lambda > 3.0$ μm.   This means that such short-wave radiation can penetrate

deep into the individual grains.   The resulting intense heating leads to the forma-

tion of steam which causes the splitting of the particles.

<u>Optical properties of flour, dough and bread.</u>   For purposes of designing in-

frared units for the disinfection of flour an investigation has been carried out of

the optical properties of flour, dough and bread-dough during baking. Samples
of flour of durum and soft wheat were used. A comparison of the obtained results
[39] with data reported in literature [99] is shown in Fig. 6.6.

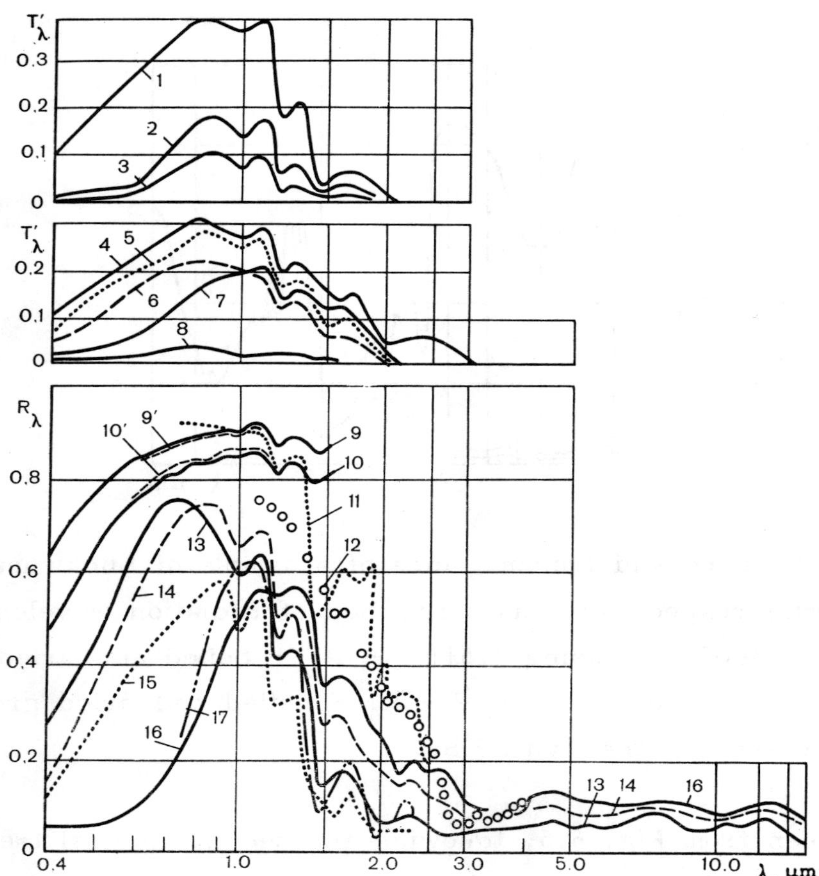

Figure 6.6 Spectral hemispherical transmittance and reflectance of flour, dough
and bread as a function of the wavelength of incident radiation [39, 99, 107]:
1,2,3 - durum wheat flour dough (ℓ=1.3, 3.2, 4.4 mm, respectively); 4 - baked
soft part of bread (ℓ=7 mm); 5 - half-baked soft part of bread (ℓ=7 mm); 6 - half-
baked white bread crust with the soft part of bread (ℓ=6 mm); 7 - baked brown
bread crust (ℓ=6 mm); 8 - wheat flour dough (ℓ=11 mm); 9 - flour from soft wheat
(W=8.2%); 9' - flour from durum wheat (W=8.2%); 10 - flour from soft wheat; 10' -
flour from durum wheat; 11 - wheat flour (grade l) [1]; 12 - wheat flour (the highest grade) [1]; 13 - wheat flour dough (ℓ=40 mm); 14 - half-baked white bread
crust with soft part of bread (ℓ=40 mm); 15 - durum wheat dough (ℓ=40 mm); 16 -
baked brown bread crust with soft part of bread (ℓ=40 mm); 17 - Ukranian bread
crust [1].

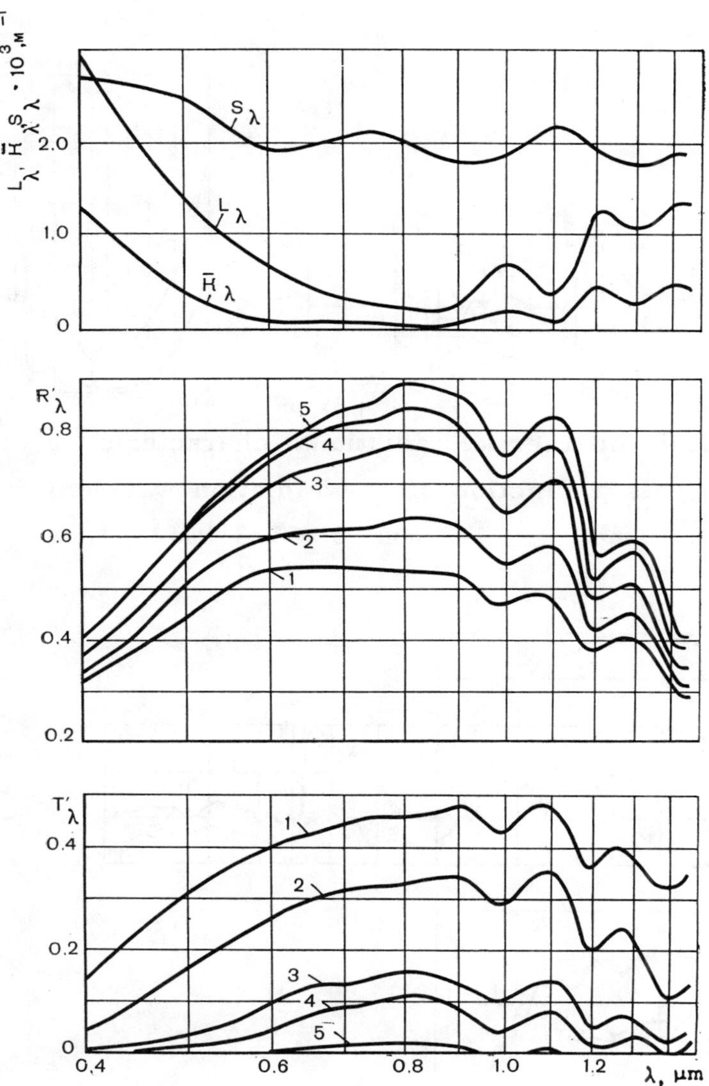

Figure 6.7 Spectral optical characteristics and hemispherical reflectance and transmittance of wheat flour dough (grade 1, W=50%) as a function of the radiation wavelength [39]: 1 - ℓ=0.65 mm; 2 - ℓ=1.0 mm; 3 - ℓ=2.6 mm; 4 - ℓ=4.6 mm; 5 - ℓ= 12.0 mm.

In terms of its structure dough is a colloidal substance; its isotropic structure changes with time. Separate measurements of the $R_\lambda(\theta; 2\pi)$ and $T_\lambda(\theta; 2\pi)$ of wheat dough, bread (grade 1), and durum wheat dough (Fig. 6.6) [1, 28, 29] make it possible to establish the similarity as well as the difference between the optical properties of flour, dough, and bread and to follow up their changes while the bread is being baked.

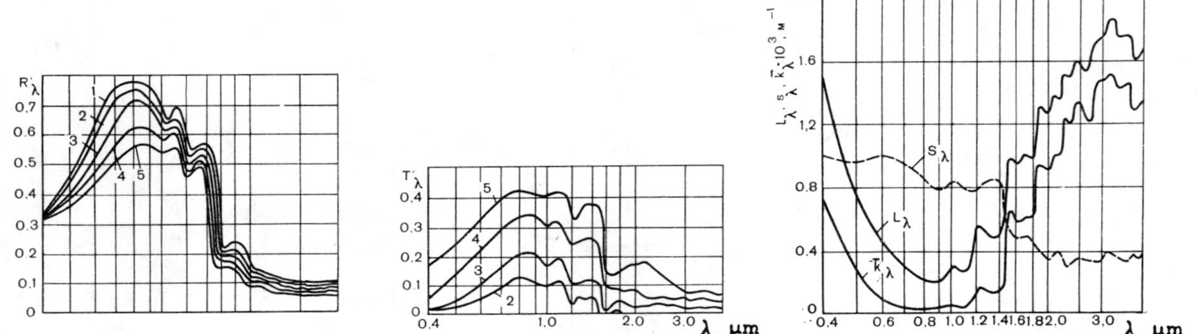

Figure 6.8 Spectral optical and thermal radiational characteristics of bread broken into small pieces (W=42.8%) as a function of the radiation wavelength [16]: 1 - ℓ= 12 mm; 2 - ℓ=7 mm; 3 - ℓ=5 mm; 4 - ℓ=2 mm; 5 - ℓ=1 mm.

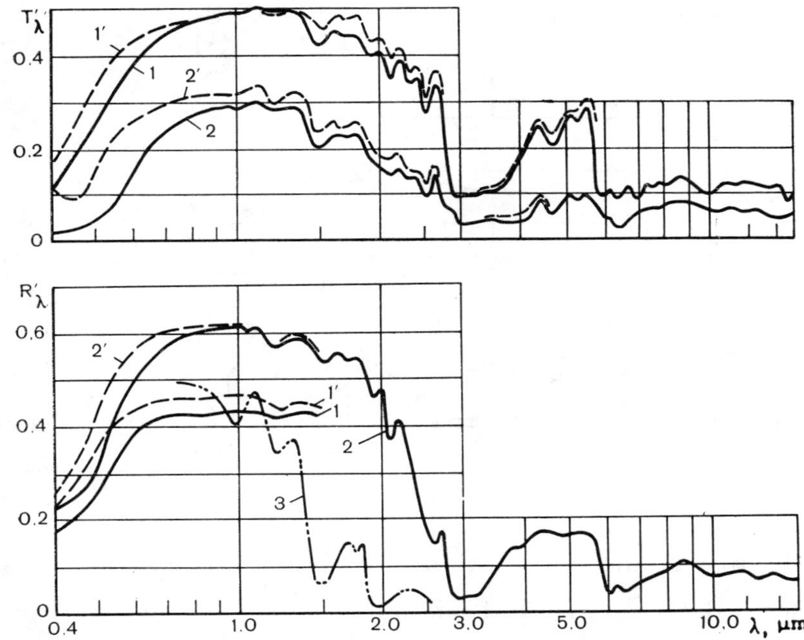

Figure 6.9   Dependence of the spectral hemispherical transmittance and reflectance of waffles (1,2) and dough (3) on the radiation wavelength at different thicknesses of the layer of sample (ℓ in mm) [1, 39]: 1,1' -1.05; 2,2' - 2.10.

A study has been carried out of the $R_\lambda(\theta; 2\pi)$ and $T_\lambda(\theta; 2\pi)$ of wheat flour dough (grade I; W=50%) just before it is baked (Fig. 6.7) [107]. The spectral coefficients of attenuation L_λ, and absorption k_λ, and backword scattering s_λ, which determine the optical properties of wheat flour dough, have been calculated by using the experimental values of R_λ and T_λ. As can be seen from Fig. 6.7, the dependence of L_λ on λ is similar in nature to the dependence of $\bar{k}_\lambda$ on the radiation wavelength, but the former is more clearly defined. The backward scattering coefficient is only slightly dependent on the radiation wavelength. The dependence of s_λ on λ is similar in nature to the dependence of the refractive index n_λ on the radiation wavelength, and represents the total dependence of the n_λ of all the components of the medium. The absorption bands near 0.98, 1.20, and 1.49 μm are due to the presence of water in the dough. In the 0.6-1.1 μm region the dough strongly scatters incident radiation ($s_\lambda = 1800 \div 2060$ m^{-1}) and only slightly absorbs it ($\bar{k}_\lambda = 14 \div 18$ m^{-1}).

Optical properties of bread broken into small pieces. To design multilayered thermal radiational units for making dried bread crumbs it is necessary to know the optical and thermal radiational characteristics of bread broken into small pieces (i.e.. particles of bread 1.0 to 4.0 mm in size). We have determined these characteristics under different irradiation conditions, using the method of simultaneous measurements of R_λ and T_λ [16].

As can be seen from Fig. 6.8, in the short-wave region bread particles are characterized by the smallest attenuation and absorption coefficients and the largest scattering coefficient. In spectral region $\lambda > 1.4$ μm the coefficients $\bar{k}_\lambda$ and L_λ increase up to $(1.5 \div 1.8) \cdot 10^3$ m^{-1}, while the coefficient s_λ decreases to $(0.3 \div 0.4) \cdot 10^3$ m^{-1}. In this region bread particles strongly absorb radiation, which causes an attenuation of radiation.

Optical properties of waffles (Fig. 6.9). In the 6.0-15.0 μm spectral range waffles reflect and transmit up to 5-10% of the infrared radiation; in the 3.0-6.0 μm range the amount of reflected radiation reaches 15-18%, and that of transmitted radiation - 5-10%. At the beginning. when the batter contains much moisture, there is strong absorption of infrared radiation at $\lambda > 3.0$ μm. and so the batter can be considered to be a "grey" body whose optical properties are close to those of a blackbody ($A_\lambda \cong 0.95$, Fig. 6.9).

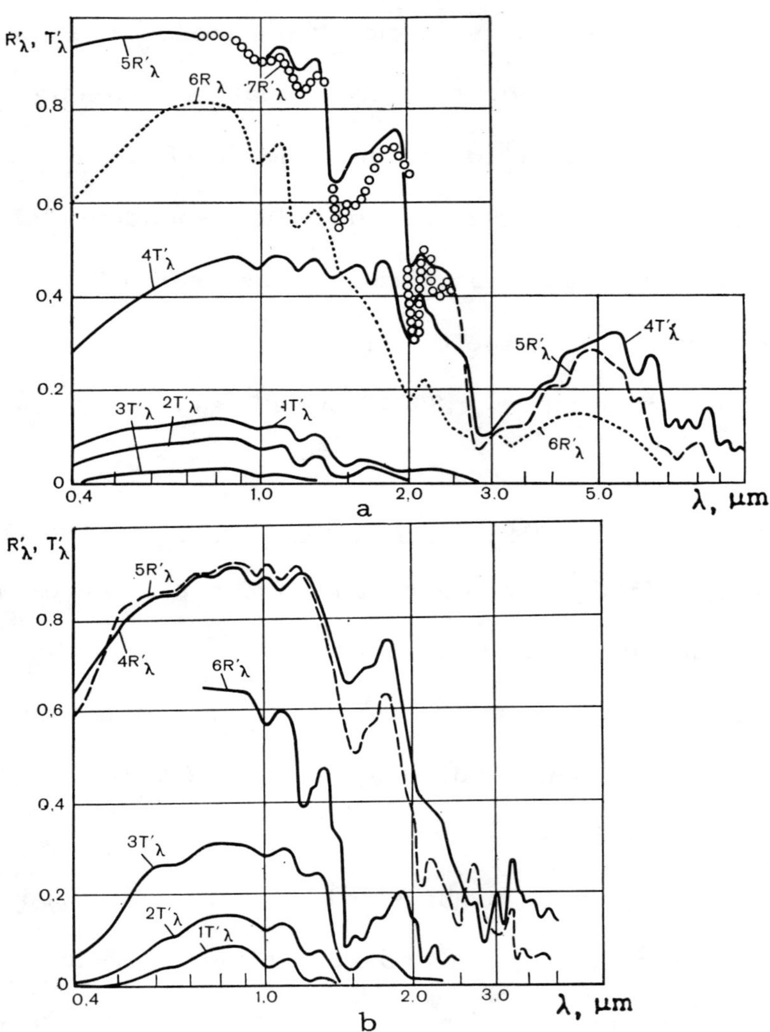

Figure 6.10 Dependence of the spectral thermal radiational characteristics of the following on the radiation wavelength [107]: a - confectioner's sugar (W=0.06%): 1 - ℓ=0.5 mm; 2 - ℓ=1.0 mm; 3 - ℓ=3.0 mm; 4 - polyethylene film (ℓ≅0.1 mm) coated with confectioner's sugar; 5 - ℓ=7.5 mm; 6 - ℓ=7.5 mm (W=6.76%); 7 - ℓ=7.5 mm (W=0.5%) [1]; b - fruit candy after it has gelled (W=30.0%): 1 - ℓ=1.0 mm; 2 - ℓ= 2.0 mm; 3 - ℓ=5.0 mm; 4 - without confectioner's sugar; 5 - dusted with confectioner's sugar; 6 - granulated sugar (W=0.2%) [1].

Optical properties of fruit candies, orange- and lemon-flavored gumdrops, and confectioner's sugar. Pastes for making fruit candies are gels which are difficult to dry. The process of drying can be considerably intensified by means of infrared irradiation. The results of experimental determination of the $R_\lambda(\theta; 2\pi)$ and $T_\lambda(\theta; 2\pi)$ of samples of fruit candy of different thicknesses after they have gelled, of confectioner's sugar, of orange- and lemon-flavored gumdrops [107] and of granulated sugar [1] are shown in Figs. 6.10 and 6.11.

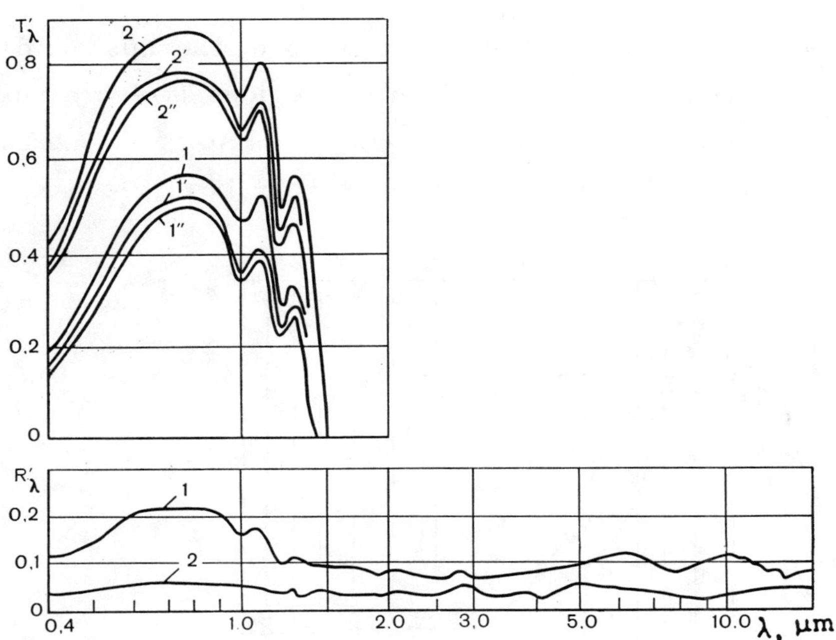

Figure 6.11 Dependence of the thermal radiational characteristics of 5 mm-thick orange-flavored gumdrops on the radiation wavelength [107]: 1,1',1" - gumdrops coated with sugar crystals; 2,2',2" - gumdrops not coated with sugar crystals.

The nature of the dependence of R_λ on the radiation wavelength is the same both for fruit candy which is powdered and that which is not powdered with confectioner's sugar. The absorption bands near the 1.0, 1.2, 1.5. 2.1, and 3.0 μm are present in all the reflection and transmission spectra. When the moisture content of confectioner's sugar is raised (from $W_1 = 0.06\%$ to $W_2 = 6.76\%$), its reflectance decreases by 20-35% (Fig. 6.10,a). The magnitude of $R_\lambda(\theta; 2\pi)$ for granulated sugar is much smaller than that for confectioner's sugar. This is due to the fact that the large crystals of granulated sugar scatter more radiation in comparison with the fine particles of confectioner's sugar.

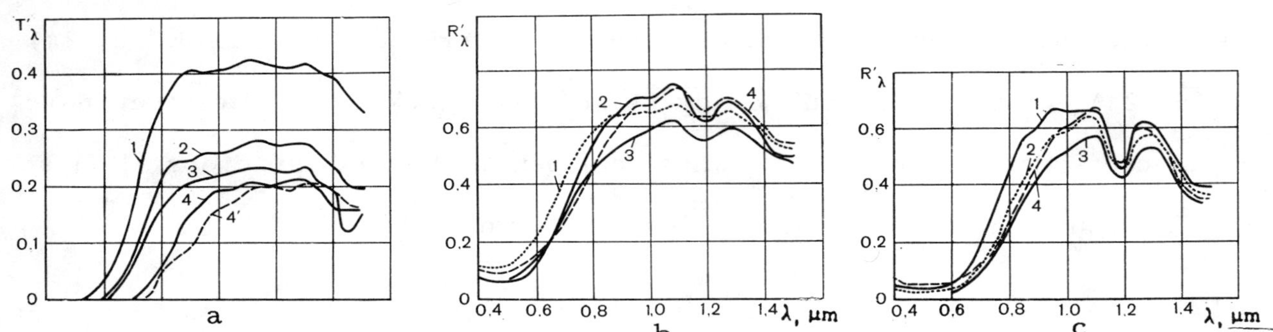

Figure 6.12 Spectral thermal radiational characteristics of the husk and kernel of cocoa beans and of whole cocoa beans from different regions (fresh samples) as a function of the radiation wavelength [39]: 1 - Bahia; 2 - Accra; 3 - Cameroon; 4,4' - Nigeria; 4' - inner side of the hull; a - husk; b - beans; c - kernel.

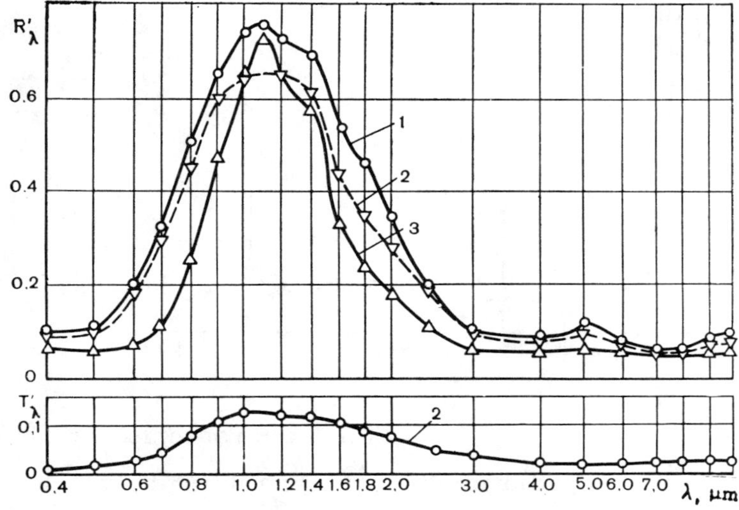

Figure 6.13 Dependence of spectral thermal radiational characteristics of the husk and the kernel of cocoa beans and of whole cocoa beans (5 years old) on the radiation wavelength [99, 118]: 1 - beans; 2 - husk; 3 - kernel.

The reflectance of orange-flavored gumdrops (ℓ=5 mm) which are not coated with sugar crystals is small in the 0.4-15.0 μm spectral region and is within the range of 5-10% (Fig. 6.11). After they are coated with sugar crystals the magnitude of $R_\lambda(\theta; 2\pi)$ increases to 20-22% in the short-wave spectral region from 0.6 to 0.9 μm. The reflectance of different samples varies within ±8%. A similar dependence of R_λ on λ is observed in the case of lemon-flavored gumdrops.

The transmittance of lemon- and orange-flavored gumdrops (ℓ=5 mm) which are not coated with sugar crystals differed from that of sugar-coated ones by 15-30%. The transmittance of different samples varies within 5-10%. The nature of the dependence of T_λ on the radiation wavelength is the same for both sugar-coated samples and samples not coated with sugar.   Absorption near 1.0 and 1.2 μm and complete infrared absorption at λ>1.5 μm are observed for all the samples. Therefore, when processing orange- and lemon-flavored gumdrops it is best to use high-temperature infrared generators emitting radiation at λ>1.5 μm.

<u>Optical properties of cocoa beans.</u> The thermal radiational characteristics of cocoa beans have been investigated in order to obtain the necessary data for selecting an infrared thermal processing unit [118]. The $R_\lambda(\theta; 2\pi)$ and $T_\lambda(\theta; 2\pi)$ of the husk and kernels of cocoa beans and of whole cocoa beans from different countries (Figs. 6.12 and 6.13) have been determined in the 0.4-10.0 μm range. Prolonged storage of cocoa beans results in an increase in the reflectance of the husks and the kernel by 10-20%. The transmittance of the husk of same beans decreases to 10-14% in the region of maximal infrared transmission (1.0-1.4 μm). These changes are due to biochemical reactions taking place in the kernel and the husk and to an increase in the density of their structure under conditions of prolonged natural drying.

Data on the maximal transmittance of the husk of cocoa beans ($\Delta\lambda(T_{\lambda max})$= 1.0÷1.4 μm) are needed in order to select an infrared radiation generator (e.g., a KG-1000 quartz lamp) for roasting cocoa beans and to determine its optimal operating temperature. The energy transmitted through the husk is absorbed by the kernel and is consumed in the heating process and in biochemical processes in which the tannins (they give a bitter and stringent taste to the untreated beans) are oxidized and the legumins become dehydrated (and can be more readily crushed).

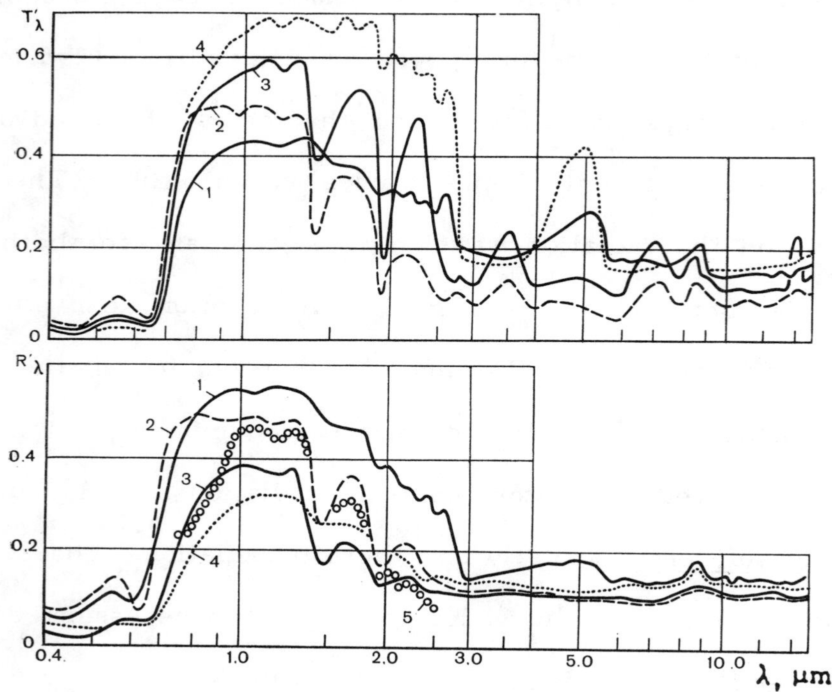

Figure 6.14   Dependence of the spectral thermal radiational characteristics of Georgian tea leaves on the radiation wavelength [55, 117]: 1 - dry green leaf; 2 - moisture-containing green leaf (W=72.4%); 3 - moisture-containing fermented leaf (W= 70.5%); 4 - dry fermented leaf (W=5%); 5 - Georgian tea [1].

<u>Optical properties of tea leaves.</u>   The thermal radiational characteristics of tea leaves have been investigated in order to obtain data necessary for designing tea-drying units [39].   As shown in Fig. 6.15, with an increase in the thickness of the layer (curves 1 and 2), $R_\lambda(\theta; 2\pi)$ in the 0.8-1.4 µm region increases by 10-12%, while $T_\lambda(\theta; 2\pi)$ decreases by 15-17%.   Measurements have also been carried out of the $R_{\lambda\infty}(\theta; 2\pi)$ of an optically infinitely thick layer ($T_\lambda < 0.01$) of green tea leaves (ℓ=30 mm).   Curves 3, 4, and 5 correspond to the different arrangements of the tea leaves with respect to the incident flux.   The orientation of the leaves strongly affects their reflectance in the 0.8-1.4 µm region, but only slightly in the 0.4-0.7 µm region.   Changes in $R_\lambda(\theta; 2\pi)$ with different arrange-

ments of the leaves at $\lambda=0.8\div1.4$ μm reach 18-20%.

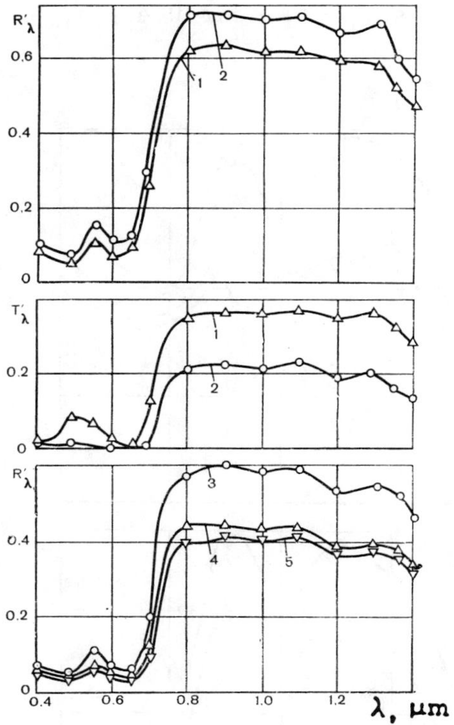

Figure 6.15 Dependence of the thermal radiational characteristics of green tea leaves on the radiation wavelength [39]: 1 - one leaf; 2 - two moisture-containing leaves on top of each other; 3,4,5 - a 30 mm-thick layer of tea leaves stacked in different ways.

The dependence of R_λ and T_λ on the radiation wavelength in the 1.0-15.0 μm region is the same for both fermented and unfermented tea leaves with the same moisture content (Fig. 6.14). Intense absorption bands near $\lambda=1.0$, 1.2, 1.4, and 1.92 μm are observed in the spectrum of moist leaves, but are practically absent in the dry ones. Fermentation and drying cause the leaves to turn black-green with a brownish tinge. That is why the magnitude of R_λ and T_λ of green leaves at $\lambda=0.55$ μm decreases.

Optical properties of fruits and vegetables. For the purpose of designing infrared drying units measurements have been made of the spectral $R_\lambda(\theta;\ 2\pi)$ and $T_\lambda(\theta;\ 2\pi)$ of carrots and beets (Figs. 6.16 and 6.17) [39, 118]. As can be seen, their spectral thermal radiational characteristics are similar to one another, and the difference in their color is revealed only in the visible spectrum owing to the different degrees of intensity of absorption near 0.4 μm.

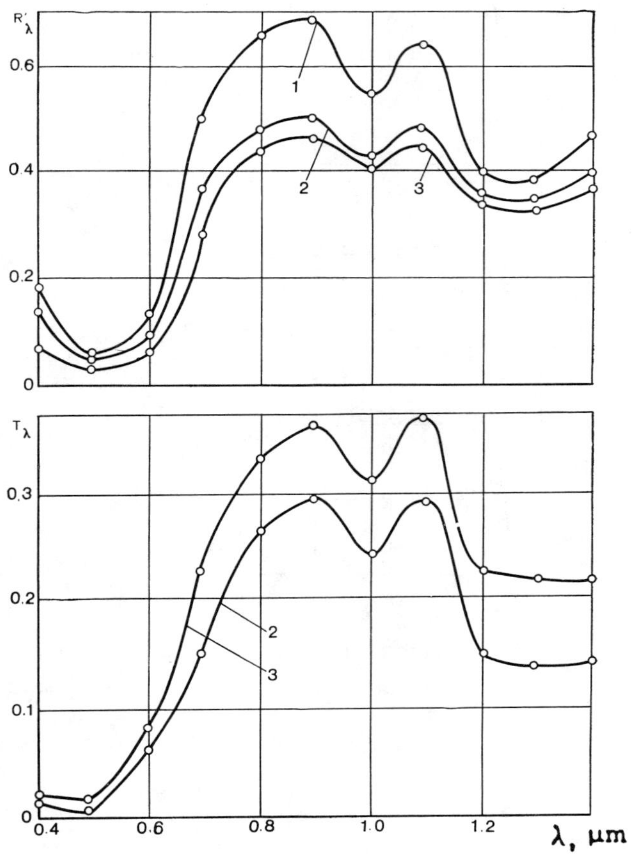

Figure 6.16 Dependence on the radiation wavelength of the spectral thermal radiational characteristics of beets [107]: 1 – ℓ=8.0 mm; 2 – ℓ=3.0 mm; 3 – ℓ=1.7 mm.

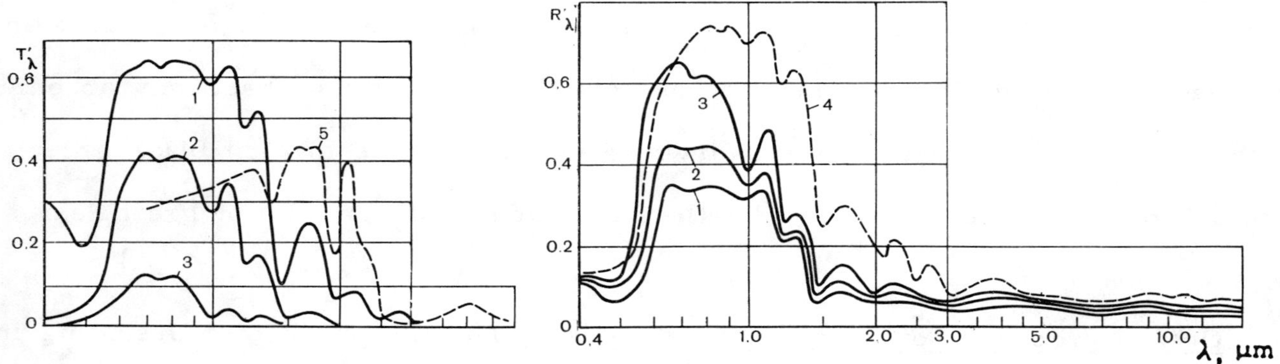

Figure 6.17 Dependence on the radiation wavelength of the spectral hemispherical transmittance and reflectance of carrots (W=86.5%) [107]: 1 – ℓ=1.0 mm; 2 – ℓ=3.0 mm; 3 – ℓ=10.0 mm; 4 – sample of dried carrot ($\ell_{undried}$=10 mm); 5 – sample of raw carrot (ℓ=0.1 mm [39, 118].

Optical properties of foodstuffs subjected to infrared pasteurization: milk, yeast, jams, tomato paste, squash paste. In Fig. 6.18.a are shown the reflectance and transmittance of milk (GOST 13277-67; 3.2% fat) measured simultaneously, with account taken of the scattering and spreading of the radiation flux [118].

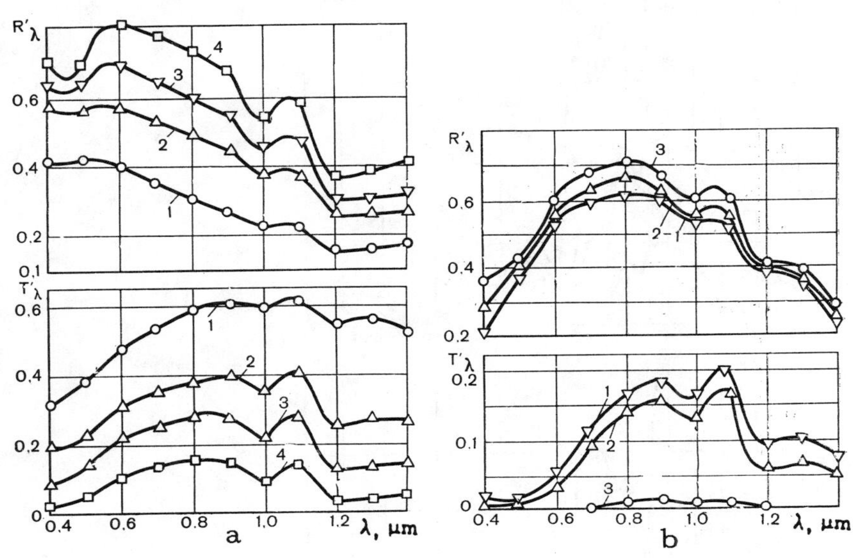

Figure 6.18 Dependence on the radiation wavelength of the spectral hemispherical reflectance and transmittance of foodstuffs subjected to infrared pasteurization [118]: a - milk (ℓ in mm): 1 - 1.0; 2 - 2.0; 3 - 3.0; 4 - 5.0; b - pressed yeat (ℓ in mm): 1 - 1.0; 2 - 2.0; 3 - 8.0.

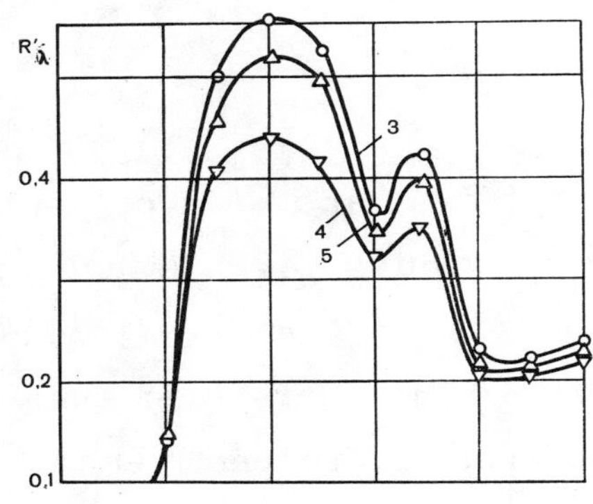

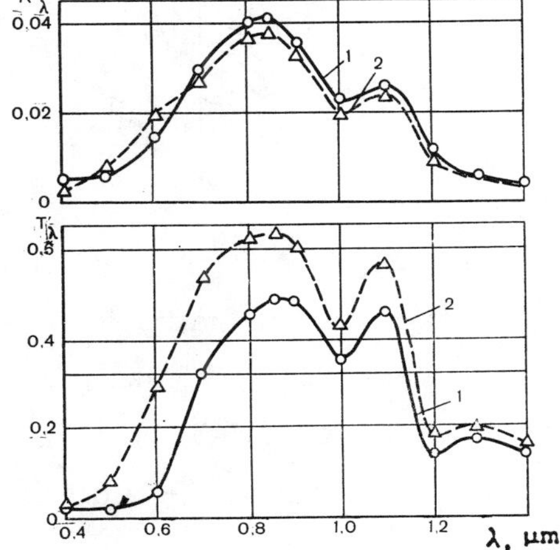

Figure 6.19 Spectral hemispherical thermal radiational characteristics of different foodstuffs as a function of the radiation wavelength [118]: 1 - plum jam (ℓ=15.0mm) 2 - apple jam (ℓ=15.0 mm); 3,4,5 - tomato paste (ℓ=8.0, 1.0, and 5.0 mm, respectively).

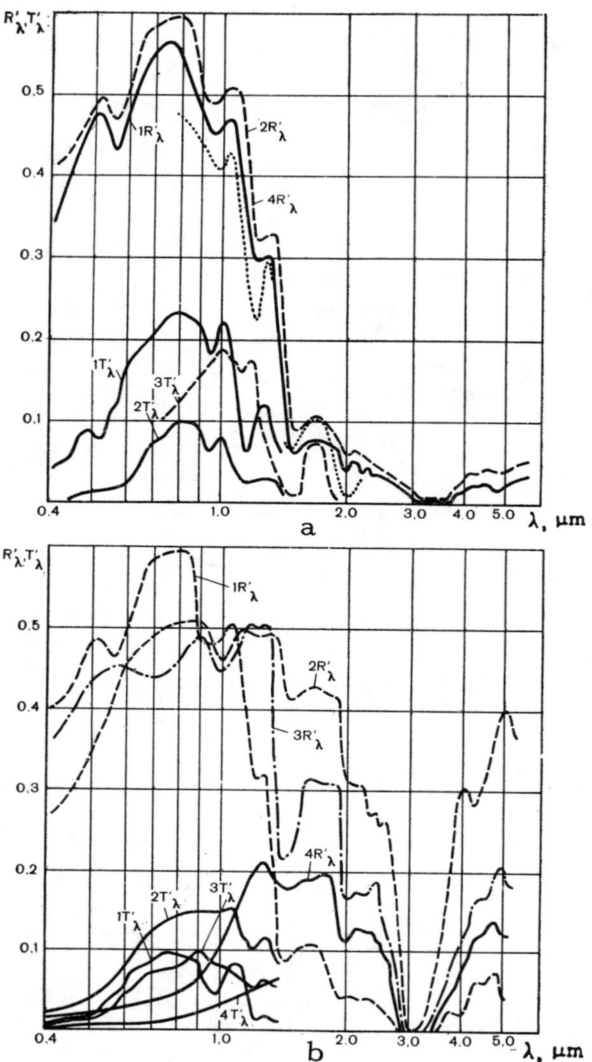

Figure 6.20 Spectral thermal radiational characteristics of meat and skin of young animals as a function of the radiation wavelength [46]: a - young pig's skin with adjoining fat and meat: 1 - ℓ=3 mm; 2 - ℓ=6 mm; 3 - pork (ℓ=2.0 mm) [97]; 4 - salted pork with some fat (W=35%) [1]; b - skin with hair or feathers (ℓ=8 mm): 1 - young pig; 2 - chick; 3,4 - white- and black-haired calves, respectively.

The reflectance of milk in the infrared region decreases at $\lambda > 0.8$ μm, while its transmittance is relatively high. Thus, infrared generators of the KG type can be used for pasteurizing milk. And since pasteurization in the ultraviolet region destroys vitamin B_1 in the milk [19, 97], the infrared generators used should emit radiation at $\lambda > 0.8$ μm.

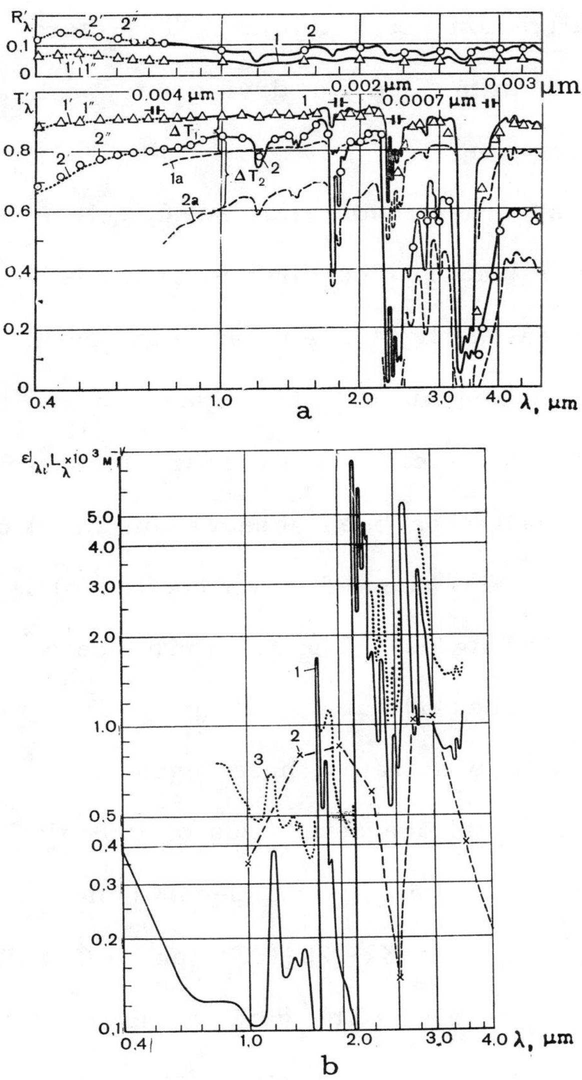

Figure 6.21 Dependence on the radiation wavelength of the spectral characteristics of polyethylene obtained by different methods [39]: a - thermal radiational characteristics: 1,2 - double-beam specular hemispherical method; 1',2' - integrating sphere method with the use of an SF-10 spectrometer; 1",2" - complex method at $\lambda = 0.4-1.4$ μm (cf. Fig. 5.11) and the single-beam specular hemisphere method at $\lambda = 1.0-5.0$ μm; 1a,2a - standard double-beam method with the use of an IKS-14 and a UR-20 spectrometer; b - optical properties: 1 - L_λ with account taken of scattering according to a method described in Chapter 3 in this work; 2 - ε_λ according to published data [94] with account taken of scattering; 3 - ε_λ with account taken of scattering.

The thermal radiational characteristics of pressed yeast are shown in Fig. 6.18,b. Its maximal reflectance lies at 0.8 μm, while its maximal transmittance is in the long-wave spectral region near 1.05 μm. Therefore, the maximum radiation intensity of the infrared units used for drying and pasteurizing yeast should lie in the 1.0-1.4 μm spectral range.

Jams and pastes, like all other foodstuffs, selectively reflect and transmit infrared radiation (Fig. 6.19) [118]. The directional hemispherical reflection coefficient of tomato paste increases slightly with an increase in the thickness of a layer of sample in the spectral regions of 0.5-0.6 μm and 1.2-1.4 μm. With an increase in the thickness of a layer of sample a sharp increase in $R_\lambda(\theta;\ 2\pi)$ in the case of jams and tomato paste has been observed in the 0.6-1.2 μm region. For apple and plum jams (ℓ=15 mm) the reflection coefficient is only 4% at λ=0.85μm, and even smaller at other wavelengths. The transmittance of apple jam is larger than that of plum jam, and reaches 60% at λ=0.85 μm. This indicates that during pasteurization infrared radiation penetrates these foodstuffs to a considerable depth.

<u>Optical and thermal radiational characteristics of biological substances.</u> The use of infrared radiation has proved highly efficient when raising young animals. The IKZK-220250, IKZ-220-500 (ZS-3) KI-220-1000, and KG-1000 infrared generators are widely used for this purpose. The data in Fig. 6.20 have been used for determining the optimal conditions of irradiation. For exposing the skin of young animals to infrared radiation it is best to use generators (Table 7.5) operating with a maximum emittance in the 0.8-1.3 μm range. For skins 3-7 mm thick these generators provide the maximal T_λ, which is 20-25%. The magnitude of R_λ depends much on the color of the hair. For the skins of white and black-haired calves it varies within 1-5% and 36-45%, respectively, in the visible spectral region, and up to 20% and 50%, respectively, in the infrared region. Furthermore, R_λ increases with an increase in the thickness of the skin. Therefore, in order to achieve maximum absorptance the skin should be at least 8-15 mm thick.

Optical properties of polyethylene as a packing material for foodstuffs. Various foodstuffs, especially meat, are processed while being wrapped in polyethylene films. Since polyethylene strongly scatters infrared radiation, we need to determine its optical properties, with account taken of the scattering and spreading of a narrow radiation flux, in order to know how much infrared radiation passes through the polyethylene film.

The maximal transmittance of polyethylene lies in the region from 0.4 to 2.2μm when there is little reflection in this region (R_λ=0.03÷0.14) (Fig. 6.21,a). This means that the thermal processing of foodstuffs packed in polyethylene films should be carried out with high-temperature infrared lamps.

A comparison of the calculated values of L_λ and ε_λ (Fig. 21,b) shows that radiation losses due to reflection and scattering are responsible for the fact that the values for ε_λ are too high (curve 3). The values for ε_λ (curve 2) [94] at λ=1.0, 1.5, and 2.0 μm are 3.35, 4.37, and 5.90 times, respectively, too high. In the 2.2-5.0 μm region they are 2.7-5.6 times too low. This is due to the large error involved in determining the $T_{\lambda 1}/T_{\lambda 2}$ ratio by the usual method [94].

6.2 Dependence of the optical and thermal radiational characteristics of foodstuffs under different irradiation conditions on their structure, density and temperature

To determine the relationship between the thermal radiational characteristics of irradiated foodstuffs and the irradiation conditions we have investigated the R_λ and T_λ for samples subjected to directional radiation at various incidence angles and to diffuse radiation. The dependence of $R_\lambda(\theta;\ 2\pi)$ and $T_\lambda(\theta;\ 2\pi)$ on the incidence angle is shown in Fig. 6.29,a in the case of irradiation at λ=0.9 μm of the soft part of bread and of bread crumbs with different thicknesses. With an increase in the incidence angle the dependence of $R_\lambda(\theta;\ 2\pi)$ increases, while that of $T_\lambda(\theta;\ 2\pi)$ decreases. For a layer of the soft part of bread 1.8 mm thick R_λ increases approximately by 10% when the incidence angle is increased from 10 to 60°.

Under the same conditions T_λ decreases by 15%. For thicker samples the increase in R_λ and the decrease in T_λ slow down with an increase in the incidence angle. The nature of the dependence of R_λ on the incidence angle is the same as that of T_λ, while the absolute value of R_λ for bread crumbs is greater, and that of T_λ is smaller than that for the soft part of bread. This difference is due to the presence in the bread crumbs of dense particles from the upper and lower crust.

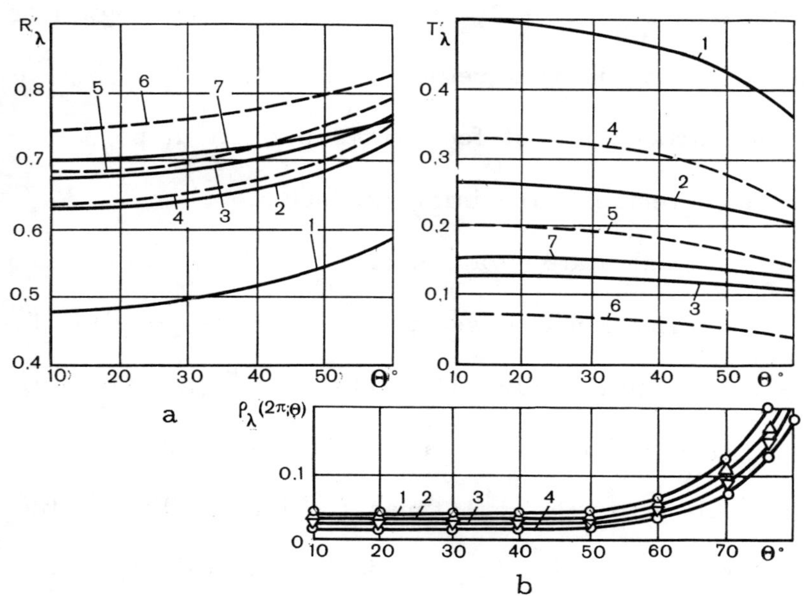

Figure 6.22 Thermal radiational characteristics of different foodstuffs as a function of the incidence angle at different radiation wavelengths [39, 118]: a - λ= 0.9 µm, soft part of bread (ℓ in mm): 1 - 1.8; 2 - 5.0; 3 - 8.0; bread crumbs (of the soft part of bread): 4 - 2.0; 5 - 5.0; 6 - 8.0; bread crust: 7 - 3.8; b - λ=2.0 µm: 1 - apple (W=76.4%, ℓ=4.0 mm); 2 - pear (W=75.7%, ℓ=4.0 mm); 3 - potato (W=78.2%, ℓ=4.0 mm); 4 - carrot (W=73.2%, ℓ=4.0 mm).

Under conditions of hemispherical irradiation the reflectance of vegetables and fruits as a function of the observation angle is of a similar nature (Fig. 6.22,b). All the measurements were carried out with the aid of the attachment for the IKS-21 spectrometer (Fig. 5.22).

Under hemispherical irradiation conditions, when the incidence angle is increased to 50°, the reflectance changes very little (by 2-3%) (Fig. 6.22,b). A further increase in the angle to 80° results in a sharp increase (5-7 times) in the reflectance of the sample.   Therefore, the bihemispherical $R_\lambda(2\pi;\ 2\pi)$ will be greater than $R_\lambda(\theta;\ 2\pi)$ in the case of directional irradiation.

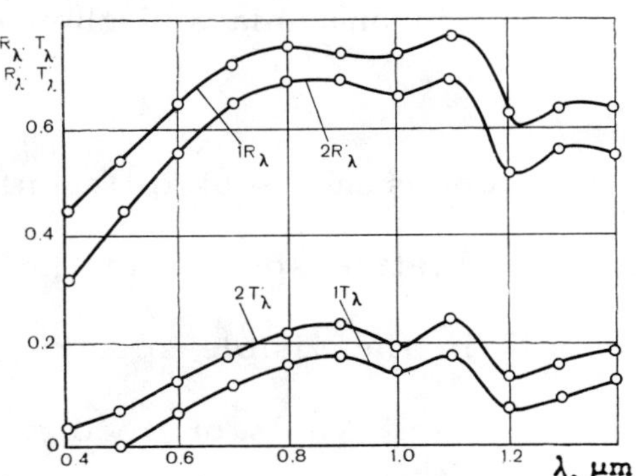

Figure 6.23   Thermal radiational characteristics of bread crumbs ($\ell=5$ mm) as a function of the radiation wavelength under different conditions of irradiation [118]: 1 - hemispherical irradiation; 2 - directional irradiation at $\theta=10°$.

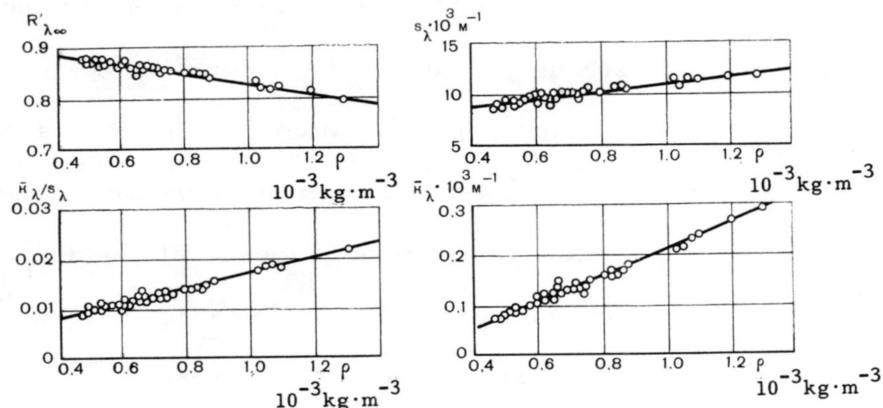

Figure 6.24   Relationship between the main characteristics of the radiation field and the density of pinewood at $\lambda=1.3$ μm [26].

To verify these conclusions we have measured the bihemispherical $R_\lambda(2\pi;\ 2\pi)$ and $T_\lambda(2\pi;\ 2\pi)$ of bread crumbs with the aid of an attachment described in Chapter 5.   The results for $R_\lambda(2\pi;\ 2\pi)$, $T_\lambda(2\pi;\ 2\pi)$, $R_\lambda(\theta;\ 2\pi)$, and $T_\lambda(\theta;\ 2\pi)$ ($\ell=2.0$ mm, W=39.1%) are shown in Fig. 6.23.   As can be seen, the bihemispherical reflectance of bread crumbs is larger than the directional-hemispherical reflectance by 10-12%; and $T_\lambda(2\pi;\ 2\pi)$ is smaller than $T_\lambda(\theta;\ 2\pi)$ by approximately 8-10%. These results confirm the theoretical conclusions made in Chapter 3.

Wood is a   suitable material for investigating the dependence of the R_λ, T_λ, and A_λ of a substance on its density.   The absorption spectra of wood are very similar to those of foodstuffs (Fig. 1.2). The density of different kinds of woods varies from 220 to 1570 kg/m^3. In our study we used high-density wood samples (880-1290 kg/m^3) obtained by saturating them in ammonia and then compressing them [26].

The dependence of the R_λ and T_λ of wood samples of different densities on the radiation wavelength shows that their qualitative spectra in the infrared region are identical. A difference is observed only in the visible spectrum, 0.4-0.7 μm. This means that the optical properties of different kinds of wood will have no effect on the relationship between the R_λ and T_λ of the sample and its density.

A decrease in $R_{\lambda\infty}$ with an increase in density at $\lambda=1.3$ μm is due to the increase in the backward scattering coefficient s_λ (Fig. 6.24) (since a denser medium scatters radiation more strongly) and in the absorption coefficient $\bar{k}_\lambda$. The decrease is also due to the fact that the absorption coefficient, as compared to the scattering coefficient, increases more rapidly with an increase in the density of the sample. Experimental results show that the dependence of the $R_{\lambda\infty}$ and $A_{\lambda\infty}$ of an optically infinitely thick layer on its density is linear in nature; and $R_{\lambda\infty}$ decreases with an increase in the density, while $A_{\lambda\infty}$ increases.

The dependence of the optical properties R_λ and T_λ of the endosperm of farinaceous and translucent wheat on the structure of the endosperm at different radiation wavelengths is shown in Fig. 6.3. Throughout the range of wavelengths employed the reflectance of the farinaceous endosperm is larger than that of the translucent one. The structural difference between the endosperms in reflecting light is most clearly manifested in the 0.5-0.6 μm range where the difference between the R_λ of farinaceous endosperm and that of transluscent endosperm is the largest. The translucent endosperm of durum wheat has a smaller R_λ and a larger T_λ and A_λ as compared to that of soft wheat. This is explained by the difference between the structures of the two translucent endosperms and by the difference between their densities.

Calculations of the absorption coefficient $\bar{k}_\lambda$, the scattering coefficient s_λ, and the effective attenuation coefficient L_λ show that the optical properties of the translucent endosperm differ greatly from those of the farinaceous endosperm in the 0.5-1.4 µm range (Tables 6.1, 6.2).

When translucent wheat (as well as other grades and kinds of wheat) is moistened the structure of the endosperm and consequently its optical properties undergo changes (Fig. 6.30). Changes in the thermal radiational characteristics of translucent wheat are brought about by many factors, the most important of which are:

- appearance of minute cracks in the endosperm and a decrease in its translucency; this leads to more intense scattering owing to the greater optical heterogeneity and structural changes of the endosperm;

- a change in the density of the endosperm following its cracking;

- a change in the supermolecular structure of the biopolymers that make up the grain;

- structural modifications of moisture due to a change in the bond energy between the tissues in the grain;

- irreversible changes in the original structure of the endosperm owing to biochemical processes.

In wheat grains containing 15-20% moisture intense structural changes take place owing to colloidal-biochemical processes. When the moisture content is above 20%, the translucent structure of the endosperm turns into a farinaceous one. At this point the endosperm begins to reflect more than the whole grain, which is characteristic of farinaceous wheat. When the moisture content is 21-25% the increase in R_λ and the decrease in T_λ are due to the complete going over from the translucent structure to a farinaceous one; this is accompanied by a rapid increase in s_λ and a decrease in $\bar{k}_\lambda$.

If a decrease in the moisture content of a grain is accompanied by its shriveling (and hence a decrease in the size of the pores), its transmittance in the 0.4-1.2 μm range decreases, while its reflectance increases (Figs. 1.2, 6.14, 6.17). A sample that has been dried but has not shriveled completely have a much larger transmittance as compared with the T_λ of a sample that has been dried and has shriveled completely. This is due to an increase in the size of the pores caused by local shriveling. The reflectance of a sample that has been dried and has shriveled in the spectral region of 1.2 μm increases with a decrease in the moisture content, but to a lesser degree than in the region <1.2 μm.

An investigation of the optical properties of foodstuffs in relation to their temperature shows that with an increase in temperature the spectral $\varepsilon_\lambda(T)$ and the integral ε_T of the emissivity factor decrease, while R_λ and R increase [190]. It has been established that for fatty beef the average value of ε_T at 65 and 100°C decreases from 0.780 to 0.776, and for lean beef - from 0.742 to 0.729, respectively. The average monochromatic emissivity $\varepsilon_\lambda(T)$ of meat dried by sublimation changes by 1-5% in the spectral region from 3.0 to 25.0 μm. For capillary-porous substances (refractory materials, calcined clay and gypsum), whose optical properties are similar to those of foodstuffs (Fig. 6.25), the reflectance increases with a rise in temperature from 24 to 100°C in the 1-5 μm region by 2-15%, depending on the radiation wavelength [187]. The maximum increase in R_λ when the sample is heated lies at wavelengths corresponding to the maximum reflectance near λ=1.6 and 3.8 μm. In the spectral region of λ>5.0 μm R_λ changes very little (1-2%) with an increase in temperature.

Surface roughness and the moisture content of the surface layer strongly affect the thermal radiational characteristics of the sample. Thus, in the case of a roughened surface R_λ can either increase or decrease, depending on the absorption coefficient and the refractive index of the medium. And moistening of the surface layer in the case of two samples of identical thickness and having the same moisture content leads to a decrease in R_λ.

6.3 Classification of foodstuffs and other radiation-scattering substances according to their optical properties

On the basis of an analysis of experimental data on the spectral thermal radiational characteristics of various moisture-containing absorbing-scattering substances we have classified these substances according to their optical properties in the 0.4–15.0 μm spectral region [40].

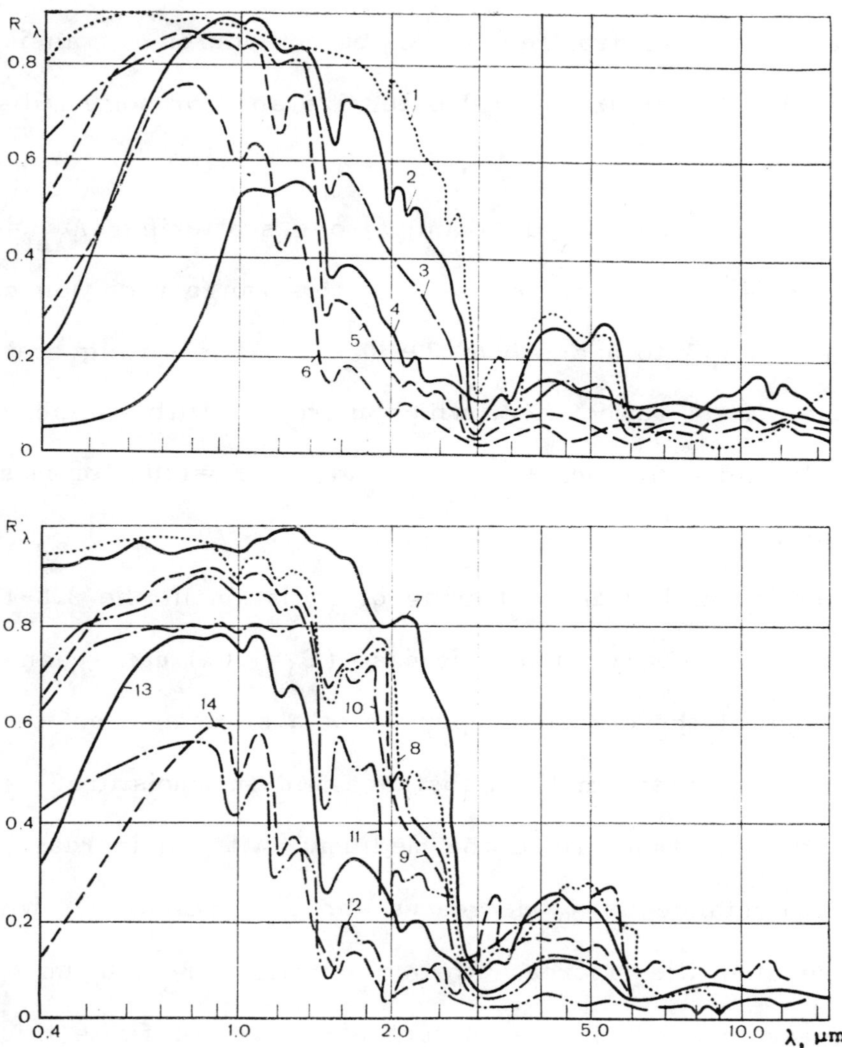

Figure 6.25 Reflection spectra of different substances [40]: 1 – enamel VL-55; 2 – pinewood (W=6.2%); 3 – flour (grade 1, W=8.1%); 4 – baked brown bread crust together with soft part of bread (ℓ=40 mm); 5 – dried potato (ℓ_{raw}=10 mm); 6 – dough made of wheat flour (ℓ=40 mm); 7 – MgO; 8 – confectioner's sugar (W=0.06%); 9 – fruit candy after it has gelled (W=30.0%); 10 – outer layer of a silk cocoon; 11 – potato starch (W=11.8%); 12 – potato starch (W=76.5%); 13 – dried pulp of pear (ℓ_{raw}==10 mm); 14 – durum wheat dough (W=31.2%).

In this region the optical properties of substances of vegetable origin are similar to one another. The spectral thermal radiational characteristics of many foodstuffs of vegetable and animal origin are similar to those shown in Fig. 6.25 [40]. To sum up, the main optical properties of the substances investigated are as follows.

1. A change from weak to average and even strong absorption ($0.001 < \beta_{\lambda ef} < 10.0$), and from strong to average and even weak scattering ($1.0 < \Lambda_{\lambda ef} < 0.05$) in the 1.5–2.5 μm range. This spectral region can be considered a transitional one. Here radiation penetration ($T_\lambda = 1.0\%$) is still extensive and for some substances it can reach 10.0 mm.

2. Weak absorption ($\beta_{\lambda ef} = \bar{k}_\lambda / s_\lambda < 0.1$) and strong scattering ($\Lambda_{\lambda ef} = s_\lambda / \varepsilon_\lambda > 0.9$) of radiation in the 0.8–1.4 μm spectral range. In this range radiation can penetrate the substance ($T_\lambda = 1.0\%$) to a depth of 20–30 mm. The strong scattering of radiation is due to the strong reflectance, which increases with an increase in the thickness of the sample and can reach 90–98% already at $T_\lambda = 1.0\%$ for most of the substances investigated.

3. Strong absorption and weak scattering of radiation in the 2.8–15.0 μm spectral range. Radiation penetration in this case ($T_\lambda = 1.0\%$) can reach 1.0–2.0 mm.

4. The dependence of the optical properties of a substance on its moisture content and the relationship between the substance and the moisture, as well as on the density and the microstructure of the medium. With an increase in the moisture content of a substance its s_λ decreases and $\bar{k}_\lambda$ increases. With an increase in ρ both these quantities increase throughout the 0.8–15.0 μm spectra range. With an increase in the moisture content of a layer of finite thickness its R_λ and T_λ decrease in the 2.8–15.0 μm spectral range where water strongly absorbs radiation; T_λ either increases or decreases, while R_λ decreases in this spectral range; and R_λ either increases or decreases, while T_λ increases in the 0.4–1.4 μm spectral range where water absorbs little radiation.

At the same time there are sharp quantitative differences between the optical properties of the substances investigated. Thus, the absolute reflectance (or absorptance) of different substances can differ by 20-60% in the 0.4-0.8 and 1.5-2.7 μm spectral ranges and by 5-20% at 3.5-5.8 μm. At 0.6-1.4 and 6.0-15.0 μm the difference between R_λ and A_λ for different substances is small (5.0-10.0%).

The main physico-chemical and optical properties of the substances examined have been summarized [40]. They enable us to forecast the approximate optical properties of each substance on the basis of their physico-chemical properties. We have also investigated [40] changes in the optical properties of substances resulting from the removal or adsorption of moisture. All these data are useful in the preliminary selection of infrared generators for the thermal processing of foodstuffs.

6.4 Factors affecting the integral optical and thermal radiational characteristics of foodstuffs

The absorption and scattering of radiation by foodstuffs are of a selective nature. Therefore, when determining the optical and thermal radiational characteristics of foodstuffs, account must be taken of multiple scattering both inside a layer of sample and inside the infrared installation. The changes in the spectral composition of the radiation following each absorption, scattering and reflection event must be taken into consideration (Tables 4.2, 4.3).

The thermal radiational characteristics R, T, and A are determined according to Eq. 4.117. Its integration is carried out with respect to the spectrum of the incident radiation flux E. The magnitude of the flux is determined according to the method described in Chapter 7, with account taken of the following: multiple reflection inside the infrared generator; absorption of infrared radiation by the air-water medium; and irradiation conditions.

In approximate calculations of the thermal radiational characteristics according to Eq. 4.117 the irradiance $R_\lambda(T_\lambda)$ of the infrared source is substituted into the equation. This yields the values of $\overline{R}$, $\overline{T}$, and $\overline{A}$ averaged with respect to the radiation which falls upon the sample once and directly from the infrared source. However, under real conditions half of the infrared radiation emitted by the generator falls upon the sample after it has been reflected from the lamp's enclosure and its spectral composition has changed. As can be seen from data in Table 6.4 on the R_∞ and T of dough and bread obtained by using Eq. 4.117, the values of R_∞ and T that are averaged with respect to the irradiance of the infrared source $R_\lambda(T)$ differ considerably from those averaged with respect to the spectrum of the radiation reflected from the lamp's enclosure $R_{\lambda refl}R_\lambda(T)$. R_∞, A, and T also depend on the temperature (T) of the filament inside the infrared lamp. For, according to Planck's rule, a change in T causes changes in the spectral composition of the radiation [108].

Table 6.4 Thermal radiational characteristics (T/R_∞) for dough and bread calculated according to Eq. 4.117 [108].

Type of radiation	Sample	T/R_∞ (temperature of the filament, K)			
		1700	2100	2500	2800
Infrared lamp (KG-1000)	dough (ℓ=11 mm)	0.003/0.224	0.006/0.301	0.008/0.370	0.009/0.407
	dough-bread (ℓ=6 mm)	0.068/0.309	0.091/0.363	0.115/0.445	0.120/0.457
	baked bread (ℓ=6 mm)	0.096/0.357	0.125/0.399	0.137/0.432	0.119/0.343
Reflected radiation	dough (ℓ=11 mm)	0.002/0.207	0.005/0.294	0.007/0.344	0.008/0.388
	dough-bread (ℓ=6 mm)	0.062/0.290	0.088/0.356	0.108/0.423	0.119/0.454
	baked bread (ℓ=6 mm)	0.059/0.281	0.093/0.319	0.104/0.334	0.108/0.334

As has been shown in Table 4.3, a disregard for changes in the spectral composition of incident radiation due to multiple reflection results in large errors (up to 31%) in calculating the absorptance. Changes in the operating conditions of the infrared generator also affect the spectral composition of radiation and thus the thermal radiational characteristics of foodstuffs.

Table 6.5 Thermal radiational characteristics of potato starch [39]

| Filament temperature (K) | With compensation for reflection | | Without compensation for reflection | | Relative deviation (%) | |
	R_∞	$L \cdot 10^{-3} (m^{-1})$	$L \cdot 10^{-3} (m^{-1})$	R_∞	δR_∞	δL
1600	0.616	2.064	2.715	0.445	27.8	24.0
2000	0.703	1.670	2.407	0.534	24.0	30.6
2400	0.751	1.442	2.164	0.602	19.8	33.4
2800	0.779	1.306	1.973	0.651	16.4	33.8
3373	0.806	1.185	1.749	0.707	12.3	32.2

Data in Table 6.5 have been averaged with respect to the spectrum of incident radiation flux at different temperatures of the filament with and without account being taken of multiple reflection in the oven.

Table 6.6 Optical properties of different substances [39]

Sample	W (%)	T(ℓ, mm)	R_∞	$L \cdot 10^{-3} (m^{-1})$	$\bar{k} \cdot 10^{-3} (m^{-1})$	$s \cdot 10^{-3} (m^{-1})$
Dough	48.0	0.008(11.0)	0.370	0.424	0.195	0.363
Dough-bread	46.0	0.115(6.0)	0.445	0.321	0.123	0.356
Baked bread	38.0	0.137(6.0)	0.432	0.295	0.117	0.313
Durum dough	31.2	0.168(1.3)	0.283	1.310	0.732	0.806
Potato starch	11.8	0.066(1.0)	0.615	2.264	0.540	4.730
Pinewood	6.2	0.169(1.0)	0.740	1.060	0.158	3.468
Cornmeal	14.3	0.005(10.0)	0.400	0.514	0.215	0.489
Fruit candy	30.0	0.157(1.0)	0.748	1.100	0.155	3.336

In approximate calculations the averaged optical characteristics L, $\bar{k}$, and s are obtained according to Eqs. 3.10-3.12, and the averaged characteristics $\bar{R}$, $\bar{T}$, and $\bar{R}_\infty$ - according to Eq. 4.117. These characteristics are independent of the coordinate x and are considered to be constant inside a layer of sample.   At the same time, because of changes in the spectral composition of the radiation inside the layer (Fig. 4.7) the optical characteristics L, $\bar{k}$, and s depend on x (Fig. 4.8) and should be determined by using Eqs. 4.104, 4.109, and 4.111.

The optical and thermal radiational characteristics of different foodstuffs averaged with respect to the spectrum of a KG-1000 infrared lamp (filament temperature 2250 K), without compensation for multiple reflections in the oven, are shown in Table 6.6.

Table 6.7   Thermal radiational characteristics with account taken of multiple reflection with the aid of Eq. 4.117 [16]

Type of spectrum	Reflector (Al) $\bar{R}$	$\bar{R}_{\infty}$	Bread crumbs (ℓ=5.0 mm) $\bar{R}$	$\bar{A}$	Air-vapor medium (ℓ=100 mm) T
Filament T=2600 K	0.850	0.471	0.417	0.484	0.988
Filament T=2400 K	0.859	0.426	0.391	0.517	0.986
Radiation E_1	0.845	0.520	0.486	0.395	0.991
Radiation E_2	0.840	0.523	0.484	0.398	0.992

Note: E_1 - radiation spectrum with compensation for multiple reflection;
E_2 - radiation spectrum with compensation for reflection and absorption by the medium.

In Fig. 6.26 are shown the dependence of the optical and thermal radiational characteristics of different kinds of starch,   with account taken of multiple reflection, on the filament temperature in a closed oven made of polished aluminum [61]. As can be seen, the R_{∞} of starch increases with an increase in the filament temperature, while the R of the reflector decreases.   This is due to a shift of the maximum of infrared radiation into the short-wave region; there the starch reflects radiation more strongly (Fig. 6.2), while aluminum absorbs radiation near 0.8 μm (Table 5.3).   Potato starch reflects less radiation since it scatters radiation more strongly than the rice starch and   cornstarch.

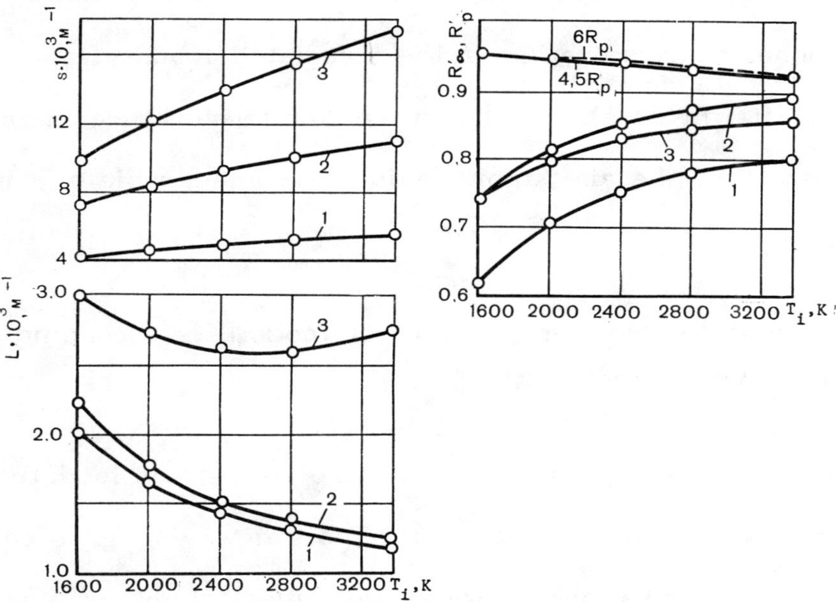

Figure 6.26 Optical and thermal radiational characteristics of aluminum and different kinds of starch as a function of the filament temperature (KG-1000 infrared lamp) [39]: 1 - potato starch; 2 - cornstarch; 3 - rice starch; 4,5,6 - polished aluminum.

The coefficient of effective attenuation for potato starch and cornstarch decreases with a decrease in the filament temperature from 1600 to 3200 K; for rice starch there is a well-defined minimum at 2500 K.

The presence of an air-vapor medium in the oven also affects the incident flux and consequently the thermal radiational characteristics of the sample. In Table 6.7 are shown the values of $\overline{R}$ for the reflector, of $\overline{R}$ and $\overline{A}$ for bread crumbs, and of T for the air-vapor medium, with account taken of multiple relection in the oven equipped with a KG-1000 infrared lamp. When a layer of the air-vapor medium is relatively thin and when its moisture concentration corresponds to that encountered in drying bread crumbs, the transmittance T of the medium is close to unity. In this case the absorption of radiation by the medium can be disregarded when carrying out engineering design calculations. However, changes in the spectral composition of the incident infrared radiation due to multiple reflections and variations in the filament temperature must be taken into consideration in such calculations. Failure to do so will lead to large errors in calculating the optical and thermal radiational characteristics of the irradiated foodstuffs.

In the next chapter it will be shown that when low-temperature infrared generators are used the effect of the air-vapor medium is considerable and must be taken into account.

6.5 Methods for determining the moisture content of foodstuffs according to their optical and thermal radiational characteristics

To work out methods for determining the moisture content of foodstuffs it is necessary to know their optical properties in the region of wavelengths corresponding to the frequencies of the oscillations of water molecules, their combination and overtones (Figs. 1.3, 1.4). The sensitivity of such methods depends on the absorption coefficients of the foodstuffs being investigated and of water at the maximum in the absorption band in the infrared spectrum, as well as on several spectrophotometric factors [1, 39, 196]. Any changes in the structure of foodstuffs (such as swelling, formation of minute fissures, the bonding of water with the substance in question, etc) also affect the optical properties of foodstuffs.

In our investigation of the dependence of R_λ, T_λ, $R_{\lambda\infty}$, L_λ, $\overline{k}_\lambda$, and s_λ on the changes induced by the presence of moisture, we have chosen foodstuffs that are different in structure: starch, grain, groats, wood, bread crumbs, and so on. The first three of the above-mentioned characteristics have been determined at λ_i corresponding to the maxima in the absorption bands and at λ_j, which is outside the absorption range of water.

It has been found [19, 44, 106] that for a majority of capillary-porous colloidal substances an increase in their moisture content leads to a decrease in their transmittance and reflectance. This is due to the fact that moisture absorbs considerable amount of infrared radiation. However, in some experiments only when a high-temperature radiation source was used did the transmittance of the sample increase   with an increase in its moisture content.

Investigations have also shown [39, 107] that in the case of such capillary-porous colloidal foodstuffs as fruits, vegetables, tea leaves and so on, with an increase in their moisture content their spectral transmittance increased in the 0.4-1.2 μm range and decreased in the 1.4-15.0 μm range.   In Fig. 6.14 is shown the dependence of T_λ on λ for tea leaves with different moisture content.

For some substances their reflectance can increase with an increase in their moisture content in the range from 0.4 to 0.8 μm [39, 107], and for tea leaves – even in the range up to 1.4 μm (Fig. 6.14).   Such behavior is characteristic only of foodstuffs exhibiting intense absorption near 0,40 μm (beets, pears, apples, carrots, tea leaves).   For other foodstuffs, with an increase in their moisture content R_λ decreases throughout the spectral region from 0.4 to 15.0 μm.   At the same time, a decrease in R_λ in the 0.4-1.2 μm region is accompanied by an increase in T_λ.   In the same spectral region with a decrease in the moisture concentration of the sample R_λ increases and T_λ decreases, while in the 2.5-15.0 μm region T_λ increases.   Even the presence of small amounts of structural or bonded water (in the substances being investigated) or the presence of HO-groups in the molecule of the substance leads to the appearance of intense wide absorption bands near 2.92, 6.12, and 15.80 μm and less well-defined ones at 0.75, 0.85, 0.98, 1.45, 1.93, and 4.74 μm (Figs. 1.2, 6.14, 6.25).

Because of the low reflectance of water (1.6-4.8%) in the 1.0-15.0 μm region (Fig. 6.13) the presence of moisture in the sample, especially on its surface, will lead to a decrease in its reflectance.   This has been confirmed experimentally (figs. 6.14, 6.25) [39, 107].   In this case the presence of moisture leads to a decrease in the reflection from the surface of the walls, pores and particles, while the moisture itself practically does not absorb radiation.

As can be seen from Fig. 6.27, the optical properties of cornmeal are in linear dependence on moisture concentration.   An increase in transmittance with an increase in the moisture content is due to a decrease in the attenuation coefficient L_λ, despite the fact that the absorption coefficient $\bar{k}_\lambda$ increases. The decrease in L_λ is due to a sharp decrease in the scattering coefficient s_λ with an increase in the moisture concentration. In this case the absorption coefficient increases only slightly. The spectral transmittance T_λ increases with an increase in W only in the 0.4-1.4 μm region, where the absorption coefficient of water is small. A similar dependence of R_λ and T_λ on W has been reported in the case of bread crumbs (Fig. 6.28) [16].

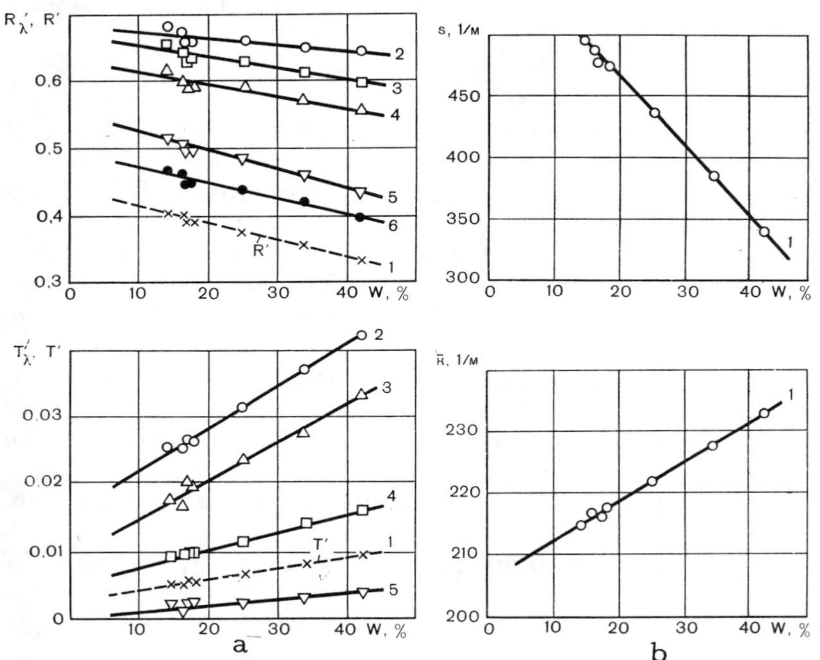

Figure 6.27 Integral and spectral thermal radiational characteristics of cornmeal as a function of its moisture content [39]: a - integral (curve 1) and spectral (curves 2-6) characteristics: 1 - averaged with respect to the spectrum of the infrared lamp (filament temperature 2500 K); 2 - λ=.9 μm; 3 - λ=1.0 μm; 4 - λ=1.1 μm; 5 - λ=1.2 μm; 6 - λ=1.3 μm; b - optical properties averaged with respect to the spectrum of the infrared lamp (filament temperature 2500 K).

The reflectance of an optically infinitely thick layer $R_{\lambda\infty}$ and the transmittance of a layer of finite thickness T_λ as a function of W are shown in Fig. 6.29: for potato starch at $W_1 \cong 12\%$ and $W_2 \cong 8\%$, for cornstarch at $W_1 = 8 \div 9\%$ and $W_2 \cong 4\%$, and for rice starch at $W_1 \cong 3\%$. For potato and cornstarch $R_{\lambda\infty}$ in the W_1-W_2 range remains approximately constant. The complex nature of the dependence of R_λ and T_λ on W is due to the different forms of bonds formed by the water in the sample and to the swelling of starch grains as they are being moistened.

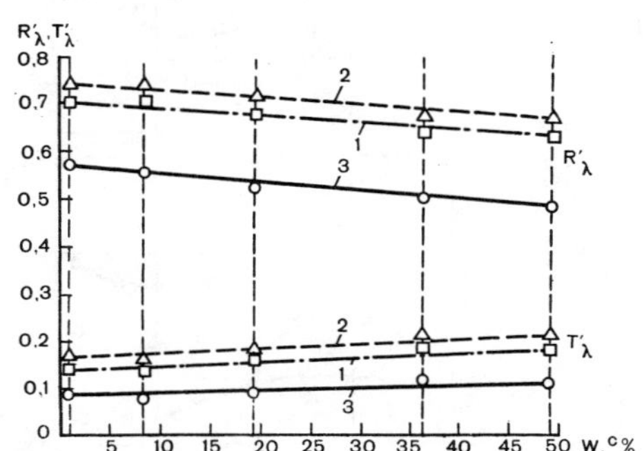

Figure 6.28 Spectral thermal radiational characteristics of bread crumbs as a function of their moisture content at different wavelengths [16]: 1 - 1.0 µm; 2 - 1.1µm; 3 - 1.2 µm.

It has been shown [23] that an increase in the moisture content of translucent wheat grains from 11 to 18% results in an increase in the reflection by a layer of sample (Fig. 6.30) and in a larger scattering coefficient (Fig. 6.31). The fact that the changes in the optical properties of translucent and farinaceous wheat grains and of corn (Fig. 6.30) at W=15-25% are of the same nature indicates that these changes are due to structural changes resulting from an increase in the moisture content, to the special mechanism of interaction between the grain and the moisture, and to the biochemical properties of the grain as a living organism.

Thus, an analysis of the changes in the reflectance and transmittance of foodstuffs shows that the presence of water strongly affects their thermal radiational and optical properties. For some substances the dependence of R_λ and T_λ on W can be considered to be approximately linear. However, in the range of the equilibrium concentrations of water such dependence is of a complex nature (Fig. 6.30) and therefore cannot be used in determining the moisture content. The linear relationship between R_λ and W can be used in designing new infrared contact-free moisture meters for foodstuffs.

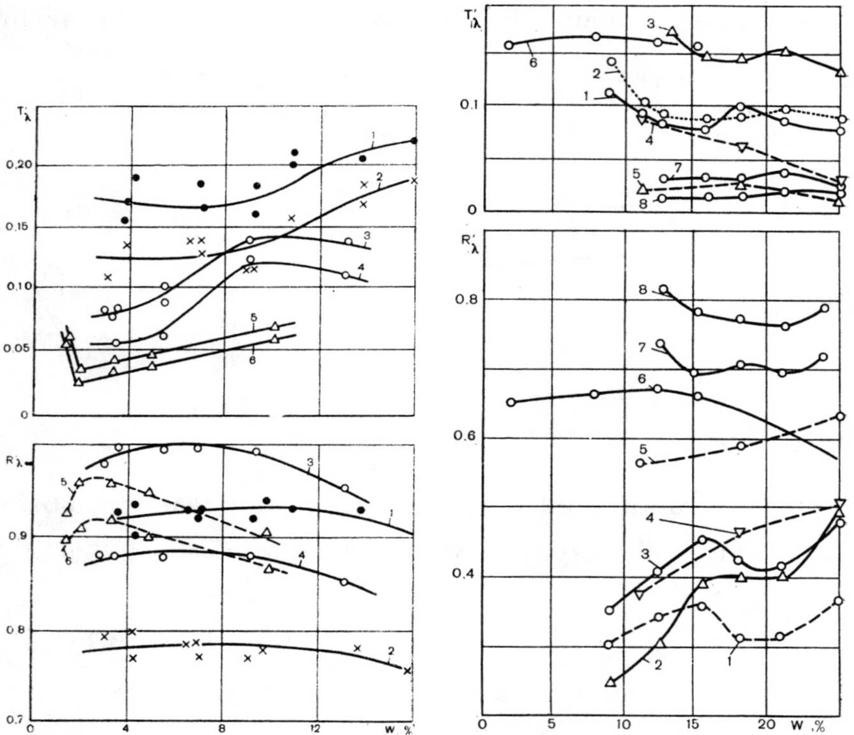

Figure 6.29 Spectral thermal radiational characteristics of different kinds of starch as a function of its moisture content at different wavelengths (T_λ values are given for a layer of sample ℓ=1.0 mm) [39, 61]: 1,2 - potato starch; 3,4 - cornstarch; 5,6 - rice starch; 1,3,5 - 0.8 µm; 2,4,6 - 1.2 µm.

Figure 6.30 Spectral thermal radiational characteristics of different substances as a function of their moisture content λ=1.1 µm [23, 26]: 1 - corn; 2 - endosperm of translucent wheat (<u>Hordeum</u> genus) (ℓ=2.9 mm); 3 -grain of translucent wheat (<u>Hordeum</u> genus); 4 - grain of translucent wheat (Bezostaya-I); 5 - grain of farinaceous wheat (Vesna); 6 - pinewood; 7 - endosperm of farinaceous wheat (PPG-269); 8 - grain of farinaceous wheat (PPG-269).

6.6 Methods for estimating the quality of foodstuffs according to their optical properties

The quality of foodstuffs can often be evaluated on the basis of their color. In this country the use of color as an indicator of the quality of foodstuffs must conform to the technical standards set by the All-Union State Standards (GOST).

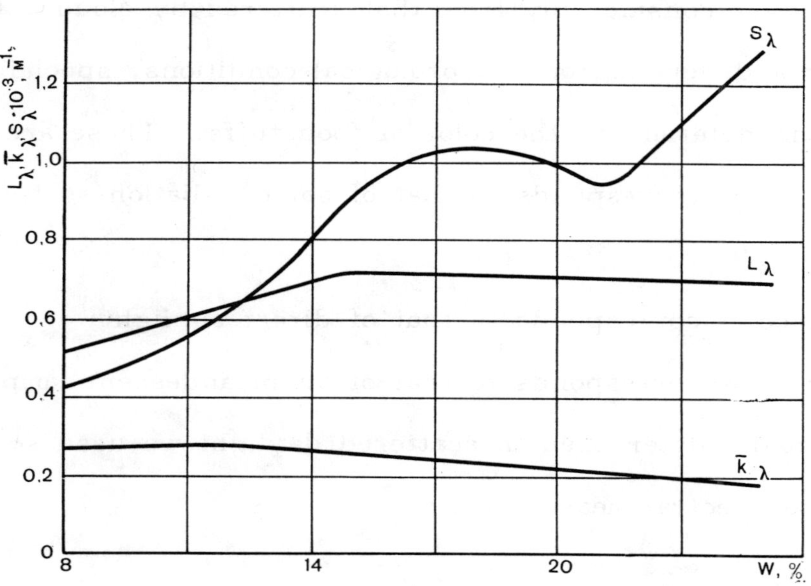

Figure 6.31 Dependence of the optical properties of the endosperm of wheat grain on its moisture content at $\lambda=0.9$ μm [23].

Usually the color of a substance is determined visually by comparing it with that of a standard. This is a subjective method, however, and is not reliable. The human eye perceives the color of scattered radiation reflected and transmitted by a sample at a given spectrum of the illumination. Thus, the color of a substance depends on the spectral composition of the source of illumination. For example, starch and flour appear white in daylight, but become yellowish under incandescent illumination. The eye perceives visible light selectively. In Table 6.8 are summarized the average approximate spectral ranges of wavelengths corresponding to the commonest colors.

For a quantitative evaluation of the color of foodstuffs spectral and colorimetric methods are used.

The absolute spectral method consists of an evaluation of color based on calculations with the aid of measured spectral coefficients of reflection and transmission. Data are also needed on the spectral distribution of energy in the spectrum of the light source employed. Owing to the fact that the spectral composition of direct solar radiation differs considerably from that scattered by clouds, that it changes during the day and depends on meteorological conditions, special standards of white light are used in determining the color of foodstuffs. These are:

- source C, whose color corresponds to that of solar radiation scattered by clouds;

- source B, whose color corresponds to that of direct sunlight;

- source A, whose color corresponds to that of an incandescent lamp.

Since colors are usually determined in scattered daylight we used source C for carrying out quantitative spectral measurements.

Table 6.8 Spectral range of different colors [9]

Color	Wavelength (μm)	Color	Wavelength (μm)
Red	0.770-0.620	Green	0.550-0.510
Orange	0.620-0.585	Blue	0.510-0.480
Yellow	0.585-0.575	Dark blue	0.480-0.450
Yellow-green	0.575-0.550	Purple	0.450-0.380

The relative spectral method for determining colors is based on a comparison of the reflection spectrum of a sample with that of an ideally white reflecting surface, by which is meant a surface which reflects radiation at all wavelengths: $R_\lambda = 1.0 = const$. The color of the sample is determined on the basis of the difference between the reflection spectrum of the sample and that of the standard in the visible spectral region.

For light sources having a continuous radiation spectrum the calculations involved in determining colors are laborous and time-consuming. For this it is necessary to measure the spectral distribution of the radiation energy of the source and the R_λ of the sample in the visible spectral range. The methods and the equip ment used for measuring the R_λ of foodstuffs have been described in Chapter 5.

The color of a given foodstuff is determined by the relationship between its reflectance and the radiation wavelength in the visible spectral range. In the case of some foodstuffs the difference in their color is due to the presence of an additional absorption band near 0.45 μm. This explains the widening of the absorption band in the ultraviolet spectral region at $\lambda < 0.40$ μm, which is characteristic of many foodstuffs. Thus, for a dough-bread mixture (Fig. 6.6) at $\lambda = 0.40$ μm the spectrum contains only a wing of the absorption band, which in the case of half-baked bread crust shifts into the long-wave region. For baked bread crust together with the soft part of bread there is a broad maximum in the absorption band at 0.4-0.6 μm. This accounts for the brown color of the bread crust, since most of the radiation from the violet to the red spectral region is absorbed.

Data in Table 6.9 show that vividly colored foodstuffs absorb more strongly blue-violet radiation at $\lambda = 0.4-0.5$ μm. Thus, the absorption of brown bread crust at $\lambda = 0.45$ μm is: $A_\lambda = 0.95$, i.e. approximately twice as large as the absorptance of dough ($A_\lambda = 0.49$), which is light-yellow in color with grayish hues. Red-purple beets absorb red light at $\lambda = 0.70$ μm, just like potato of a light-yellow color. However, at $\lambda = 0.45$ μm the beets absorb violet light 1.5 times more strongly than potato. The spectral reflectance and absorptance of a given foodstuff can be used for the qualitative and quantitative determination of its color. The smaller its A_λ in the visible spectrum, the less vivid is its color; often it is light in color, approaching white.

The colors of foodstuffs can be determined quantitatively only on the basis of color mixture. It is well-known that all existing shades of color can be obtained by mixing three colors: red (R), green (G), and blue (B), taken in certain amounts. In the first international three-color colorimetric system red color ($\lambda = 0.700$ μm) was obtained with the aid of an incandescent lamp and a red filter; and green color ($\lambda = 0.5461$ μm) and blue color ($\lambda = 0.4358$ μm) were obtained by separat-

Table 6.9 Colors of different foodstuffs determined by different methods (λ in μm) [39, 99]

Foodstuffs	W (%)	Whiteness (Eq. 6.13) $R_{\lambda\infty}\overline{W}$(%)	Yellow color (Eq. 6.10) Y(%)	$A_\lambda(\theta;2\pi)$ at different λ			
				0.45	0.52	0.70	0.90
Confectioner's sugar	0.06	88	4	0.06	0.04	0.02	0.06
Potato starch	13.2	73	6	0.15	0.12	0.10	0.08
Wheat flour (white)	8.1	44	18	0.32	0.24	0.17	0.10
Wheat flour (yellow)	8.3	23	31	0.46	0.35	0.22	0.15
Fruit candy uncoated with sugar	30.0	48	18	0.28	0.20	0.12	0.09
Wheat flour dough	50.0	12	39	0.49	0.36	0.16	0.13
Potato	74.5	7	33	0.60	0.49	0.40	0.47
Waffles	2.1	-	39	0.72	0.60	0.54	0.53
Cornmeal	14.3	-	75	0.83	0.65	0.33	0.32
Carrots	86.5	-	-	0.86	0.82	0.36	0.47
Beets	85.5	-	-	0.91	0.92	0.40	0.32
Bread crust	10.1	-	-	0.95	0.94	0.85	0.58

ing the corresponding spectral lines from the emission spectrum of a mercury lamp.

By using the three main colors, R, G, and B, the color C of a sample can be expressed quantitatively and qualitatively with the aid of a color equation:

$$C \equiv r'R + g'G + b'B; \qquad (6.1)$$

where, r', g', and b' - chromaticity coordinates; these coefficients indicate how many units of each of the three colors, R, G, and B, should be used in order to obtain a given color C. Sign $\equiv$ indicates that the color of the radiation reflected by the sample (or standard) and the color of the incident radiation have identical chromaticity: hue, color purity, and brightness. The products $r'R$, $g'G$, and $b'B$ are called color components.

For a qualitative estimation of the color of a luminous flux reflected by a given foodstuff it is sufficient to use the relative values of chromaticity coordinates $(r, g,$ and $b)$. These values are determined from the following relationship:

$$r=r'/(r' + g' +b'). \qquad (6.2)$$

And their sum equals unity:

$$r + g + b = 1. \qquad (6.3)$$

This equality can be used to represent any color graphically as a point located inside or outside an equilateral color triangle RGB whose altitude equals unity (Fig. 6.32,a). In this case the sum of the perpendiculars from any point to the sides of the triangle is also unity and can represent the sum of the chromaticity coordinates.

The points corresponding to all the colors which cannot be obtained by directly mixing the basic colors are located outside the color triangle. In this case one of the chromaticity coordinates is negative (Fig. 6.32, b). For such foodstuffs the equation can be written as follows:

$$C + r'R \equiv g'G + b'B, \text{ or } C \equiv g'G + b'B - r'R. \tag{6.4}$$

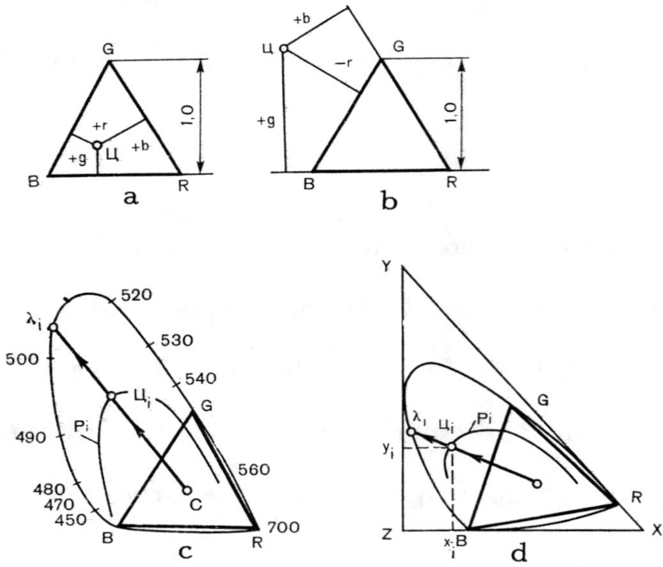

Figure 6.32 Location of the basic colors in different three-color systems [9]: a,b - color represented with the aid of a color triangle; c - color according to the RGB system; d - color according to the XYZ system.

The color graph in Fig. 6.32 can be used to represent all the existing shades of color parallel to the line of spectral color, λ_i. Point C (or B and A) in the center of the triangle corresponds to the color having chromaticity coordinates r = g = b = 1/3.

For calculating the color C_i of a sample it is necessary to determine the chromaticity coordinates and locate point C_i on the graph (Fig. 6.32,c). By drawing a line linking point C corresponding to the white color with the point for a given color C_i and continuing the line until it intersects with the line corresponding to the spectral colors λ, we obtain the hue λ_i. All the colors on the $CC_i\lambda_i$ line are of identical hue (λ_i), and differ only in their color purity (p_i); i.e., they differ by the extent to which they are diluted by the white color. The purity of color at C point equals zero and is 100% on the line for the spectral colors (purely monochromatic color).

The color purity p_i is determined from the ratio of the luminous flux of a given spectral color to the sum of the luminous fluxes of the white and the spectral color, which must be mixed in order to obtain the given color. The closed lines p_i=const includes point C corresponding to the white color.

The use of such a color system (Fig. 6.32,c) for the quantitative determination of a color is rather complicated owing to the presence of the negative chromaticity coordinates r, g, and b. This problem was solved in 1931 when the International Commision on Illumination (C.I.E.) adopted a new colorimetric system. In this system (Figs. 6.32,d and 6.33) the chromaticity coordinates x. y, z for all the colors are positive. The provisionally chosen basic colors X, Y, Z, which do not exist in reality, can be mathematically expressed through the basic colors R, G, and B. Thus, any color can be determined with the aid of the following relationship [135, 147]:

$$C \equiv x'X + y'Y + z'Z; \qquad\qquad (6.5)$$

where, x', y', and z' – chromaticity coordinates in the C.I.E system.

In a quantitative evaluation of the color of a given foodstuff, the color being dependent on the spectral composition of the radiation E_λ and R_λ emitted by the foodstuffs, we first determine the monochromatic components of the conditional colors X_λ, Y_λ, Z_λ. Then the total color characteristics X, Y, Z are determined by integration with respect to all the wavelengths in the visible spectrum:

$$X = \int_{\lambda_1}^{\lambda_2} E_\lambda R_\lambda X_\lambda d\lambda. \qquad\qquad (6.6)$$

The quantities of X, Y, Z are obtained experimentally with the aid of special spectrometers and three-color colorimeters. The spectral sensitivity curves of the photocells are converted to the internationally established functions of specific coordinates $\bar{y}(\lambda)$, $\bar{z}(\lambda)$, and $\bar{x}(\lambda)$. The normalized function $\bar{x}(\lambda)$ is reproducible experimentally, and for the C source its relationship with the above-mentioned coordinates can be expressed as follows:

$$\overline{x}_n(\lambda)=0.833\overline{x}(\lambda) + 0.333\overline{y}(\lambda) - 0.167\overline{z}(\lambda). \tag{6.7}$$

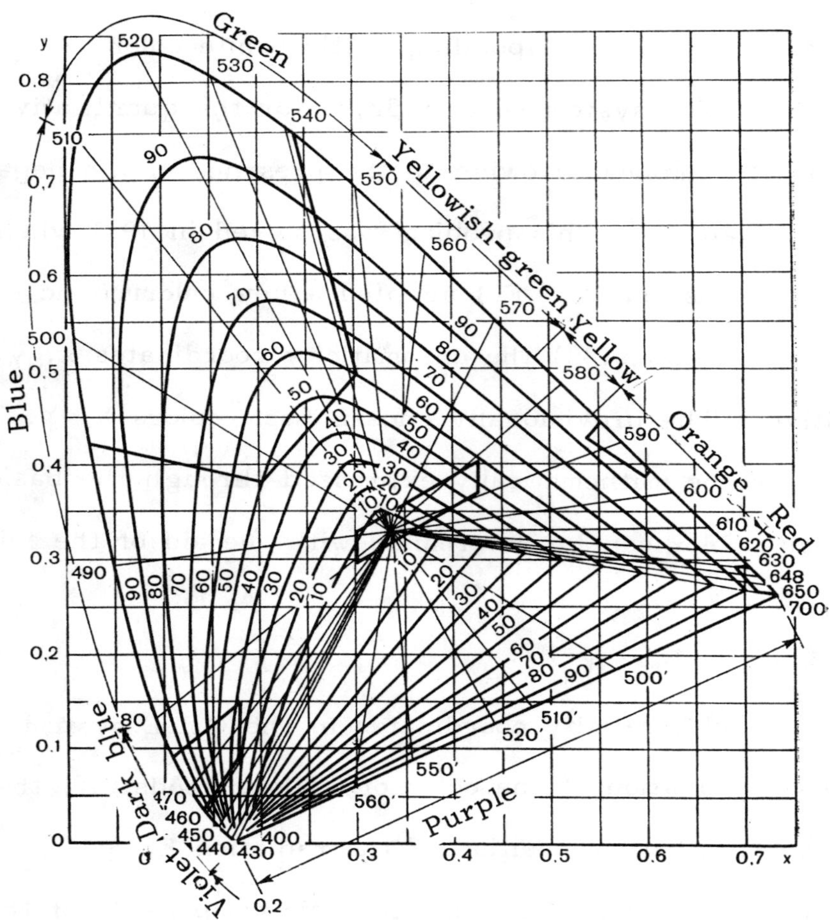

Figure 6.33 International colorimetric system XYZ (the C.I.E system) [9].

The chromaticity coordinates are determined with the aid of a relative method, which consists in a comparison of the characteristics of the sample with that of the standard. The ratio of the measured luminous fluxes is proportional to the ratio of the chromaticity coordinates of the samples being compared:

$$n_x=x'_n/x'_{n\,(st)}, \quad n_y=y'/y'_{st}, \quad n_z=z'/z'_{st}. \tag{6.8}$$

When we know the chromaticity coordinates of the standard, $x'_{n(st)}$, y'_{st}, and z'_{st}, we can calculate the chromaticity coordinates of the sample, x'_{st}, y'. and z'.

The color of a sample of foodstuff is established by first determining the absolute color coordinates x', y', and z' with the aid of Eq. 6.8, and then by calculating the absolute chromaticity coordinates x, y, and z as follows:

$$x = x'/(x' + y' + z'). \qquad (6.9)$$

By plotting the values of x_i and y_i on the coordinate axes (Figs. 6.32,d; 6.33) we have point C_i which corresponds to the chromaticity of a given color. Its hue λ_i is represented by the point where the lines of the spectral color intersect with the straight line leading through point C_i. The color purity p_i is determined from the curve of identical purity which passes through point C_i and encompasses point C.

The color of starch, one of the main indicators of its quality, is usually determined visually. For an objective determination we have investigated [30] the optical properties of starch in the visible (0.4-0.76 μm) and near-infrared (0.76-5.0 μm) regions (Figs. 6.1, 6.2). The chromaticity of starch depends on the relationship between $R_{\lambda\infty}$ and the radiation wavelength in the visible range. As can be seen from Fig. 6.2, the $R_{\lambda\infty}$ of rice starch in the 0.4-0.6 μm range is smaller than that of potato starch and of cornstarch. This confirms the color of the given sample of rice starch which is white with a grayish and a yellowish tinge.

The grayish tinge, which is characteristic of all kinds of starch, is due to the fact that their reflection depends only slightly on λ and is less than one at 0.50-0.76 μm: $R_{\lambda\infty} = 0.8 \div 0.9$. The relatively greater or lesser absorption on radiation at 0.4-0.5 μm determines the intensity of the yellowish tinge, which is also characteristic of all kinds of starch. In this spectral range R_λ noticeably decreases: at $\lambda_z = 0.4$ μm $R_{\lambda z}$ is less than 10-20% compared with R_λ for the remaining part of the visible spectral range. This is due to the presence of an intense absorption band in the ultraviolet region at $\lambda < 0.4$ μm.

The intensity of the yellowish tinge (Y) can be determined according to the relationship below, following measurements of $R_{\lambda z}$ at $\lambda_z = 0.457$ μm and at $\lambda_x =$

0.614 µm:

$$Y = (1 - R_{\lambda z}/R_{\lambda x})100\%; \qquad\qquad (6.10)$$

where, R_λ - spectral reflectance of the sample relative to that of the standard which is white in color (MS-14 glass, etc).

Photometric measurements have shown [66] that potato starch is white in color with yellowish and grayish tinges and is a nonluminous substance. The chromaticity coordinates of different grades of potato starch irradiated by a C source vary within the limits: $x = 0.313 \div 0.322$, $y = 0.321 \div 0.329$.

By comparing the chromaticity coordinates of starch with those of the white color $C(x_C = 0.31006$ and $y_C = 0.31616)$, it can be established with the aid of the colorimetric system (Fig. 6.33) that all the samples of starch are white in color.

Spectral investigations [66] suggest that the degree of whitness ($\overline{W}$) of starch can be determined quantitatively according to the approximate equation:

$$\overline{W} = 4B - 3Y; \qquad\qquad (6.11)$$

where, $B = 0.847z$; Z,Y - chromaticity coordinates according to the C.I.E. colorimetric system. The degree of whiteness can be approximately calculated by using the spectral reflectance of an optically infinitely thick layer at $\lambda_z = 0.459$ µm $(R_{\lambda z})$ and at $\lambda_y = 0.522$ µm $(R_{\lambda y})$ according to the equation:

$$\overline{W} = 4R_{\lambda z} - 3R_{\lambda y}. \qquad\qquad (6.12)$$

To simplify calculations only two chromaticity coordinates are used in Eqs. 6.11 and 6.12: z - at $\lambda_z = 0.45$ µm for the yellow tinge, and y - at $\lambda_y = 0.52$ µm for the grayish tinge. The magnitude of the third coordinate x at $\lambda_x = 0.70$ µm is close to that of y and thus also characterizes the grayish tinge. The combination of x, y, and z coordinates defines the overall color of the sample and its difference from an absolutely white body.

It has been proposed [66] that in relative measurements of the whiteness of starch, i.e., in comparing the light reflected by a sample of starch and the reference sample with that of an MS-standard, the following simplified relationship be used with account taken of the color characteristics of the MS-standard [66]:

$$\overline{W} = 3.795 n_5 - 2.88\, n_6;$$

(6.13)

where, $n_6 = y/y_{st}$ - the ratio of the chromaticity coordinates of the sample to those of the standard with the use of a green light filter No.6 ($\lambda_y = 0.52$ μm) in diffuse daylight; $n_5 = z/z_{st}$ - the ratio of the chromaticity coordinates of the sample to those of the standard with the use of a blue light filter No.5 ($\lambda_3 = 0.45$ μm) in diffuse daylight.

The n_5 coordinate characterizes the intensity of the yellowish tinge, and the n_6 coordinate - the grayish tinge. Instead of n_5 and n_6 in Eq. 6.13 we used the values of $R_{\lambda\infty}$ at the same wavelengths (0.45 and 0.52 μm) since the dependence of $R_{\lambda\infty}$ on the wavelength in the visible spectrum determines the chromaticity of starch.

A comparison of the data on the whiteness of starch obtained by the above-mentioned method with those obtained by the approximate spectral method for different kinds of starch is shown in Table 6.10. The calculated values of whiteness obtained by various methods differ by 1-2.2%, thus confirming the validity of Eq. 6.13.

Data on the whiteness and the yellow tinge of different foodstuffs calculated by the method proposed above with the use of $R_{\lambda\infty}$ values are summarized in Table 6.9.

Thus, the chromaticity of starch and other foodstuffs can be determined from the dependence of the reflectance of an optically infinitely thick layer of sample ($R_{\lambda\infty}$) on the radiation wavelength in the visible spectrum.

The optical and thermal radiational properties of foodstuffs, as it has been shown (Fig. 6.30), are affected by their moisture content.

Table 6.10 Whiteness of starch obtained by different methods [39, 61]

Sample	n_6	n_5	$\overline{W}$ (%) (n_6 and n_5)	$R_{\lambda\infty}$ (0.45 μm)	(0.52 μm)	$\overline{W}$ (%) $R_{\lambda\infty}$
Potatoe starch	84.3	87.3	68.5	81.5	84.0	67.5
Cornstarch	90.4	94.9	69.8	86.5	90.5	67.7
Rice starch	81.8	86.9	60.2	79.5	83.0	62.7

The color index (e.g., whiteness), as determined on the basis of $R_{\lambda\infty}$ or the chromaticity coordinates x, y, z, also depends on the moisture content of the sample. Many foodstuffs become lighter in color when moistened, and turn darker in color when they are dried. This is due to the selective change in the spectral reflection of foodstuffs when they are moistened (Figs. 1.2, 6.14, 6.25). Thus, for cornmeal [39], with an increase in its moisture content up to W=40%, R_λ increases in the 0.45-0.50 μm region and decreases at λ>0.70 μm. For capillary-porous colloidal systems the dependence of R_λ on W is of a complex nature (Fig. 6.30). But within a narrow range of water concentration approaching the equilibrium state, R_λ only slightly depends on W. As has been shown [66], the whiteness of two kinds of starch (W012-20%) varies by 1.5 to 2.0%. This is a very important factor when the optical method is used for determining the chromaticity of different kinds of starch under industrial conditions.

When spectral methods are used for determining the chromaticity of sugar it is necessary to measure the light reflected by sugar with the aid of integrating spheres as attachments for spectrometers. Similar methods have been developed at the Braunschweig Institute of the Sugar Industry (FRG) and are used in France [146]. We have worked out a method for measuring the chromaticity of crystalline sugar with the aid of an FMSh-56M spectrometer [31]. So that the dimensions and forms of the sugar crystals would not affect the reflection coefficient, R_λ is measured at two wavelengths: $\lambda=0.368$ and $\lambda_2=0.758$ μm. The chromaticity is calculated according to the following equations:

- for granulated sugar at $(\lg R_{\lambda 2} - \lg R_{\lambda 1}) > 0.1050$

$$C = K_1 (\lg R_{\lambda 2} - \lg R_{\lambda 1}) + K_2 (\lg R_{\lambda 2} - \lg R_{\lambda 1})^2 \qquad (6.14)$$

- for refined granulated sugar at $(\lg R_{\lambda 2} - \lg R_{\lambda 1}) < 0.0.650$

$$C = K_0 (\lg R_{\lambda 2} - \lg R_{\lambda 1}); \qquad (6.15)$$

where, K_1, K_2, and K0 - callibration coefficients. The fact that R_λ is dependent on several factors (the density, moisture content, and dimensions of the sugar crystals, the thickness of a layer of sample, and the state of the surface) prevents this method from being widely used for color control in the sugar industry.

It has been shown [104] that the reflectance of flour can be used for determining its quality. A relationship has been established between the ash content Z of flour and its R_λ, and between the whiteness $\overline{W}$ of the flour and its bran content. The relationship between the ash content and the whiteness of different kinds of translucent wheat can be described as follows:

$$Z = a_0 + a_1 \overline{W} + a_2 \overline{W}^2; \qquad (6.16)$$

where, a_0, a_1, and a_2 - parameters determined on the basis of graphic representation of Z and $\overline{W}$; at $Z = 0.35 \div 1.20\%$ and $\overline{W} = 12 \div 72$, the arbitrary units are: $a_0 = 0.5052$, $a_1 = -0.0066$, $a_2 = 0.00028$. The relatively high value of the coefficient of correlation ($\eta = 0.96$) between Z and $\overline{W}$ indicates that the data on whiteness of the flour can be used for determining its ash content.

To improve the quality of dried foodstuffs and to find out how the different methods of preliminary drying affect their quality it is important to choose an effective color indicator [41]. Photocolorimetric determinations of the chromaticity of dried fruits and raisins based on the optical density of their aqueous-alcohol extracts do not coincide with the visual assessment of their colors [108]. On the other hand, the spectrophotometric method of determining the chromaticity of a substance, which consists in measurements of its spectral reflectance (R_λ), makes it possible to carry out automatic color control.

The reflectance of samples of dried apples and quince and raisins of identical thickness and moisture content, but with different shades of color, has been measured at 0.40–0.75 μm with the aid of an SF-26 spectrometer with an integrating sphere attachment [41]. The absolute values of the directional-hemispherical coefficient of reflection (R_λ) has been determined (without the use of reference standards) with a precision of ±1.5% [39]. It has been found that the greater the values of R_λ, the lighter (approaching white) the color of the dried fruits.

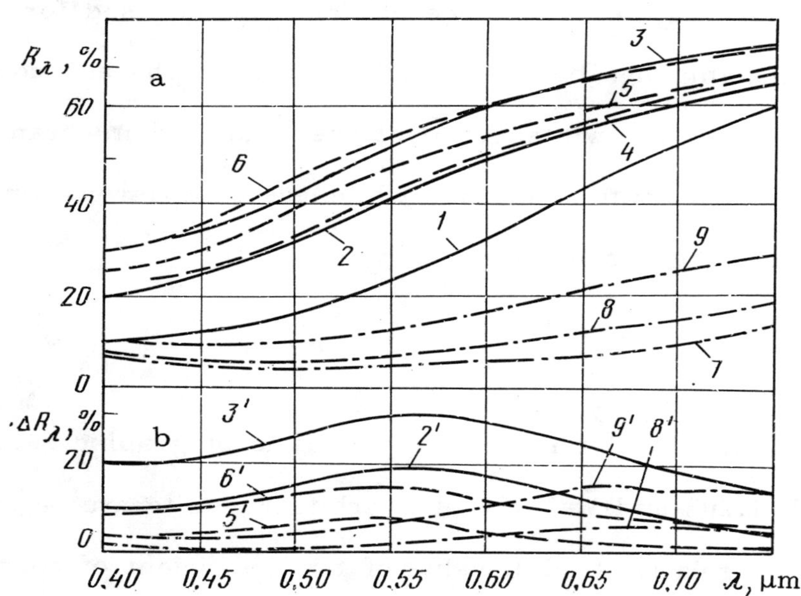

Figure 6.34 Chromaticity of dried fruits [41]: a – reflection spectra of dried quince (curves 1, 2, 3), dried apples (curves 4, 5, 6), and raisins (curves 7, 8, 9) with different shades of color: 1,4,7 – dark (reference); 2,5,8 – light; 3,6,9 – lighter; b – the difference ΔR_λ between the R_λ of light-colored samples and that of the reference sample: 2′,3′ – quince; 5′,6′ – apple; 8′,9′ – raisins.

In Fig. 6.34 are shown the reflection spectra of dried fruits. The color of samples 3, 6, and 9 is close to the natural color; their qualitative characteristics correspond to first-grade foodstuffs. Samples 1, 4, and 7 are marked by a darkening of the surface due to fermentation. The reflectance of light-colored samples is greater than the dark ones throughout the visible spectral range; this accords with the organoleptic examination. For an objective estimation of the fraction of light reflected by different colors (blue, $\lambda=0.51$ μm; green, $\lambda=0.55$ μm; yellow, $\lambda=0.585$ μm; orange, $\lambda=0.62$ μm) with respect to that reflected by the red color ($\lambda=0.75$ μm) the $R(\lambda_i)/R(\lambda=0.75$ μm) ratio has been determined for each sample (the red color is characteristic of foodstuffs undergoing fermentative oxidation). The data obtained show that the lighter the color of the samples, the larger the fraction of light reflected by different colors with respect to that reflected by the red color.

If we take the darkest samples as the reference samples, the greatest difference ΔR_λ between the R_λ of light-colored samples and that of the reference samples will indicate the spectral region in which the difference in color between the dark- and light-colored samples is the greatest. Thus, for dried quince the largest value for ΔR_λ lies at 0.54-0.58 μm (which corresponds to the spectrum of greenish-yellow to yellow; for dried apples - at 0.50-0.57 μm (green to yellowish green); and for raisins - at 0.60-0.70 μm (orange-red).

On the basis of the reflectance of the dried fruits measured at the wavelengths corresponding to the maximum values for ΔR_λ we can determine the degree of darkening of the samples since an increase in the reflectance at a certain λ in the maxΔR_λ region corresponds to an increase in the reflectance throughout the visible range.

Spectral methods have also been used for determining the composition and thus the quality of tobacco [36]. These include: spectrophotometric measurements of tobacco extracts in the ultraviolet and visible range; determination of the reflection spectra of chopped tobacco leaves in the visible spectral range with the aid of an

SF-10 and an SF-18 spectrometer; and determination of the absorption spectra of the tissues of a whole tobacco leaf by measuring its transmittance with the aid of an SF-10 and SF-18 spectrometer.

The quality of tobacco (A - aroma, B - flavor; in number of points) has been determined on the basis of degustation and of the transmittance of tobacco extracts at $\lambda=0.665$ µm:

$$A=13.3 + 0.063T_\lambda \tag{6.17}$$

$$B=13.8 + 0.052T_\lambda. \tag{6.18}$$

The results of degustation and the calculated values are in accord with each other ($\delta_A=3.1\%$; $\delta_B=3.0\%$).

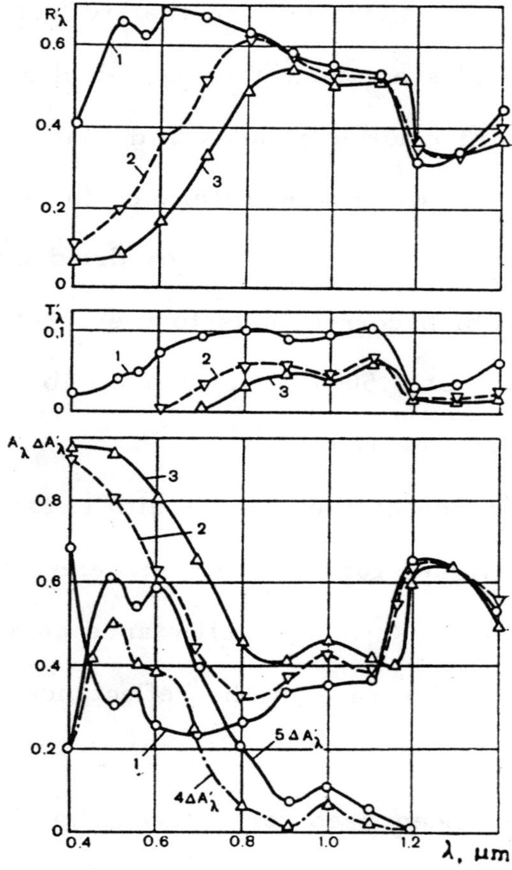

Figure 6.35 Thermal radiational characteristics of fatty pork subjected to various degrees of thermal processing as a function of the radiation wavelength [39. 118]: 1 - not subjected to thermal processing; 2 - with normal thermal processing; 3 - with intense thermal processing; 4,5 - ΔA_λ as a function of λ ($A_{\lambda(2-1)}$ and $A_{\lambda(3-1)}$.

We have not been able to derive mathematical relationships describing the aroma and taste of tobacco in terms of R_λ and T_λ on the basis of its reflection and transmission spectra.

A method for controlling the thermal processing of meat products has been worked out [39, 99, 118] which consists in a comparison of the spectral absorptance A_λ of a given meat product at two different wavelengths. For example, an increase in the degree of thermal treatment of fatty pork at 0.4-1.0 μm results in an increase in its absorptance (Fig. 6.35). The maximum difference ΔA_λ between thermally processed and unprocessed pork has been observed at 0.5 and 0.6 μm. In the 1.0-1.4 μm region this difference decreases, and at 1.3 μm the absorptance of fatty pork subjected to various degrees of thermal treatment is close to the absolute value. The reflectance of fatty pork also varies, depending on the degree of thermal processing.

The degree of thermal processing of pork can be determined on the basis of its thermal radiational characteristics:

- absorptance at $\lambda_1 = 0.6$ μm and $\lambda_2 = 1.3$ μm; $A_\lambda^* = A_{\lambda 1}/A_{\lambda 2}$;

- absorptance of thermally processed ($A_\lambda(T)$) and unprocessed pork ($A_\lambda(0)$) at $\lambda_1 = 0.6$ μm ($\Delta A_\lambda = A_\lambda(T) - A_\lambda(0)$);

- reflectance at $\lambda_1 = 0.6$ μm and $\lambda_2 = 1.3$ μm ($R_\lambda^* = R_{\lambda 1}/R_{\lambda 2}$).

A comparison of the results obtained with visual examination (Table 6.11) shows that the values for A_λ^*, ΔA_λ, and R_λ^* depend on the degree of thermal treatment and can therefore be regarded as objective indicators of the extent of such processing.

The degree of thermal processing of crushed rice and pearl barley can be determined on the basis of a comparison of their R_λ values at two different wavelengths [23].

Table 6.11   Comparative data on the thermal processing of pork [118]

Samples subjected to different degrees of thermal treatment	Visual assessment of the color of pork	A^{*}_{λ}	ΔA_{λ}	R^{*}_{λ}
Unprocessed	white with yellow tinge	0.38	0.00	1.98
With normal processing	golden	0.98	0.39	1.05
With intense processing	dark-brown	1.31	0.59	0.47

The ratio $R_{\lambda 1}/R_{\lambda 2}$ increases with an increase in the degree of thermal treatment (here, $R_{\lambda 1}$ – reflectance at a wavelength corresponding to the maximum change in the optical properties in the course of thermal treatment; $R_{\lambda 2}$ – reflectance at λ_2; $R_{\lambda 1} \cong R_{\lambda 2}$; $\lambda_1 = 0.50$ and $\lambda_2 = 0.85$ μm). The measuring process takes less than 5 minutes, and no preliminary grinding of the grain is required.

The values of R_λ and T_λ of wheat grains depend on the translucency of the grains and can therefore be used for determining the latter [23]. These values are determined for individual grains according to the method of simultaneous measurements (Chapter 5). The R_λ/T_λ ratio is taken at 0.9–1.3 μm where the difference between the reflectance and transmittance of translucent grains ($R_\lambda = 0.48 \div 0.54$) and farinaceous grains ($R_\lambda = 0.68 \div 0.73$) is the greatest and where the effect of color is excluded (Fig. 6.3). The maximum transmittance of farinaceous wheat grains at $\lambda = 1.1$ μm is $0.02 \div 0.03$, while that of translucent ones – $0.11 \div 0.12$ (for durum wheat) and $0.08 \div 0.10$ (for soft wheat). Therefore, the R_λ/T_λ ratios for translucent soft wheat grains ($5.0 \div 7.0$) and for farinaceous wheat grains ($22 \div 37$) can be used for determining the translucency of wheat grains.

Wheat grains may lose their natural color partially or completely during their growth or harvesting as a result of periodic rainfalls and sunshine or during their processing and storage. The process of decolorization is acccompanied by changes in the physical properties of the grain [23]. It leads to an increase in the reflectance of the grain (regardless of the type of wheat) at all the wavelengths. The difference between the values of R_λ of normal wheat grain and those of the decolorized grain is greater in the infrared region than in the visible spectrum.

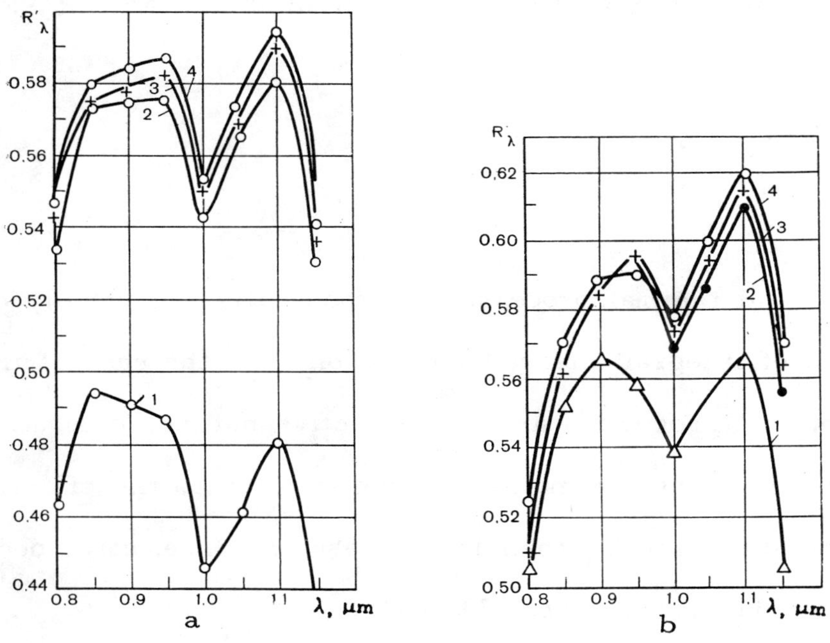

Figure 6.36 Reflectance of decolorized wheat (a - Melyanopus-26; b - Bezostaya-1) as a function of the radiation wavelength [23]: 1 - not decolorized; 2 - 1st degree of decolorization; 3 - 2nd degree of decolorization; 4 - 3rd degree of decolorization.

In Fig. 6.36 are plotted the values of R_λ in the 0.80-1.15 μm region for decolorized and non-decolorized wheat. A well-defined reflection maximum appears at 0.85 μm for the Melyanopus-26 wheat variety, and at 0.9 μm for the Bezostaya-1 variety. The form of the curve of the spectral reflectance in the 0.8-1.1 μm region changes as decolorization intensifies, with the reflection maximum shifting slightly into the long-wave region. Upon further decolorization the difference between the reflectance at 0.95 μm and the reflectance at 1.1 μm increases. In the case of Bezostaya-1 wheat, which is not decolorized, the reflection spectrum shows a maximum at 0.85 μm, and the magnitude of R_λ is greater than at 1.10 μm. The spectrum of completely decolorized wheat shows a maximum in R_λ at 0.95 μm, and its magnitude is less than at 1.10 μm.

EFFECT OF THE SELECTIVE NATURE OF THE OPTICAL PROPERTIES OF FOODSTUFFS EXPOSED TO INFRARED RADIATION ON HEAT TRANSFER PROCESSES

The desiccation and thermal processing of foodstuffs are usually carried out under the conditions of integral infrared irradiation. In the case of irradiated foodstuffs, whose optical properties are of a selective nature, it is more difficult to determine the magnitude of the incident radiation flux E, the effective radiation E_{ef}, and the net radiation flux E_n than in the case of nonselective bodies subjected to infrared irradiation [101, 123, 157, 170, 203].

We have developed methods for estimating quantitatively the effect of the radiation penetrating separate layers of a sample on heat and mass transfer and the absorption of radiation by the layers. We have also derived the basic equations for heat and mass transfer and defined their corresponding boundary conditions. An explanation is given of the anomalous distribution of temperature in the upper layers of the sample under different irradiation conditions.

7.1 Effect of the selective nature of the thermal radiational characteristics of foodstuffs on heat transfer in infrared units

The amount of energy absorbed by a substance per unit time is determined by AES (here, A - absorptance; E - incident radiation flux; S - surface area of the sample). The incident radiation flux should be determined with account taken of multiple reflection inside the oven and of the absorption of infrared radiation by the vapor-air medium. The amount of energy received by the substance per unit time during heat transfer equals the difference between the absorbed and emitted radiation: $E_n S = (AE - E_e)S$; where, $E_e = \varepsilon E(T)$; E_n – emitted radiation flux. The sum of the radiation emitted by the sample and the reflected radiation equals the effective radiation: $E_{ef} = E_e + RE$.

In simple terms the problem of heat transfer between an infrared source and the irradiated substance can be regarded as a problem of heat transfer between two parallel planes having selective spectral thermal radiational characteristics.

When a directional or diffuse monochromatic radiation flux E_λ falls upon the surface of a layer of sample, it is reflected by the layer, and after being repeatedly reflected by the surface of the radiation source and the walls of the oven, it again falls upon the surface of the layer. After each reflection event a part of the radiation, $A_\lambda E_\lambda$, is absorbed by the surface of the radiation source, the walls of the oven and the layer of sample. The spectral region with a greater A_λ of the sample, the walls of the oven, the radiation source, and the vapor-air medium is the first to disappear from the spectral composition of the incident radiation flux. When taking into account multiple reflection in the case of unidirectional irradiation the monochromatic radiation flux incident upon the sample (if we disregard the absorption and scattering or radiation by the vapor-air medium) is defined as follows:

$$E_\lambda = E_\lambda^* / (1 - R_\lambda R_{\lambda ef}); \qquad (7.1)$$

where, R_λ – spectral bihemispherical reflectance of the layer of sample; $R_{\lambda ef}$ – spectral effective reflection coefficient of the oven, which is determined from the following equation:

$$R_{\lambda ef} = R_\lambda^* \phi + \sum_{i=1}^{n} R_{\lambda i} \phi i; \qquad (7.2)$$

where, ϕ – coefficient of irradiance indicating the fraction of reflected radiation that takes part in multiple reflection; i – index characteristic of the walls of the oven.

In our model of linear generators, with the reflector being in the form of a conditional radiation-emitting plane, E_λ^* in Eq. 7.1 stands for the sum of the fractions of incident radiation flux coming directly from the infrared generator, $E_\lambda = 0.5F_\lambda S^{-1}$, and the radiation coming from the reflector, $E_{\lambda_r} = 0.5F_\lambda R_{\lambda r} S^{-1}$. Thus,

$$E_\lambda^* = E_\lambda / 2S(1 + R_{\lambda r}). \qquad\qquad (7.3)$$

When calculating radiation heat transfer in the thermal processing of foodstuffs, the absorption of radiation by air is usually ignored [19]. However, during the drying, searing and other forms of thermal treatment of foodstuffs a considerable amount of moisture and other volatile components escapes through the surface of the substance in question.

Such an air-vapor medium as has been described [26] strongly absorbs and scatters infrared radiation. However, this medium also has spectral windows through which infrared radiation (especially radiation at $\lambda < 2.4$ μm) passes almost without attenuation.

Table 7.1   Absorption of infrared radiation by air

Radiation source	Average transmittance of air T_λ (at $\phi=0.7$, $\ell=0.6$ m)	Radiation flux E (W/m^2)
Infrared lamp T=2080 K	0.908	~ $3 \cdot 10^4$
Ni-Cr heating coil T=1033 K	0.875	~ $8 \cdot 10^4$
Electrical heater	0.862	~ $3 \cdot 10^4$

As can be seen from data in Table 7.1, even when $\phi=0.7$, air strongly absorbs infrared radiation emitted by both low-temperature and high-temperature generators. In industrial thermal radiation installations the distance from the generator to the surface of the sample varies within 5.0–20.0 cm, and the intensity of desiccation can reach 10 kg/m$^2\cdot$hr [20]. Therefore, the air over the sample becomes increasingly humid, and more infrared radiation is absorbed and scattered as a result.

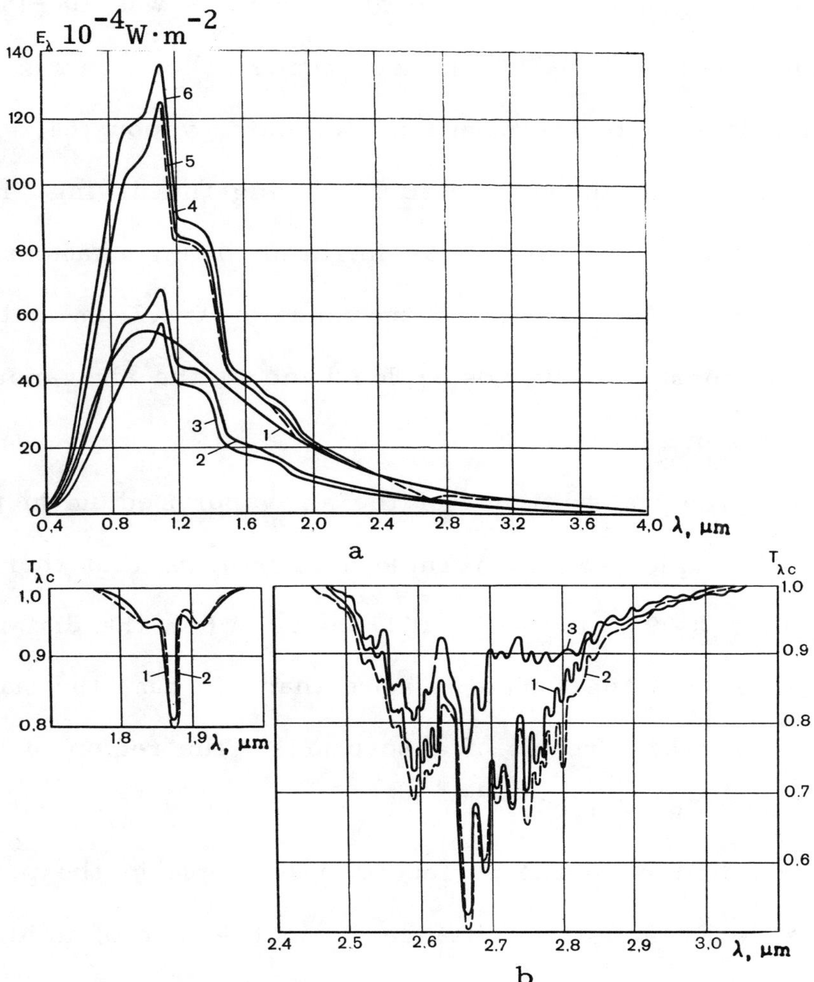

Figure 7.1 Radiation flux (a) and the transmittance of the air-vapor medium (b) as a function of the radiation wavelength [16, 32, 39]; a: 1 - radiation from a quartz KG-1000 generator at 2600 K; 2 - incident radiation from an infrared generator; 3 - incident radiation from the reflector, with account of multiple reflection; 4,5 - incident radiation, with account taken of the absorption of radiation by the air-vapor medium and of multiple reflection; 6 - incident radiation from the generator, each being equipped with a reflector; b: 1 - water vapor (condensed water layer ℓ=0.01 mm) [64]; 2 - saturated air-vapor medium in a drying oven (ℓ=110mm); 3 - air-vapor medium in a drying oven (ℓ=110 mm).

The effect of the absorption of radiation by the air-vapor medium on heat transfer should be considered in each particular case following the experimental determination of $T_\lambda(\theta, \theta_T)$. Measurements have been carried out of the bidirectional T_λ of the air-vapor medium in an oven for drying bread crumbs [16]. The distance between the infrared generator and the bread crumbs was 100-150 mm. The results of spectrophotometric (IKS-14A) measurements, shown in Fig. 7.1 b, accord well with reported data on the transmittance of water vapor [64, 130]. To determine the transmittance of the air-vapor medium during the drying process a special attachment for IKS-14A in the form of an infrared drying chamber was used. During the period of drying $\tau=2\div4$ min the transmittance at $\lambda=2.7$ μm in the absorption band of water remains constant (the absorption band of the air-vapor medium inside the chamber is shown in Fig. 7.1,b).

The average values of the transmittance of the air-vapor medium in the case of a 110 mm thick layer of sample, shown in Table 6.7, indicate that there is little absorption, with $T_\lambda=0.98\div0.99$ K for a KG-1000 lamp. When the distance between the infrared generator and the sample is more than 110 mm, the magnitude of T_λ will apparently decrease as a result of absorption in the region of wavelengths of 1.87 and 2.7 μm (Fig. 7.1,b).

The radiation flux incident upon the surface of the sample in the presence of a medium having reflectance R_λ and transmittance T_λ in the case of unidirectional irradiation, with account taken of multiple reflection, is determined from the following relationship:

$$E_\lambda=E_\lambda^* T_\lambda/(1 - R_\lambda R_{\lambda ef});\qquad(7.4)$$

where, $R_{\lambda ef\ medium}=R_{\lambda ef}T_\lambda^2 + R_\lambda$ (E_λ^* and $R_{\lambda ef}$ are determined $\qquad(7.5)$ from Eqs 7.3 and 7.2, respectively).

In the case of bilateral irradiation $E_{\lambda 1}=E_{\lambda 2}$ if the emissive and thermal radiational characteristics of the generator, of the walls of the oven, and of the air-vapor medium equal one another:

$$E_\lambda=E_\lambda^* T_\lambda[1-(R_\lambda + T_\lambda)R_{ef}].\qquad(7.6)$$

This equation can be used in determining the effect of the air-vapor medium and multiple reflection on the radiation flux. In Fig. 7.2,a is shown the dependence of the dimensionless incident radiation flux on the reflectance of the sample R_λ and the effective reflection coefficient $R_{\lambda ef}$ of the oven, with no account being taken of the medium $(T=1.0)$. From this it follows that at $R_\lambda > 0.5$ and $R_{\lambda ef} > 0.5$ the incident radiation flux can exceed E_λ^* more than twofold. The attenuation of radiation by the medium has a strong effect on E_λ (Fig. 7.2,b). Under real irradiation conditions, when different radiators are used, $(T_\lambda + R_\lambda)R_{\lambda ef}$ can reach values of $0.5 \div -.8$, and the transmittance of the medium $- 0.8 \div 0.9$. Therefore, the real magnitude of E_λ for the medium will be 40-50% below that calculated with the aid of Eq. 7.6

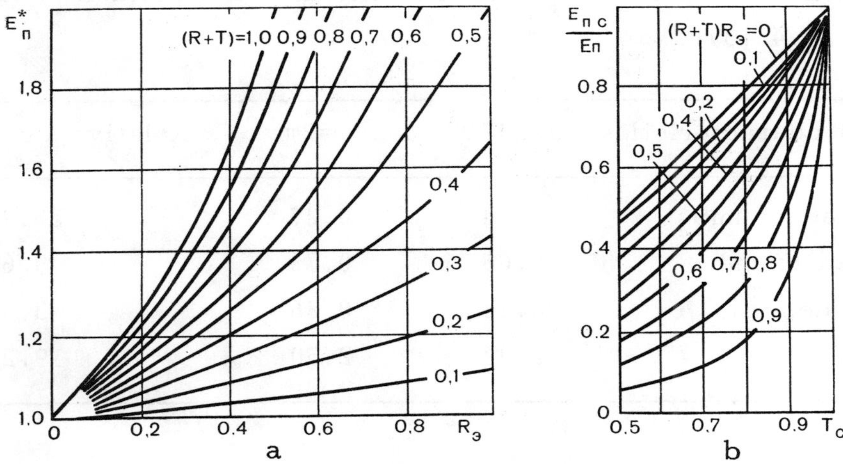

Figure 7.2 Dimensionless incident radiation flux as a function of multiple reflection (a) and of absorption by the air-vapor medium (b) [39].

The calculated values for the radiation fluxes shown in Fig. 7.1,a clearly show the effect of the absorption by the medium, the multiple reflection in the oven, and the selective nature of the thermal radiational characteristics of the sample on E_λ. Thus, the curve representing the spectral radiation flux emitted by a KG-1000 infrared generator (Fig. 7.1,a; curve 1) differs from the dependence of E_λ on the wavelength, if compensation is made for multiple reflection and for the spectral thermal radiational characteristics of the sample. This is evidenced by the appearance of absorption bands near 1.0, 1.2, 1.4, 1.9, and 3.0 μm. The effect of the air-vapor medium on E_λ (shown by dashed line) at λ=1.87 and 2.7 μm is insignificant. This is due to the fact that the air-vapor layer in our experiment was fairly thin (110 mm).

A reflecting surface in the upper section of a quartz lamp makes it possible to raise E_λ by an additional 8-10% (Fig. 7.1, a; curve 6). The total incident radiation flux is determined by integrating Eqs. 7.1, 7.4-7.6 according to the spectrum of the infrared generator:

$$E= \int_{\lambda_1}^{\lambda_2} E_\lambda \, d\lambda . \tag{7.7}$$

Table 7.2 Thermal radiational characteristics obtained under the conditions of directional ($\theta \cong 5 \div 10°$) and diffuse ($\omega = 2\pi$) irradiation, and the incident radiation flux calculated from Eq. 7.4 [39].

Thermal radiational characteristics	$\theta \cong 5 \div 10°$	$\omega = 2\pi$	Relative deviation δ (%)
Absorptance of bread crumbs	0.34	0.27	25.9
Reflectance of bread crumbs	0.66	0.73	9.6
Reflectance of polished Al	0.83	0.86	3.5
Incident radiation flux	2.02	2.50	19.2

The incident radiation flux obtained according to Eq. 7.7 takes into account the absorption by the medium and multiple reflection in the oven. However, owing to the absence of data on the bihemispherical R_λ and T_λ of the irradiated sample, the directional hemispherical thermal radiational characteristics (R_λ of the sample and of the reflector and T_λ of the medium measured at a small angle $\theta = 5 \div 10°$) should not be used when calculating the radiation flux according to Eq.7.7.

In real thermal radiational installations there is diffuse radiation; and the quantities $A_\lambda(2\pi; 2\pi)$, $R_\lambda(2\pi; 2\pi)$, and $T_\lambda(2\pi; 2\pi)$ differ from $A_\lambda(\theta; 2\pi)$, $R_\lambda(\theta;$

2π), and $T_\lambda(\theta; 2\pi)$ by 24.3-24.7% (Table 3.6) or 32.1-42.5% (Table 5.8), depending on the thickness of the sample and the scattering indicatrix. Therefore, in the calculations of E an error is possible owing to different thermal radiational characteristics resulting from different irradiation conditions.

To estimate the magnitude of this error we have calculated the incident radiation flux E_λ, using the directional hemispherical and bihemispherical characteristics of some foodstuffs. In Table 7.2 are shown the values of E_λ for irradiated bread crumbs (ℓ=12 mm, W=42.8%) at 1.1 μm (the wavelength which corresponds to the maximum radiation emitted by a KG-1000 infrared quartz lamp). The data obtained show that the error is quite large. The value of E_λ is lower than the true value by 19.2%. The integration of E_λ according to the spectrum of the infrared generator gives the precise values for the incident radiation flux.

For engineering calculations we have developed a simpler method based on averaged characteristics [26, 44]. It involves a systematic analysis of the heat transfer between the sample, on the one hand, and the infrared source, the air-vapor medium, the reflector, and the walls of the oven, on the other. And instead of the spectral characteristics of the sample use is made of R, T, and A averaged with respect to the spectrum of the sample, the infrared source, the medium, etc:

$$\bar{A}_i = \left[\sum_{\lambda_1}^{\lambda_2} A_\lambda \varepsilon_{\lambda i} E_{\lambda a}(T_i)\Delta\lambda\right]\left[\sum_{\lambda_1}^{\lambda_2} \varepsilon_{\lambda i} E_{\lambda a}(T_i)\Delta\lambda\right]^{-1}. \tag{7.8}$$

For substances having highly selective thermal radiational characteristics the results of the calculation of the radiation flux according to the method of averaged characteristics may contain a large error.

To estimate the magnitude of the error we have determined the dimensionless value of E^* for potato starch (W=11.8%) according to two methods (Table 7.3):

- by summing up the spectral E_λ^* according to the emission spectrum:

$$E^* = \sum_{\lambda_1}^{\lambda_2}\left[E_\lambda/(1 - R_\lambda R_{\lambda ef})\Delta\lambda\right]/\sum_{\lambda_1}^{\lambda_2} E_\lambda \Delta\lambda ; \tag{7.9}$$

- by using Eq.7.7 and the values of R and R_{ef} averaged according to Eq.7.8:

$$\overline{E}^* = 1/(1 - RR_{ef}). \tag{7.10}$$

Table 7.3 Dimensionless value of the incident radiation flux for potato starch (W= 11.8%) calculated according to different equations [39]

Infrared source	E^*			Absorptance		
	Eq.7.9	Eq.7.10	Error(%)	Eq.7.8	Averaged with respect to E_λ	Error(%)
KG-1000 lamp with a reflector (T=2250K, R=0.9)	2.75	2.24	18.8	0.294	0.385	31.0
Ni-Cr spiral with a reflector (T=1033 K, R=0.5)	1.11	1.10	0.9	0.802	0.820	2.2
Metal plate (T=570 K, R=0.2)	1.020	1.018	0.2	0.900	0.912	1.3

For the purpose of simplifying the calculations the reflectors were considered to be nonselective bodies (R_{ef}=const), and the effect of the medium (air) was disregarded.

To determine the exact magnitude of the total incident radiation flux it is necessary to integrate Eqs. 7.1-7.6 with respect to the entire radiation spectrum of the infrared source.

The total absorptance of starch, as determined with the aid of Eq. 7.8 with and without compensation for multiple reflection (Table 7.3), indicates that it strongly depends on the spectral composition of the incident flux and on the irradiation conditions.

The magnitude of the total incident radiation has been calculated on the basis of numerical integration of Eqs. 7.1 and 7.4 according to the infrared source spectrum (T=2600 K) in the case of irradiated bread crumbs. The results show that the method of averaged characteristics can be used in calculating the relevant engineering parameters. A comparison of the results obtained by the integration method and those obtained by the method of the averaged characteristics $\bar{R}$ and $\bar{T}$ shows that the deviation from the total E obtained on the basis of Eq. 7.4 does not exceed ±6% [39].

Most moisture-containing capillary-porous substances (including foodstuffs belonging to the 1st and 2nd groups in the classification according to their optical properties) can be considered to be nonselective bodies. And the heat transfer between them and low-temperature generators (whose temperature does not exceed 600 K) can be calculated with the aid of an equation obtained on the basis of the tensor method [3]. This is possible because the spectral reflectance of such substances is small (0.1-0.2), and their absorptance in the 2.5-15.0 μm spectral region depends slightly on the radiation wavelength, lying within the limits of 0.85-0.95. Most of the radiant energy (~90%) emitted by nonselective bodies at T>600 K is within this spectral region. On the other hand, foodstuffs of the 3rd and 4th groups can be considered to be nonselective bodies only in the 6.0-15.0 μm region and farther; i.e., the temperature of a nonselective radiation source must lie below 400 K.

Some foodstuffs (e.g., tea leaves) cannot be considered to be nonselective bodies. Therefore, the equation obtained on the basis of the tensor method [3] can be used in calculating heat transfer only in those cases where such use is justified.

The processes of heat and mass transfer taking place during the infrared treatment of foodstuffs are of an unstable nature. Therefore, when determining the radiation flux $E_{net}(\tau)$, account should be taken of the spectral thermal radiational characteristics of the infrared generator, of the walls of the oven (R_λ, ε_λ), of the substance being irradiated, and of the air-vapor medium (R_λ, T_λ, A_λ, ε_λ) as a function of time. The time factor is often ignored in calculations based on commonly used methods.

The net radiation flux is defined as the difference between the total absorbed radiation flux $E_A(\tau)$ and the radiation flux that is emitted $E(\tau)$:

$$E_{net}(\tau)=E_A(\tau) - E(\tau). \tag{7.11}$$

$E(\tau)$ is determined from the following relationship with account taken of multiple reflection:

$$E(\tau)=\varepsilon_T(\tau)\sigma_0 T^4(\tau)D(\tau); \tag{7.12}$$

where, $D(\tau)$ - coefficient which compensates for multiple reflection inside the oven and is defined as follows:

$$D(\tau)=1 - [R_{ef}(\tau)/(1 - R(\tau) + T(\tau))R_{ef}(\tau)]. \tag{7.13}$$

Into this equation are substituted the time-dependent quantities $T_{ef}(\tau)$, $T(\tau)$, and $R(\tau)$, which are averaged with respect to the radiation spectrum of the sample.

Since the optical and emissive properties of the sample, the infrared generator the air-vapor medium, and the reflecting surfaces of the oven are selective in nature, while the temperatures of the surfaces of all the components taking part in the heat transfer are different, the radiation emitted by the components is anisotropic and heterogeneous. Therefore, $E_A(\tau)$ should be determined from the following relationship:

$$E_A(\tau)=A(\tau)E(\tau)\phi + A_m(\tau)E_m(\tau)\phi_m + \sum_{i=1}^{n} A_{(i)}(\tau)E_i(\tau)\phi_i; \tag{7.14}$$

where, $A(\tau)$, $A_m(\tau)$, and $A_{(i)}(\tau)$ - absorptance of the sample averaged with respect to the spectrum of the emitter, the medium, and the reflectors, respectively (cf. Eq. 7.8); E, E_i, and E_m - radiation flux emitted by the infrared generator, the reflectors, and the medium, respectively (cf. Eqs. 7.4-7.6), with account taken of multiple reflection and absorption of radiation by the air-vapor medium.

If the temperature of the sample is close to that of the reflectors and the air-vapor medium, the radiation flux emitted by the sample, the reflectors, and the medium will be insignificantly small, and the net radiation flux is determined from following approximate relationship:

$$E_{net}(\tau) \cong A(\tau)E(\tau)\phi. \qquad\qquad (7.15)$$

Thus, when closed installations equipped with high-temperature (T>1200 K) infrared generators designed for processing all four groups of foodstuffs are used, E and E_{net} should be determined with the aid of Eqs. 7.4-7.6 and 7.11 by means of integration with respect to the spectrum of the infrared source [40]. And when low-temperature (T<1200 K) infrared generators are used, E and E_{net} are determined by a simple method: the substitution of the averaged characteristics R, T, etc into Eqs. 7.4-7.6 and 7.11. In conclusion, it should be noted that for the 1st and 2nd groups of foodstuffs the calculation of E_{net} with the aid of Eqs. 7.12 and 7.13 is possible only when low-temperature generators (T<600 K) are used, and for the 3rd and 4th groups - when T<400 K. In these cases foodstuffs can be considered to be nonselective bodies.

7.2 Calculations of the temperature fields in foodstuffs subjected to infrared radiation

Experimental investigations of the spectral thermal radiational characteristics of foodstuffs show that in the case of irradiation by high-temperature infrared generators or solar irradiation a small fraction (0.01) of the incident radiation can penetrate a sample to a depth of 5-10 to 40 mm. Intensive heating of a sample occurs at a certain depth: $x_m=1\div6$ mm (Figs. 7.3,a and 7.4). For many foodstuffs (dough, bread, starch, etc), the temperature at this depth is $1\div10°C$ and in some instances 20°C higher than the surface temperature of the sample [80].

During the initial stage of infrared processing (with temperature of the sample reaching 100°C, Fig. 7.3,a; curve 2) there is no noticeable loss of moisture. What happens is that moisture is being redistributed, and it moves deeper into the sample (Fig. 7.3,b; curve 2).

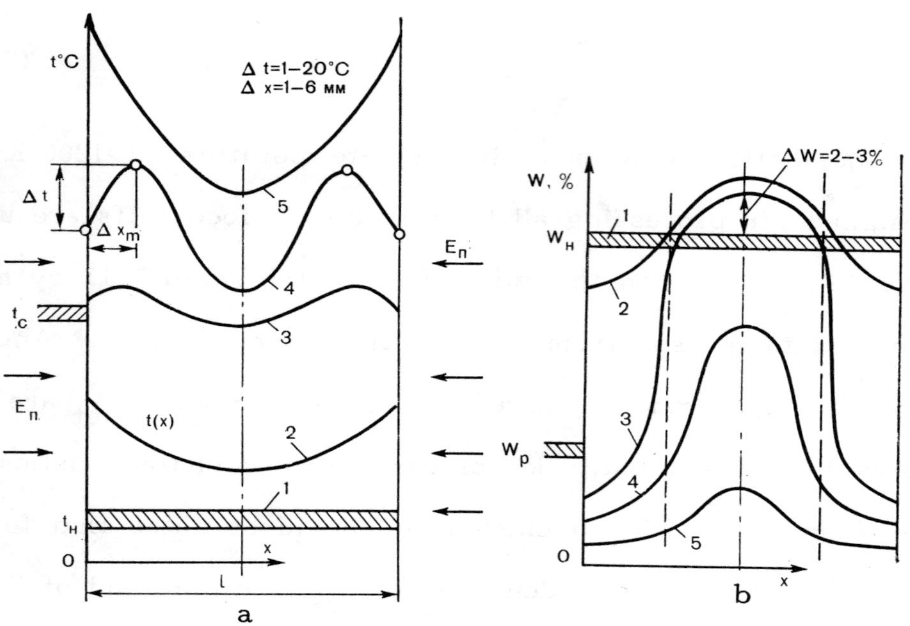

Figure 7.3 Characteristic temperature field (a) and humidity field (b) in capillary-porous substances at different stages of infrared irradiation [39, 49]: 1 - initial stage; 2,3,4 - intermediate stages; 5 - final stage.

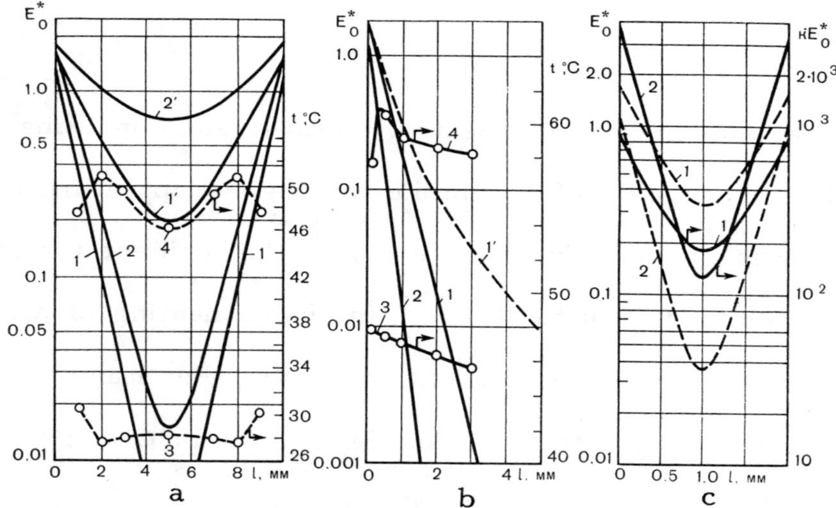

Figure 7.4 Radiation and temperature fields for different substances subjected to infrared irradiation [39, 227]: a - KG-1000 lamps: 1,1' - dough made of durum wheat flour (W=31.2%); 2,2' - fruit candy after it has gelled (W=30.0%); 3,4 - dough made of durum wheat flour after 5 and 60 min [19]; b,c - potato starch: 1 - KG-1000 lamps; 2 - hot plate (T=570K); 3,4 - Philips lamps [184].

An anomalous distribution of temperature was observed as a rule only in those cases where the temperature of the medium was below that of the sample surface. Thus, it was found that when granulated sugar was dried with a 300°C-source the temperature at the surface was 1-5°C lower than at 3-6 mm below the surface [75]. However, when bread was baked (T=300°C), the temperature at the surface of the bread was from the very start higher than at 1-2.5 mm below the surface. Meanwhile, the temperature of the medium was higher than the temperature on the surface of the bread. When a high-temperature infrared source was used, the temperature at 1-2.5 mm below the surface of the bread was higher than at the surface, while the temperature of the medium was below that of the sample surface [19, 20, 79].

The anomalous distribution of temperature with respect to the thickness of the sample can be explained by the following factors [19, 77, 78]:

- the absorption of infrared radiation by the sample in which the radiation is converted into thermal energy;

- the loss of energy (through emittance, convections, heat conduction) at the surface of the sample and its dispersion into the surrounding medium;

- the phenomenon of thermal spreading: the circulation of air inside the pores caused by the temperature gradient;

- the phenomenon of molecular flow: the effusion of gas from micro-capillaries in the direction of the temperature gradient;

- the consumption of heat upon the evaporation of moisture at the surface.

The first of the factors mentioned above requires further investigation. In order to explain the mechanism of the transfer and absorption of infrared radiation and its effect on the temperature field in the sample it is necessary to determine the radiation field inside the sample. For this purpose we can use Eqs. 4.101-4.112 and 4.117-4.123.

The distribution of the radiation energy absorbed per unit time by the volume element $w^*(x)$ can be determined with the aid of Eq. 2.26. The distribution function $E_0^* = w_{\lambda 0}^* (1 - \Lambda_{ef})^{-1}$ of the flux density of monochromatic radiation $w_{\lambda 0}^*$ and that of total radiation w_0^* inside a layer for some foodstuffs is shown in Fig. 7.4. From this it follows that under diffuse irradiation conditions the anomalous distribution of temperature (curve 4) [19, 26, 75] cannot be due to the penetration of infrared radiation into the sample.

As can be seen from Fig. 7.4, in the case of dough the volume flux density of the absorbed energy of both total and monochromatic radiation in the surface layer is several times greater than that at a depth of 2.2 mm [19] where the maximum temperature is reached. This means that the surface layer emits more heat owing to the absorption of radiation energy than the layer lying at 1-5 mm below the surface (curve 3). At a depth of 2.2 mm of a layer of sample (ℓ=10 mm) of durum wheat flour dough E_0^* (x=2.2 mm)=0.074, and on the surface - E_0^*(x=0)= 1.283; i.e., the surface layer emits 17.4 times more heat than the layer lying at 2.2 mm below it, if the absorption coefficients are the same: $\overline{k}$(x=2.2 mm)=$\overline{k}$(x=0). But if these coefficients were not the same, then in order that the same amount of heat is emitted at a depth of 2.2 mm as at the surface of the sample, one of the absorption coefficients should be 17.4 times greater than the other.

Our experimental investigations show that such an increase in the magnitude of the absorption coefficient at depth x for one and the same substance is impossible [39]. Indeed, the density of the surface layer as a result of desiccation will always be greater than that of the layer lying at 2.2 mm below the surface. Therefore, $\overline{k}$(x=2.2 mm)<$\overline{k}$(x=0). In the case of high-temperature irradiation an increase in the moisture content of the surface from 10 to 30% results only in a small (less than 1.5-fold) increase in the absorption coefficient (Fig. 6.27).

In the case of directional irradiation of strongly scattering substances such an anomalous distribution of the absorbed energy w_λ^* is possible if at a certain depth x_m the $w^*(x)$ function has a maximum. In Fig. 2.9 are shown the calculated values of w^* in the case of diffuse monochromatic irradiation and of w_λ^* in the case of directional monochromatic irradiation at $\lambda=1.1$ μm of an optically infinitely thick layer of wheat flour dough (W=48%, $\ell>40$ mm) immediately before baking. The presence of a maximum value of w_λ^* at $x_m=0.16$ mm is due to the propagation of directional and scattered radiation inside the layer subjected to directional irradiation.

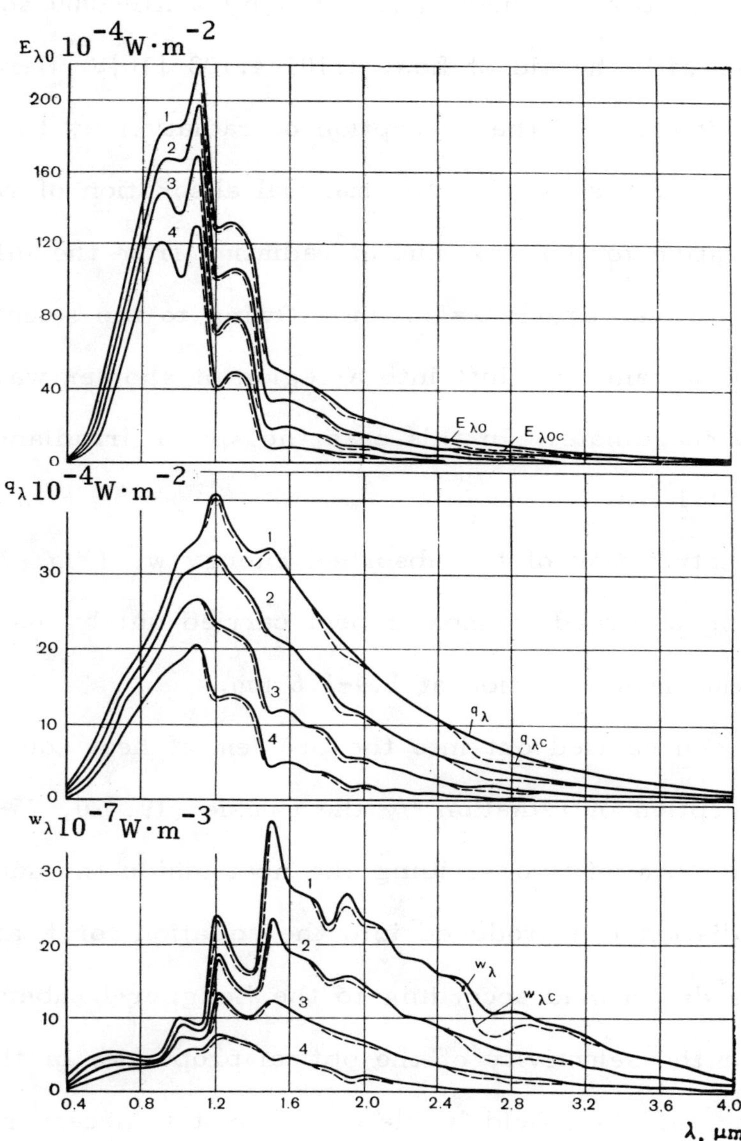

Figure 7.5 The dependence of $E_{\lambda 0}$, q_λ, and w_λ on the radiation wavelength for a semi-infinite layer of bread crumbs at different depths [16]: 1 - 0 mm; 2 - 0.4mm; 3 - 1.0 mm; 4 - 2.0 mm (the dashed lines show the effect of the air-water vapor medium on the calculated values).

The irradiance $E^*_{\lambda 0s}$ of a volume element at a certain depth x_m resulting from scattered irradiation is greater than $E^*_{\lambda 0}$ resulting from directional irradiation (Fig. 2.8,b), and the absorption coefficient $\bar{k}_\lambda$ of the sample subjected to scattered irradiation is always m times greater than k_λ (cf. Eq. 3.79).

The distribution of the radiation flux and of the absorbed radiation at different wavelengths in a layer of bread crumbs irradiated by a KG-1000 source (T= 2600 K) has been analyzed with the aid of Eqs. 4.101-4.103 [16]. Here account is taken of multiple reflection and of the absorption of radiation by the air-water vapor medium. Data in Fig. 7.5 show that the maximal absorption of radiation w_λ at x<1.0 mm is shifted relative to the maximum of radiation from the infrared source (λ_m=1.1 μm) into the long-wave region, λ=1.5 μm. Owing to the selective absorption the maximal values of w_λ and q_λ shift into a region of shorter wavelengths with an increase in the x coordinate. In this case the spatial irradiance $E_{\lambda 0}$ remains the highest near λ=1.1 μm.

An analysis of the distribution of the absorbed energy w_λ (Fig. 7.5) shows that the thermal processing of bread crumbs is best carried out by using an infrared generator with a maximum of radiation at 1.4-1.6 μm.

Much research has been carried out into the problem of heat conduction with account taken of the absorption of radiation by the sample [19, 20, 75-78]. For this purpose an additional term w(x) describing the heat inside the sample due to the absorbed infrared radiation is introduced into the equation for heat conduction. The magnitude of w(x) is determined according to the Bougner-Lambert law, with no account being taken of the selectivity of the optical properties of the sample.

In order to know the radiation field inside a sample it is necessary to calculate the function of distribution of the absorbed radiation energy w(x) averaged with respect to the spectrum of the infrared source. The incident radiation flux under real conditions is 10^4-10^5 W/m^2 [19]. Therefore, even when $w(x_1)$=0.01 at a certain depth x_1, the flux density of the absorbed energy $w(x_1)$ per unit volume is large and should not be disregarded in calculations.

According to our method for calculating the heat conduction of the infrared source, it is necessary to introduce a conditional minimal flux density of the absorbed energy $w_{min}(x_1)=10^m$ W/m^3, which can be disregarded at x_1 depth. The choice of the magnitude of x_1 and index m depends on the conditions of the particular problem concerned. If at depth x_1 $w(x_1)<w_{min}(x_1)=10^m$ W/m^3, $w(x)$ of the radiation source is introduced into the boundary condition; i.e., it is assumed that $w(x)=w(0)$. On the other hand, if $w(x_1)>w_{min}(x_1)$, $w(x)$ is taken into account in the heat conduction equation.

Let us suppose that a plate having ℓ thickness is irradiated on both sides by two infrared generators in the form of two theoretically boundless plates separated by a weakly-absorbing, scattering air-water vapor medium. The heat conduction equation for a unidimensional symmetrical problem, with account taken of the penetration of infrared radiation into the sample to a certain depth, has the following form:

$$c\rho_0(\partial t/\partial \tau)=\lambda(\partial^2 t/\partial x^2) + w(x,\tau). \tag{7.16}$$

And the boundary conditions are:

$$(x,0)=f(x), \tag{7.17}$$

$$\partial t(0,\tau)/\partial x=0, \quad -B<x<B \quad (\ell=2B), \tag{7.18}$$

$$-\lambda[\partial t(B,\tau)/\partial x] + q(\tau)=0. \tag{7.19}$$

Eqs. 7.16-7.19 can be solved by a known method [77, 80, 206] if the type of $w(x,\tau)$ function and the total flux density of thermal radiation at the $q(\tau)$ boundaries are known. The boundaries depend on the thermal physical and thermal radiational properties of the substance concerned, the radiation source, the medium, and the reflectors, and are functions of time and the x coordinate.

For a quantitative description of heat and mass transfer in foodstuffs subjected to infrared irradiation we can use conjugated boundary heat conduction equations [77] which contain terms for additional heat sources, w(x) and Aq. The conjugated problems are reduced to a simultaneous solution of differential heat conduction equations for the desiccated and moist parts of the sample. These regions differ from each other in their moisture content, thermal physical properties, and optical and thermal radiational characteristics, and in terms of the presence of heat sources and the outflow of heat. The conjugation of neighboring radiation fields is realized under the conditions defined by Eqs. 4.7-4.12, 4.31, and 4.32, while that of neighboring temperature fields - by Eqs. 7.21 and 7.22.

In order to simplify the calculations of heat and mass transfer the equation of heat conduction, which includes all the above-mentioned internal heat sources for a multilayered system, should be replaced by a system of n equations for i=1,2,3,....,n layers. It should be borne in mind that short-wave radiation ($\lambda < 2.8$ μm) penetrates both dry and moist capillary-porous substances to a considerable depth (up to 40 mm), and should therefore be taken into account in the heat conduction equation with the aid of the $w_i(x_i, \tau)$ function. On the other hand, long-wave radiation ($\lambda > 2.8$ μm), which partially penetrates dry substances, is practically completely absorbed by moist substances when $\ell > 0.1$ mm, and can therefore be considered to be an additional heat source A_q at the interface where the dry and moist portions of the sample meet. The heat sources and the heat outflow due to phase transitions are also included in the boundary conditions. As a result, the equation of heat conduction is simplified:

$$c_i \rho_i (\partial T_i / \partial \tau) = \partial \{ [\lambda_i (\partial T_i / \partial x)] + w(x_i, \tau) \} / \partial x. \tag{7.20}$$

The conjugation conditions at the mobile boundaries for the i-th and (i + 1)-layers ($x_i = \ell_i$, $x_{i+1} = 0$) will be the conditions of the temperature equality:

$$T_i(\ell_i, \tau) = T_{i+1}(0, \tau) = T_{outflow} f(\tau). \tag{7.21}$$

And the equality of heat fluxes, with account taken of the heat outflow $(T_{outflow})$ at the interface due to changes in the enthalpy of the dried layer, phase transitions, and the absorption of long-wave infrared radiation, can be defined as follows:

$$-\lambda_i(\partial T_i(\ell_i,\tau)/\partial x) + \lambda_{i+1}(\partial T_{i+1}(0,\tau)/\partial x)=A_i{+}_1 q(\ell_i) + \varepsilon rm(\tau) +$$
$$+ [c_{i+1}\rho_{i+1}T_{i+1}(0,\tau) - c_i\rho_i(T_i(\ell_i,\tau)]dx/d\tau. \tag{7.22}$$

Here, $q(\ell_i)=E_{i+} - E_{i-}$ is determined by Eqs. 4.16 and 4.17 at $x_i=\ell_i$ in the case of unidirectional irradiation $(E^*_{i,2}=0)$:

$$q(\ell_i)=\{E_{i,1}[(1 - R_{i\infty})(1 - B_{i+1,i}R_{i\infty})c_{i,i+1}]/\{1 - B_{i-1,i}B_{i+1,i}\psi_i^2\}\exp(-L_i\ell_i) \tag{7.23}$$

Let us now consider how the methods of determining $w(x)$ and $q(\tau)$ are applied in practice. We shall take a single layer $(L\ell<10;\ \ell=10\ mm)$ of durum wheat flour dough exposed to a high-temperature radiation source $(E=10^4\div10^5\ W/m^2)$. The distribution of the absorbed energy $E^*_0(x)$ in such a layer is shown in Fig. 7.4. Under these conditions the sample has the following optical and thermal radiational characteristics as calculated according to Eq. 7.8: $R\cong R_\infty=0.283,\ T\cong 0;\ L=1310\ m^{-1}$, $\overline{k}=732\ m^{-1}$. At $x_1=0.1\ell$ (1.0 mm for durum wheat flour dough) $E^*_0(x_1)=0.342$ (Fig. 7.4,a); at $E=10^4\div10^5\ W/m^2$ and $\overline{k}=10^2\div10^3 m^{-1}$ the minimal value of $E^*_0(x_1)$ should be $10^{-2}\div10^{-3}$. By comparing the dimensionless values of the volume density of the absorbed radiation energy with the magnitude of $w^*_{min}(x_1)$, we have: $w^*(x_1)\gg w^*_{min}(x_1)=0.01$.

Therefore, it is necessary to take into account the additional heat source $w(x)$ due to the absorbed infrared radiation in the differential equation of heat conduction. The magnitude of $w(x)$ is determined by a method (cf. Chapter 4) involving the use of the spectral optical properties $\overline{k}_\lambda$, L_λ, and $R_{\lambda\infty}$, which are correlated to the spectrum of the infrared source with account taken of Eq. 2.26.

The calculations of E and $w(x)$ for bread crumbs show that in precise determinations of heat transfer it is necessary to use the bihemispherical characteristics $R(2\pi; 2\rho)$, $T(2\pi; 2\pi)$, and $A(2\pi; 2\pi)$ which correspond to the real irradiation conditions. Thus, by substituting into the equation for $w(x)$ the values of R_∞, $\bar{k}$, and L obtained for bread crumbs with the aid of directional hemispherical and bihemispherical characteristics of the layer, we get different values of $w(x)$. The magnitude of $w(x)$ at any depth x in the layer turns out to be too low: by 18% at x=0, and by 23% at x=1.5 mm.

The flux density of the radiation emitted by the air-vapor medium, by the sample, and by the reflector is less than E by 2-3 orders of magnitude. When a low-temperature radiator is used, $E^*(x_1=1.0 \text{ mm})<10^{-2}$; we can thus conclude that taking part in the external heat transfer are the surface layer of the sample up to 0.5 mm thick, the reflector, and the medium. Therefore, an additional term for the heat source or heat outflow must be introduced into the boundary condition. Then, $q(\tau)$ is defined by the following relationship, with account taken of Eqs. 7.11 and 7.12:

$$q(\tau)=\alpha[t_c - t(B,\tau) + E_{net}(\tau) \tag{7.24}$$

Next we shall consider a case of low-temperature irradiation ($E=10^3 \div 10^4$ W/m^2) of a layer $L\ell>10$. In Fig. 7.4,b (curve 2) is shown the distribution of the dimensionless radiation observed per unit time by a volume element E_0^* in a layer of starch ($\ell=20$ mm) (T=590 K, $\lambda_{max}=5.1$ μm). It can be seen that $E_0^*(x_1)$ is less than $E_{0,min}(x_1)=10^{-3}$ at $x_1=0.1$, $\ell=2.0$ mm. Therefore, the heat source due to the absorbed radiation lies practically in the surface layer, and we only need to take it into account in the boundary condition. If $w(x)=0$, the problem is then defined by Eqs. 7.16-7.19, and $q(\tau)$ is determined from Eq. 7.24 (in this equation $E_{net}(\tau)$ is determined from Eqs. 7.11-7.14). The magnitudes of A, R, and T of starch and of T_m of the air-water vapor medium, which are needed for calculating $E_{net}(\tau)$, are determined according to Eq. 7.8: A=0.9, R=0.1, T≅0, T_m=

0.86.

Now let us consider a third case involving high-temperature (L_1) and low-temperature (L_2) irradiation ($Ll<10$). For starch we have the following optical characteristics of an optically finite layer: L_1=2264 m^{-1}, L_2=4088 m^{-1}, $\bar{k}_1$=540 m^{-1}, $\bar{k}$=3426 m^{-1}. For a layer of starch to be of finite optical thickness with respect to high- and low-temperature irradiation it should be less than 2 mm thick since only then $L_2 l$=8.17<10.

In this case the differential equation should also take into account the heat source $w(x)$ (Fig. 7.4,c) due to the absorption of the infrared radiation from the generator, the reflector, and the air-vapor medium. In addition, the radiation emitted by the medium itself should be taken into account, and the equation for $w(x)$ becomes:

$$w(x)=w_1(x) + w_2(x) + w_3(x) + w_4(x). \tag{7.25}$$

Here, $w_1(x)$, $w_2(x)$, and $w_3(x)$ – the function of the distribution of the radiation energy absorbed per unit time by the volume element of the sample, radiation energy which is emitted by the infrared generator, the air-vapor medium, and the reflectior, respectively. These quantities are determined according to Eq. 2.26, with the substitution into the equation of $\bar{k}$, L, R_∞ , and E averaged with respect to the spectrum of the infrared source, the air-water vapor medium, and the reflector. The function $w_4(x)$ describes the contribution of the radiation energy from the layers of the medium having thicknesses x and (l-x) to the total radiation energy absorbed at x depth. It is determined with the aid of the following equation:

$$w_4(x)=\bar{k}/2[\sigma_0\int_0^x T^4(\xi,\tau)\Phi_1(x,\xi)d\xi] + \bar{k}/2[\sigma_0\int_x^l T^4(\xi, \tau)\Phi_2(x,\xi)d\xi -$$
$$- \bar{k}\sigma_0 T^4(x,\tau). \tag{7.26}$$

The $\Phi_1(x,\xi)$ and $\Phi_2(x,\xi)$ functions, which describe multiple scattering and reflection of radiation in a layer of sample, depend in a complex manner on the spectral, optical and thermal radiational characteristics of the sample and the infrared source.

The temperature field in a layer of sample subjected to infrared irradiation can be determined from relationships describing $w(x)$ and $q(\tau)$. A general solution of Eqs. 7.16-7.19 with any heat source $w(x,T,\tau)$ yields [77, 172]:

$$T(x,\tau)=B^{-1}[\int_0^B f(x)dx + (c\rho)^{-1}\int_0^\tau w(0,T,\phi)d\phi] + (2/B)\sum_{n=1}^\infty \cos(n\pi x/B)\cdot$$

$$\exp(-\lambda n^2\pi^2\tau/c\rho B^2)\int_0^B f(x)\cos(n\pi x/B)dx + (2/c\rho B)\sum_{n=1}^\infty \cos(n\pi x/B)\int_0^\tau \int_0^B \cdot$$

$$w(x,T,\phi)dx \, \exp[-(\lambda n^2\pi^2/c\rho B^2)(\tau-\phi)]d\phi. \tag{7.27}$$

The use of the obtained relationships for describing the temperature field in foodstuffs can be illustrated with the aid of Eqs. 7.28-7.31 and of nomograms (see Fig. 7.6 a,b,c) [16, 39, 88]. These nomograms have been used in determining the temperature fields in a layer of bread crumbs irradiated with infrared generators (KG-1000 lamps). The solution of Eq. 7.16 with respect to the dimensionless temperature yields:

$$T(x,Fo)=Po_c\Phi_I + Ki\Phi_{II}; \tag{7.28}$$

where, $\Phi_I=\{[1-\exp(-K)]/K\}Fo+\{2K\sum_{n=1}^\infty \cos(\mu_n X)[1-\exp(-\mu_n^2 Fo)]\cdot$

$$\cdot[1-(-1)^n\exp(-K)]/\mu_n^2(K^2+\mu_n^2)\}; \tag{7.29}$$

$$\Phi_{II}=Fo-[(1-3X^2)/6]-\sum_{n=1}^\infty (-1)^n[2\cos(\mu_n X)\exp(-\mu_n^2 Fo]/\mu_n^2; \tag{7.30}$$

$Po=E(1-R_\infty)LR_v^2/\lambda t_0$ - constant part of the Pomerantsev criterion; Fo, Ki - the Fourier and Kirpichev criterion, respectively; $K=LR_v$ - the average (with respect to the thickness of the layer) dimensionless coefficient of attenuation; $T(X,Fo)=\Delta T_f/T_i$ (here, T_f, T_i - the final and initial temperature of the sample, respectively). The first term in Eq. 7.28 determines the effect of the internal heat sources due to the absorption of infrared radiation on the temperature field. The second term describes the change in the temperature field as a result of convective heat transfer with the surrounding medium. The solution (Eq.7.28) holds true in the case of uniform distribution of the initial temperature (T_i) in the layer of the sample. In the case of nonuniform distribution of T_i, for example, in the case of parabolic distribution of $T_i(X)$, Eq. 7.28 should be modified by the introduction of the following term:

$$F_I(XFo)=4\sum_{n=1}^{\infty}[\cos(\mu_n X)\exp(-\mu_n^2 Fo)^2(-1)^n/\mu_n^2].\qquad(7.31)$$

The solution of Eqs. 7.29-7.31 at different X and Fo yields Φ_I, Φ_{II}, and F_I as a function of K, Fo, and Fo, respectively (Fig. 7.6). For bread crumbs the magnitudes of T_f calculated with the aid of Eq. 7.28 and those measured in the process of baking are sufficiently close to one another (they vary within ±5%).

When radiation and temperature fields are known, it becomes possible to establish the cause of the anomalous distribution of temperature in the upper layers of different substances subjected to infrared irradiation. Thus, in the case of diffuse irradiation the data in Figs. 4.3-4.9 and 7.4 show that the radiation energy absorbed per unit time by the volume element of the sample is tens of times more in the upper layers than at a depth of 1-6 mm where the maximum temperature is recorded. The presence of a maximum in temperature at a certain depth x>0 is due to the distribution of heat sources $w(x_i)$ (Eq. 7.20) under the boundary conditions: $T_{i\pm1} - T_i(0,\tau)<0$. This explains why an anomalous distribution of temperature is observed only when the temperature of the medium in the oven $T_{i\pm1}$ is lower than the surface temperature of the sample $T_i(0,\tau)$; i.e., when there is loss of heat due to the convection of radiation into the surrounding medium.

In the case of bread-dough, as it has been established experimentally (Fig. 7.7) and shown by the analytical solution of Eq. 7.16, when there is no convective

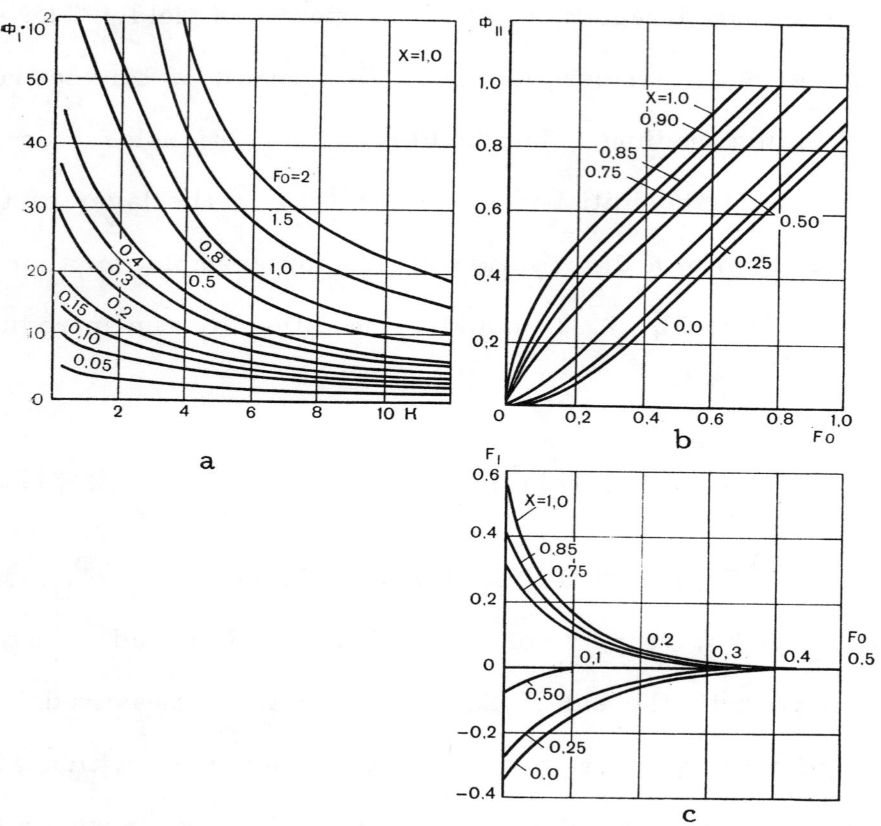

Figure 7.6 Parameters Φ_I (a) and Φ_{II} (b,c) as a function of K and the Fourier criterion, respectively, at different values of X [16].

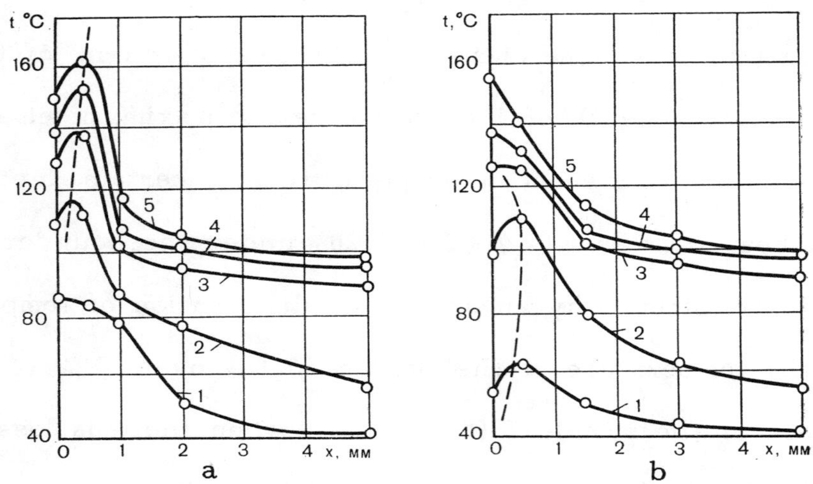

Figure 7.7 Temperature fields in bread-dough subjected to infrared processing under different boundary conditions [108]: a: 1-5 at t_c=80°C=const; b: 1,2 at t_c= 40°C; 3,4,5 - at t_c=f(τ)>t; 1-5 - τ=0.5, 1.0, 2.0, 3.0, 4.0 min, respectively.

heat transfer ($\alpha \cong 0$), the maximum temperature will be found in the surface layer ($x_m=0$). Therefore, a necessary condition for $x_m>0$ is the heat transfer from the surface at $t_c-t(0,\tau)<0$. A similar conclusion is drawn on the basis of available data on the solar heating of water [108].

However, under widely used infrared irradiation conditions the function $w(x)$ is of a more complex nature, being not of the exponential type, and Eqs. 7.29 and 7.30 are inapplicable.

When there is a need for rapid and uniform heating of foodstuffs (e.g., for baking lavash (thin, unleavened bread) and flat cakes, or for drying fruits and vegetables, etc), the substance concerned is subjected to bilateral diffuse irradiation.

The real function of the internal heat sources, $w(x)$, is determined by the method involving the integration of the function (Eq. 2.24) with respect to λ within the limits $\lambda_1-\lambda_2$. This is the most precise method for calculating the total energy transfer in irradiated substances. Here it is necessary to approximate the complex integral function $w(x)$ with the aid of simpler exponential relationships (Eqs. 7.28-7.30).

As an example let us consider a layer of fruit subjected to bilateral symmetrical diffuse irradiation [43]. In Fig. 7.8 (curve 1) is shown the real distribution of the internal heat sources calculated with account taken of the variables with respect to the coordinate of the optical characteristics. We divide the sample conditionally into two symmetrical layers and approximate the integral function with the aid of simpler functions. The approximation with the aid of the exponential function emerging from the point tangent to the curve (Fig. 7.8, curve 3) is represented as follows:

$$w_3(x)=EL_3(1 - R_\infty)\exp(-L_3x), \tag{7.32}$$

where, L_3 is defined by the following relationship:

$$L_3 = R_v^{-1}[\ln(w(R_v)/w_3(0)];$$
(7.33)

where, $w_3(0)$ - radiation energy absorbed at $\ell/2$ depth (it is determined from the plot).

The approximation with the aid of the exponential function emerging from the point $w_4(R_v)$ and intersecting curve 1 (Fig. 7.8, curve 4) is represented as follows:

$$w_4(x) = EL_4(1 - R_\infty)\exp(-L_4 x),$$
(7.34)

and L_4 is determined with account taken of the equality between the cut-off areas over and under the integral function $w(x)$ (Fig. 7.8, curve 1):

$$L_4 = R_v^{-1}[\ln(\int_0^{R_2}(w(x)dx/w(0)) + 1].$$
(7.35)

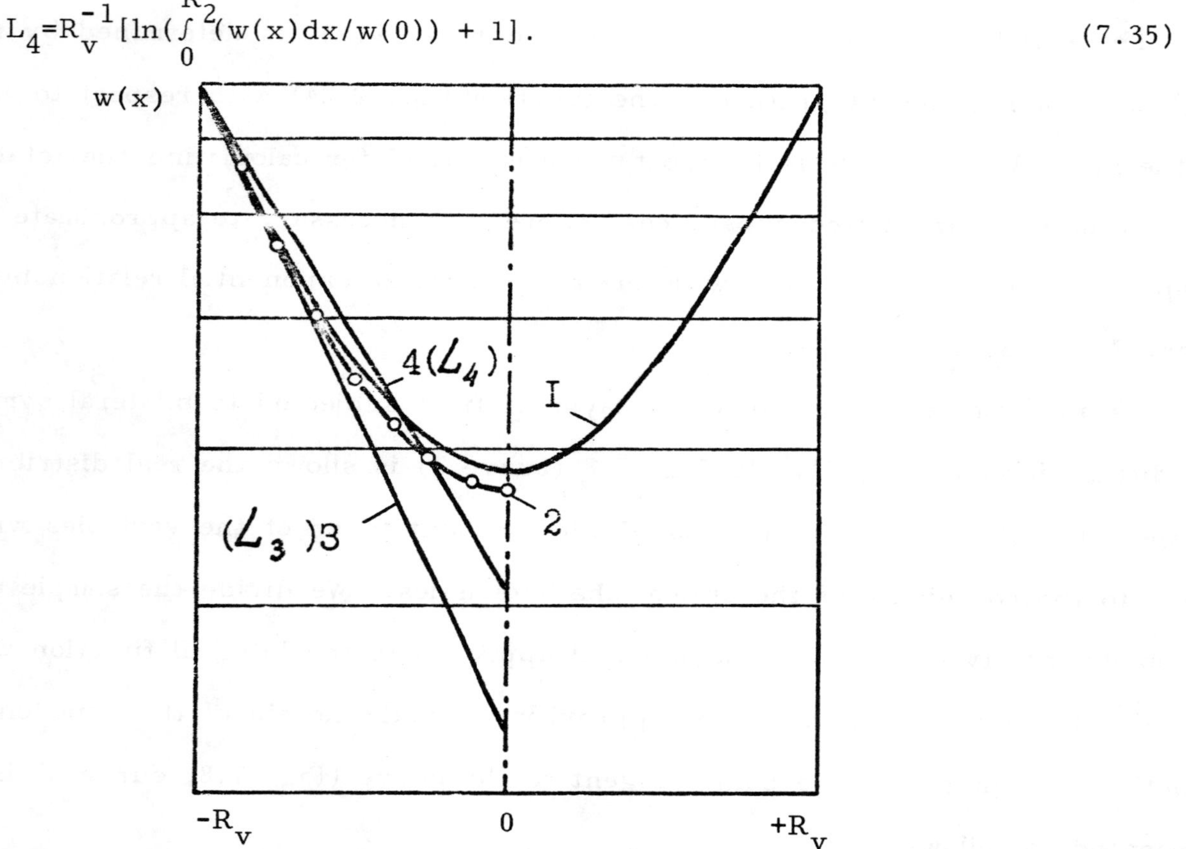

Figure 7.8 Approximation of the variable part of the integral dependence $w(x)$ (curve 1) in the case of bilateral irradiation of a layer $\ell = 2R_v$ with the aid of different functions [43]: 2,3,4 - according to Eqs. 7.36, 7.32, and 7.34, respectively.

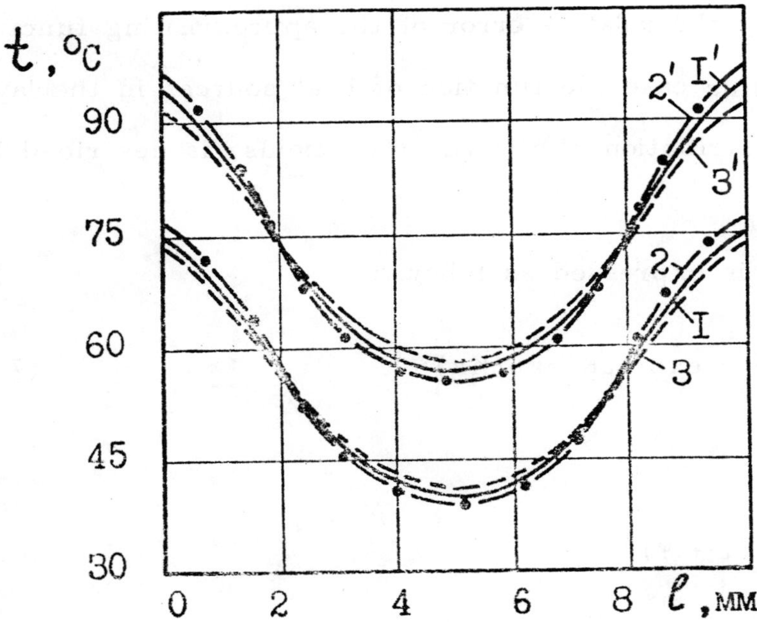

Figure 7.9 Temperature fields in the case of bilateral irradiation of a layer of quince calculated according to functions approximating the integral function w(x) at time τ (sec) [43]: 1,2,3 - 60; 1',2',3', - 90.

The approximation with the aid of the function (Eq. 7.50) is represented (curve 2) as follows:

$$w_2(x)=[EL_1(1 - R_\infty)/(1 + \Psi)][\exp(-L_3x) + \exp(-L_3(\ell - x))];\qquad(7.36)$$

where, L_3 is determined according to Eq. 7.33.

As an example, in Fig. 7.9 are shown the temperature fields in the case of bilateral diffuse irradiation of a layer of quince. They are calculated without compensation for mass transfer and convective heat transfer. The function describing the distribution of heat sources, w(x), is calculated by means of integration with respect to the spectrum of the precise function. Curves 1 and 1' in Fig. 7.9, obtained according to the approximation of the integral function w(x) (Eq. 7.36), describe the heating process more precisely. Curves 2 and 2' are obtained according to the approximation with the aid of the tangential function (Eq. 7.31) in a similar way as curve 3 (Fig. 7.8). Curves 3 and 3' are obtained by approximation of the function (Eq. 7.34). An analysis of the plots in Fig. 7.9 shows that during 90 seconds of heating the relative error of the approximating functions does not exceed ±2.6%. In this case the function of heat sources in the layer, which corresponds to the distribution of the radiation fields, is described by the sum of the exponents.

The Po(x,Fo) function is expressed as follows:

$$Po(x,Fo) = Po_c[\exp(-Kx) + \exp(-2K)\exp(Kx)]; \qquad (7.37)$$

where,

$$Po_c = E(1 - R_\infty)LR_v''/(\lambda t_c(1 + \Psi)). \qquad (7.38)$$

Integral transformations yield:

$$D(x,Fo) = Ki\Phi_{II}(x,Fo) + (Po_cFo/b)(1 - e^{-2K}) + 2\sum_{n=1}^{\infty} Po_c[K(1 - e^{-2K})/$$

$$(\mu_n^2 + b^2)\mu^2 n][\exp(\mu_n^2 Fo) - 1]\cos\mu_n\exp(-\mu_n^2 Fo). \qquad (7.39)$$

In the case of unilateral diffuse irradiation of a layer of finite thickness the criterion Po(x,Fo) becomes [52]:

$$Po(x,Fo) = Po_c[\exp(-bx) - R_\infty\exp(-2b)\exp(bx)]; \qquad (7.40)$$

where, $Po_c = E(1 - R_\infty)LR^2/\lambda t_c(1 - \Psi^2)$ - constant portion of the Po(x,Fo) criterion. Following integration we have:

$$\theta(x,Fo) = Po_c\{(Fo/K)[1 - e^{-K} + R_\infty e^{-2K}(1 - e^K)] + 2\sum_{n=1}^{\infty} [1 - (-1)^n e^{-K} -$$

$$- R_\infty e^{-2K}(e^K(-1)^n - 1)][K/(\mu_n^2 + b^2)\mu_n^2[1 - \exp(-\mu_n^2 Fo)]\} + Ki\Phi_{II}(x,Fo). \qquad (7.41)$$

Next we shall consider the question of unilateral diffuse irradiation of a plane layer on a conveyer support (in the case of drying stone fruits the stone can be considered an opaque support). If the layer is not an optically infinitely thick one, a part of the radiation reflected by the support becomes an additional heat source acting at the bottom of the sample. The criterion $Po(x,Fo)$, which corresponds to the function of the heat sources $w(x)$, is described as follows:

$$Po(x,Fo)=Po_c[exp(-bx) - [(R_\infty - R)/(1 - R_\infty R)]exp(-2b)exp(bx)]; \quad (7.42)$$

where, $Po_c=E(1 - R_\infty)LR^2/(1 - R_{ef}\Psi^2)\lambda t_c$ - constant part of the $Po(x,Fo)$ criterion. After integration we obtain [43]:

$$\theta(x,Fo)=Po_c\{(Fo/K)[1 - e^{-K} + [(R_\infty - R)/(1 - R_\infty R)]e^{-2K}(1 - e^{-K})] +$$

$$+ \Sigma [K/(\mu_n^2 + K^2)\mu_n^2][1 - (-1)^n e^{-K} - [(R_\infty - R)e^{-2K}]/(1 - R_\infty R)]\cdot$$

$$\cdot(1 - (-1)^n e^{-K}|cos\mu_n[1 - exp(-\mu_n^2 Fo)]\} +Ki\Phi_{II}(x,Fo). \quad (7.43)$$

Of great interest is the case of directional irradiation of a layer of sample where the absorbed energy is distributed anomalously in the layer. This anomalous distribution of the absorbed energy is described as follows:

$$Po(x,Fo)=Po_c[(1 + R_\infty)c_2 exp(-Kx) - (c_1 + c_2 - 0.5)exp(-\text{\ss}x/\mu)]; \quad (7.44)$$

where, $\text{\ss}=\varepsilon R$ - dimensionless attenuation coefficient; $Po_c=E(1 - R_\infty)LR^2/(1 + R_\infty)\lambda t_c$ - constant part of the $Po(x,Fo)$ criterion; $\mu =arccos\theta$ - coefficient compensating for the slope of incident flux at the angle θ. The solution of this equation yields:

$$\theta(x,Fo)=Po_c\{Fo[(c_2(1 + R_\infty)/K)(1 - e^{-K}) + (\mu(c_1 + c_2 - 0.5)/\text{\ss})(e^{-\text{\ss}/\mu} - 1) +$$

$$+ \sum_{n=1}^{\infty}(2/\mu_n^2)[(c_2 K(1 + R_\infty)/(\mu_n^2 + K^2))(1 - (-1)^n e^{-K}) - (\text{\ss}(c_1 + c_2 - 0.5)/$$

$$\mu(\mu_n^2 + (\text{\ss}/\mu)^2)(1 - (-1)^n e^{-\text{\ss}/\mu}]cos\mu_n[1 - exp(-\mu_n^2 Fo)]\} + Ki\Phi_{II}(x,Fo).$$
$$(7.45)$$

This equation can be used for analyzing the temperature field inside a layer of sample. In Fig. 7.10 is shown the potential $\theta(x,Fo)$ as a function of x/R during normal directional irradiation ($\mu=1$) of a layer (under the assumption that K=2, ß= 6, R_∞=3, c_1=0.3, c_2=1.3). At low values of Fo there is a maximum in the potential $\theta(x,Fo)$ at a certain value of x_m. With an increase in Fo this maximum shifts towards the surface of the layer and becomes less noticeable. The presence of the maximum and its shift are shown in Fig. 7.11. As can be seen, there is a change in time in the ratio of the potential $\theta(x,Fo)$ in the upper layers to the potential $\theta(1,Fo)$ on the surface of the layer (x=1). The maximum is present only when the values of Fo are small, and it disappears at Fo=0.3.

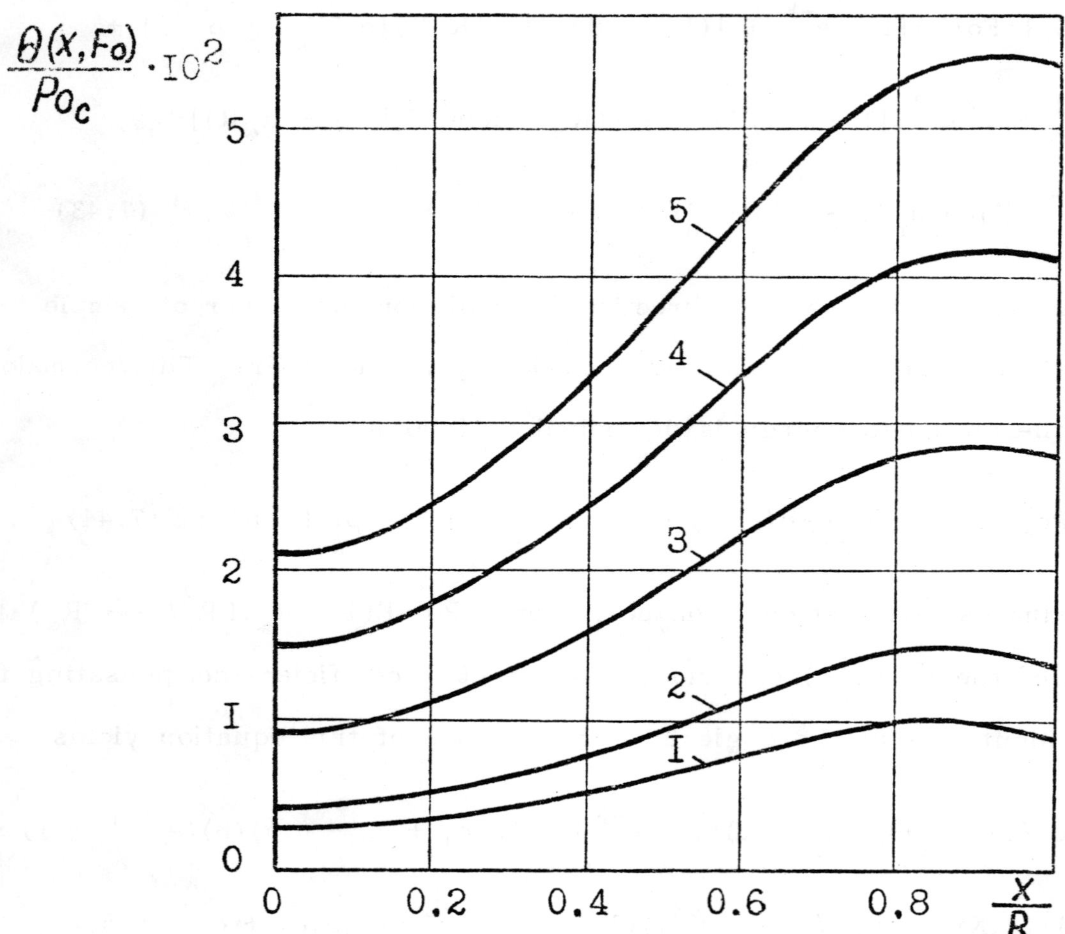

Figure 7.10   Distribution of the temperature field in a layer of sample subjected to directional irradiation at Fo [43]: 1 - 0.01; 2 - 0.015; 3 - 0.03; 4 - 0.045; 5 - 0.06.

The anomalous distribution of the temperature field under the conditions of directional irradiation helps to intensify heat and mass transfer especially in desiccation processes. The temperature gradient in the direction of the surface of the sample causes moisture to move towards the surface.

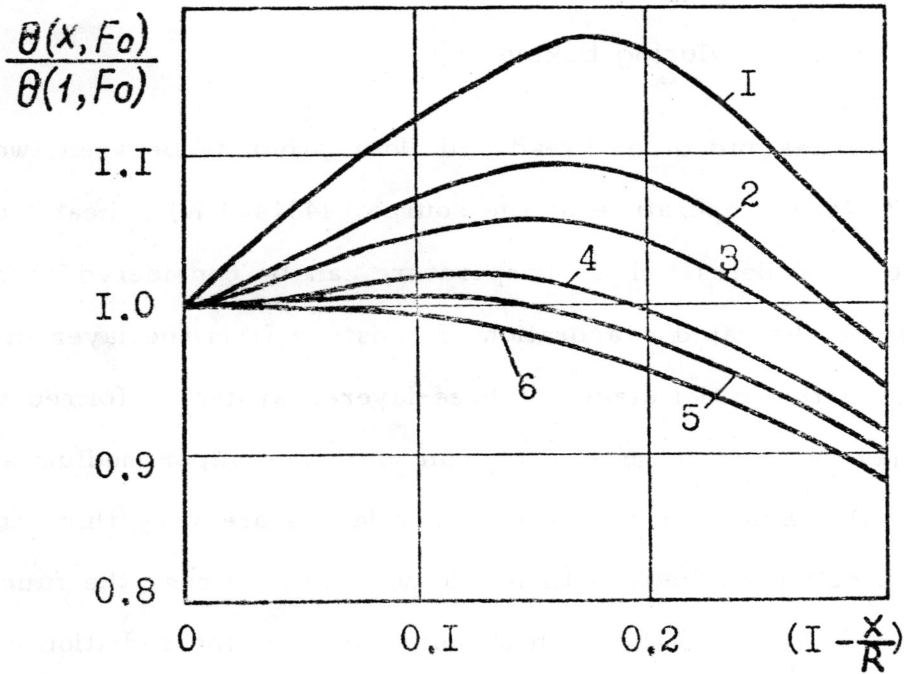

Figure 7.11   Change in the $\theta(x,Fo)/\theta(1,Fo)$ ratio in the surface layers under the conditions of directional irradiation at Fo [43]: 1 - 0.005; 2 - 0.01; 3 - 0.015; 4 - 0.03; 5 - 0.06; 6 - 0.3.

The maximum of the dimensionless temperature in a layer of sample subjected to directional infrared irradiation, with account taken of the convective heat transfer ($Ki \neq 0$), can be obtained in a general form from Eq. 7.45 by determining the local maximum $\theta(x,Fo)$.

On the basis of what has been said above the following conclusion can be drawn, which is of practical importance: the presence of a maximum in temperature at a certain depth x_m of a layer of sample due to the anomalous distribution of the absorbed radiation $w(x)$ in the case of directional irradiation is possible during the initial stages of heating (at low values of Fo). The x_m coordinate and the magnitude of the maximum $w(x)$ change with time so that x_m shifts towards the surface of the sample with an increase in Fo, and no anomaly of the temperature field is observed. In the process of heating the anomalous distribution of w(x) affects the distribution of temperature at all values of x. And the conditions of irradiation strongly affect the distribution of the temperature field in the sample.

7.3 Radiative heat transfer during baking

When baking waffles and other bread and flour products between two metallic plates at $170 \div 190°C$ (the temperature of the source: $440 \div 460$ K)   heat transfer takes place between the plates (whose temperature can be considered constant) and the dough. Owing to the rapid evaporation of moisture from the layer in contact with the plate during the initial stage, a three-layered system is formed which consists of two thin layers ($\ell_c < 0.5$ mm each) of an air-water vapor medium and a layer of waffle ($\ell \cong 2.0$ mm). Since the air-water vapor layers are very thin, their optical thickness $k_\lambda \ell_c$ will obviously be less than 0.5 mm. In this case the function $w(x, \tau)$, defined by Eqs. 7.25 and 7.26 with account taken of the radiation emitted by the waffle, the air-water vapor medium, and the infrared source, must be introduced into Eq. 7.16. Being the functional of the temperature field, $w(x,\tau)$ is determined from Eq. 7.26. The balance equation for the adjacent surface of the waffle, which is separated from the wall by a thin air-water vapor layer, has the following form:

$$q(\tau) = -\lambda dT/dx + E_{net}. \tag{7.46}$$

A general solution of Eq. 7.16-7.26 with account taken of Eq. 7.46 is shown above (see Eq. 7.27).

7.4 Principles of designing infrared irradiation plants for the food industry

The type of infrared generators needed for food processing plants is determined according to their spectrum (Figs. 7.12, 7.13). The energy absorbed by foodstuffs subjected to infrared irradiation brings about intermolecular reactions and reactions between various functional groups and atoms in the molecules the frequencies of whose vibrations overlap or are multiples of the frequency of the incident radiation. The energy of individual chemical bonds is comparable to the energy of the photons of the radiation source. Thus, at $\lambda > 1.0$ μm the energy of the photon $\varepsilon_f = h\nu \leqq 2 \cdot 10^{-19}$ J, while the energy of the C–C bond is $\sim 4 \cdot 10^{-19}$ J and that of the H–O bond – $(0.32 \div 0.48) \cdot 10^{-19}$ J. By intensifying the vibrations of certain groups in the molecule, infrared radiation accelerates various biochemical reactions. It has been reported [19, 79] that bread baked in infrared ovens has a finer and more thin-walled porosity than bread baked in ordinary ovens. And dough prepared from flour of wheat grains that have been subjected to infrared irradiation makes good-quality bread.

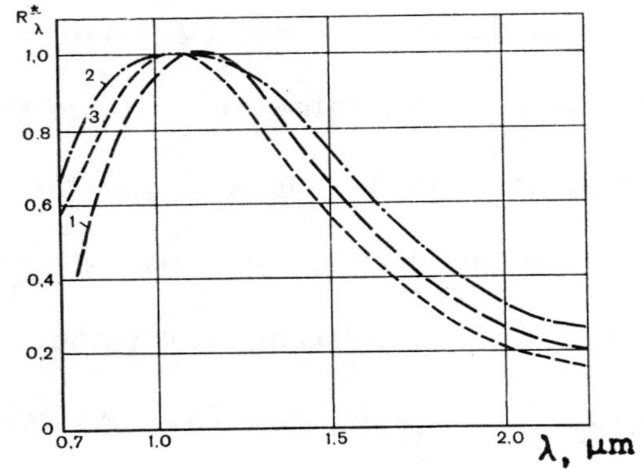

Figure 7.12 Emission spectra of different infrared lamps [8, 46]: 1 – IKZK-220-250; 2 – KG-220-1000; 3 – IKZ-220-500.

The efficiency of drying grapes has been improved [53] by their preliminary treatment for 120 s with infrared radiation (0.4–5.0 μm, 25 kW/m^2) under oscillating conditions. This cuts down the time required for drying grapes in the sun 3.2 times. The grapes subjected to preliminary infrared irradiation have better organoleptic characteristics (in terms of color and transparency). Furthermore, the preliminary treatment reduces the amount of nonspore-forming microorganisms and slows down detrimental biochemical processes in the final product.

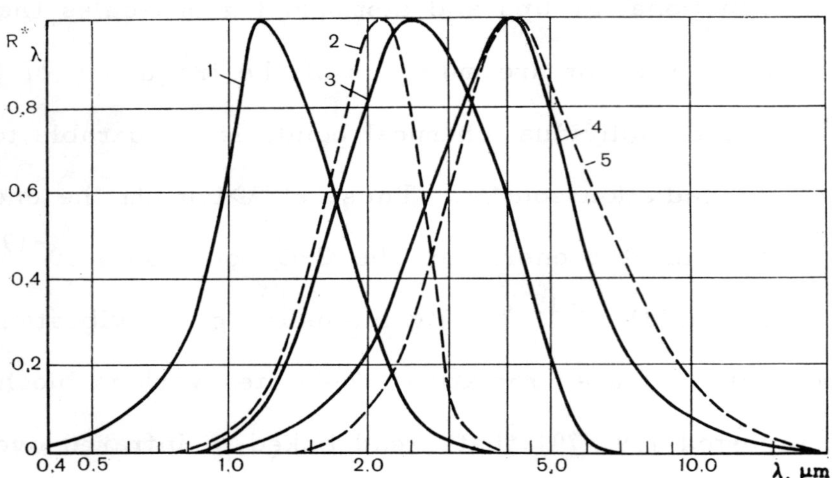

Figure 7.13 Emission spectra of different infrared generators [39]: 1 - KG-220-1000 (2250 K); 2 - hot plate GII (1093 K); 3 - Ni-Cr spiral (220 v, 6A); 4 - metal plate (220 v, 600 W); 5 - radiator TEN (220 v).

Another important factor that should be considered when selecting an infrared generator is the thickness of the layer of substance to be irradiated. Let us take, for example, two layers of starch, one 10 mm and another 2 mm thick. For the thicker layer the heating can be accomplished only by using a high-temperature infrared generator (Fig. 7.4,b). For radiation from low-temperature generators penetrates a sample to a small depth $x_1 < 2$ mm (curve 2). As for the layer 2 mm thick, either high-temperature or low-temperature generators can be used.

From Fig. 7.4c it can be seen that during irradiation with high-temperature sources (curve 1) and low-temperature ones (curve 2) the entire layer is heated through owing to the absorbed radiation. The magnitude of the dimensionless flux density E_0^* in the case of irradiation by a high-temperature source is much greater than that by a low-temperature source. However, this information is not yet sufficient for choosing an infrared generator. Calculations show that the absorption coefficient of starch irradiated by a high-temperature generator is 540 m^{-1} as compared to 3426 m^{-1} in the case where a low-temperature generator is used. This means that the radiation energy from the latter is absorbed more completely, and the heating will be more effective since the absorption coefficient in this case at a depth up to 0.6 mm (Fig. 7.4,c) will be tens of times greater than that in the case of a high-temperature source. Furthermore, the fraction of radiant energy from the low-temperature source absorbed by the entire layer (A=0.912) is 2.37 times greater than that from the high-temperature source (A=0.385).

The efficiency of an infrared processing unit can be evaluated by comparing the AE product with the energy required to produce a unit of the incident radiation flux. The minimal permissible power of an infrared source is determined on the basis of the kinetics of the process and the thermal lability of the substance. For bread crumbs, for example, the temperature of the heating should not exceed 180°C. The required radiation flux E can then be determined from Eqs. 7.4, 7.7, and 7.3, and the minimal power of the infrared source necessary for heating the substance to T(X,Fo) is expressed in the following terms:

$$[P] = P_1 / \eta_{ef} M \Phi_I [(\Delta T_1 / T_2 - (Ki\Phi_{II}))]; \qquad (7.47)$$

where, $P_1 = \lambda T_2 S / LB^2; \qquad (7.48)$

$$M = 2(1 - RR_2)/(1 - R_\infty)(1 + R_2). \qquad (7.49)$$

These relationships show that in order to decrease the power of infrared generators it is necessary to increase η_{ef}, the reflection coefficient R_2 of the reflector, and the surface area S of the irradiated sample.

On the basis of the results obtained the following principles can be formulated for the designing of infrared installation units for processing foodstuffs:

1. The physico-chemical properties of a given substance should be established (the chemical structure of the molecules and the predominant form of the bond between the substance and its moisture content); and the basic optical property $R_{\lambda\infty}$ of the substance should be determined experimentally. The results obtained form the basis for determining the group to which the substance belongs

in the classification of materials according to their optical properties.

2.   The technological operations for every stage of the thermal treatment should be determined.

3.   The optical thickness of a layer of substance $L\ell$ is determined on the basis of its spectrum in the 1.5-15.0 µm range on the following assumptions:

- that the layer is optically thick, $L\ell > 10.0$; for most of the substances belonging to all four groups in the classification a layer can be considered thick at $\ell > 10$ mm;

- that the layer is optically thin, $L\ell < 10.0$; for most of the substances of the 1st and 2nd groups a layer can be considered thin at $\ell > 5$ mm, and for the most of the substances belonging to the 3rd and 4th groups - at $\ell < 2$ mm.

4.   The type and temperature of an infrared generator are determined in accordance with the following considerations.

For an optically thick layer ($L\ell > 10.0$) of substances of the 4th group:

- for heating the surface low-temperature infrared generators ($T < 1000$ K), $\lambda_m > 2.8$ µm are the most suitable;

- for heating a substance throughout its thickness either high-temperature generators whose temperature lies within 1500 K$<T<$2000 K ($\lambda_m = 12 \div 1.6$ µm) or low-temperature generators whose temperature can be varied within 1200 K$<T<$1400 K ($\lambda_m = 1.6 \div 2.6$ µm) can be used.

For an optically thin layer ($L\ell < 10.0$) of substances belonging to the 1st and 2nd groups:

- for heating the surface low-temperature generators ($T < 1000$ K), $\lambda_m > 2.8$ µm are suitable;

- for heating a substance throughout its thickness generators whose operating temperature lies within 1200 K$<T<$1500 K ($\lambda_m = 2.0 \div 2.4$ µm) are recommended.

For substances belonging to the 3rd and 4th groups:

- for heating the surface low-temperature generators operating at 900 K$<T<$

1000 K (λ_m=2.8÷3.2 μm) and at 450 K<T<500 K (λ_m=5.5÷6.2 μm) can be used;

- for heating a substance throughout its thickness low-temperature generators operating at 500K<T<800 K (λ_m=3.5÷5.5 μm) and at 1200 K<T<1500 K (λ_m=2.0÷2.5 μm) are suitable.

5.   The thermal radiational characteristics of a layer of substance $f(\ell)$ are determined with the aid of Eq. 7.8 which includes the spectral attenuation coefficient L_λ and the reflectance of an optically infinitely thick layer $R_{\lambda\infty}$; these are determined experimentally.

6.   The flux density of infrared radiation absorbed per unit area of the surface AE is calculated as well as its distribution throughout the thickness of layer, w(x).

7.   A comparison of these two parameters for different infrared radiators provides a basis for selecting an infrared generator for which the energy absorbed by the whole layer is the greatest and the distribution of the absorbed energy throughout the thickness of the layer is the most suitable for a given technological process.

Thus, the choice of a technologically and economically optimal infrared source is determined mainly by its spectral composition, the optical thickness of the layer of substance to be irradiated, the amount of the radiation absorbed and its distribution throughout the layer, w(x).

The best location of infrared generators and the geometrical configuration of the oven are shown in Figs. 2.1 and 2.2.

7.5   Calculating the irradiance fields in thermal radiational installations

The efficiency of infrared food-processing plants depends above all on the uniform distribution of radiant power on the surface of foodstuffs. It is necessary that the distribution of the flux density and the corresponding isotherms should be determined by the same functional dependence as the surface of a sample. The irradiance fields (Fig. 7.14) on the surface are determined by the distance $h=z$ between the radiation source and the layer of foodstuffs, the spacing between the lamps, s, the distance from the reflector, h_r, and the operating conditions of the source.

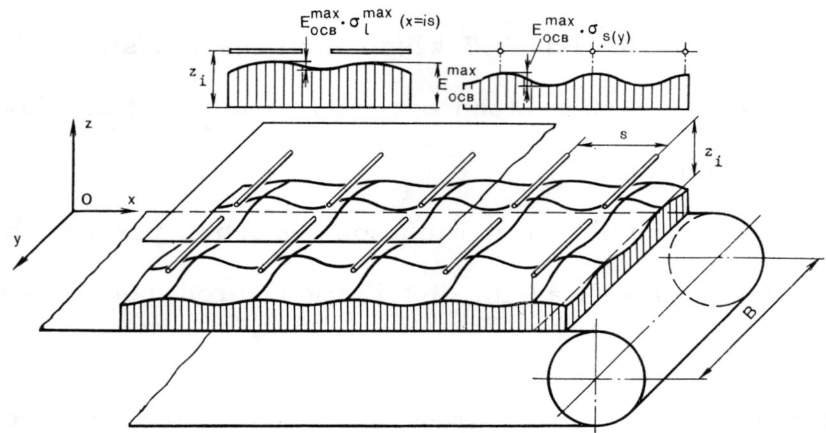

Figure 7.14 Determination of the nonuniformity of the irradiance fields produced by infrared generators in conveyor-type installations with a plane reflector [16, 68].

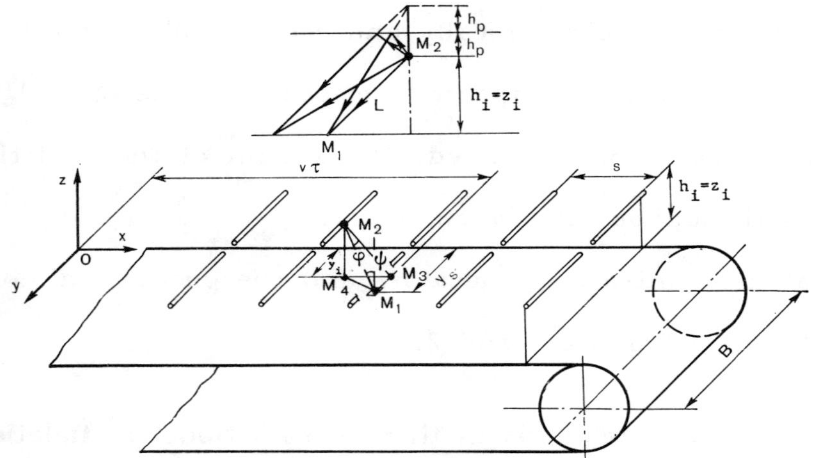

Figure 7.15 Schematic diagram for calculating the irradiance produced by assemblies of linear infrared generators equipped with a plane reflector [16, 88, 112].

Analytical methods for calculating the irradiance fields produced by an arbitrary number of generators in the case of plane, cylindrical, and spherical samples have been described in literature [16, 19, 88, 102, 108]. With their aid the relationships can be established between the flux density of the incident radiation, the parameters describing the location of the infrared sources, their emission characteristics, and the thermal radiational properties of the foodstuffs being irradiated, the reflectors, and the walls of the oven.

A linear radiator can be considered a cylinder with diameter d and consisting of an infinite number of Lambert point sources lying on one line. In this case the irradiance E_i from an i-th linear source with y_i- and x_i-coordinates on the surface of the sample is determined by the integration of the generator with ℓ_i length within the $\phi_2-\phi_1$ angles (Fig. 7.15):

$$E_i = \int_{\phi_1}^{\phi_2} 1/L^2 B_i d\cos\phi\cos\psi\, d\phi;\tag{7.50}$$

where, ϕ - angle between the normal element of a luminous cylinder $dd\phi$ and the direction of the radiation $\vec{L}$ towards the irradiated area; ψ - angle between the normal to the irradiated area and the direction of the incident radiation; L - distance between the luminous and the illuminated area elements M_1 and M_2 (Fig. 7.15):

$$L = [z^2 - (v\tau - is)^2]^{\frac{1}{2}};\tag{7.51}$$

here, B_i - radiance of an element of the i-th radiator. Integration of Eq. 7.50 yields the irradiance of a unit area by an arbitrary i-th source:

$$E_1(y,y_s) = B_{ef}d(z/2L)\{[(y-y_s)/(L^2+(y_s-y)^2)]-(1/L)\text{arc tg}(y_s-y)/L\}|$$

$$\begin{matrix}(y_s-y_1)\\(y_s-y_2)\end{matrix};\tag{7.52}$$

where, y_1 and y_2 - coordinates of the end points of the radiator.

The total radiation flux at the point under consideration on the surface of the sample coming from n-th assemblies of linear infrared generators, each containing an i-th infrared source, is defined as follows:

$$E = \sum_{n} \sum_{i} [_i(2n) - E_i(y_{2n-1})]; \qquad\qquad (7.53)$$

where, y_{2n} and y_{2n-1} - coordinates of the i-th source in an n-th assembly of linear infrared generators.

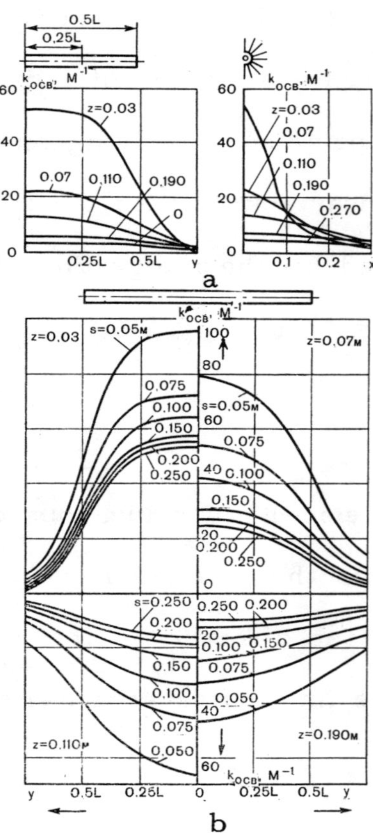

Figure 7.16 Irradiance coefficient (K) as a function of the y-coordinate on the surface of the sample at different heights of the z-radiator and at different distances between the lamps [16, 88]: a - for a single lamp; b - for an assembly of 21 lamps under a central lamp.

The presence of a plane reflector located above the radiator at distance h provides additional irradiance E_r at the point under consideration on the sample. The magnitude of E_r is determined with the aid of Eq. 7.52 as the irradiance from the virtual image of radiators located at distance $z'=z + 2h_r$. The irradiance of the imaginary source is $B'=BR_r$. By substituting z' for z and B'_{ef} for B_{ef} in Eq. 7.51 and Eq. 7.52, respectively, we obtain the irradiance due to reflected radiation.

Whether or not the irradiance will be of a nonuniform nature depends on the height (z) of the radiator located above the sample, the spacing (s) and the location of the reflector (h_r). In Fig. 7.16 are shown the calculated and experimentally obtained irradiance fields produced by a linear radiator, which indicate their considerable nonuniformity. An acceptable degree of uniformity of irradiance on the surface of a sample can be achieved by decreasing s and increasing z.

Eqs. 7.52 and 7.53 can be used for calculating the irradiance field and its nonuniformity due to spacing in relation to the location of the infrared source and the reflector.

A handy nomogram for determining the location of infrared lamps in an industrial installation for desiccating foodstuffs is shown in Fig. 7.17. In quadrant I in this nomogram is shown the irradiance coefficient:

$$K(z,s)=E/B_{ef}d_{ef}. \tag{7.54}$$

In quadrant II is shown the coefficient which compensates for the nonuniformity of irradiance due to spacing:

$$\bar{\sigma}_s=\tfrac{1}{2}(\sigma_{s,m})=f(z/s). \tag{7.55}$$

In quadrant III is shown the magnitude of B_{ef} as a function of the type of infrared source used and its operating conditions (U - voltage). And in quadrant IV is shown the magnitude of E as a function of d_{ef}. The value of E obtained with the aid of the nomogram characterizes the magnitude of the irradiance coming directly from the source.

Since foodstuffs reflect up to 90-95% of short-wave radiation an additional irradiance is produced owing to multiple reflection. The absorptance of the sample (A) and the multiple reflection in the oven are taken into account in quadrant V (Fig. 7.17):

$$AM = A(1 - RR_r); \qquad\qquad (7.56)$$

where, R, A, and R_r are averaged with respect to the spectrum of the infrared source with the aid of Eq. 7.8

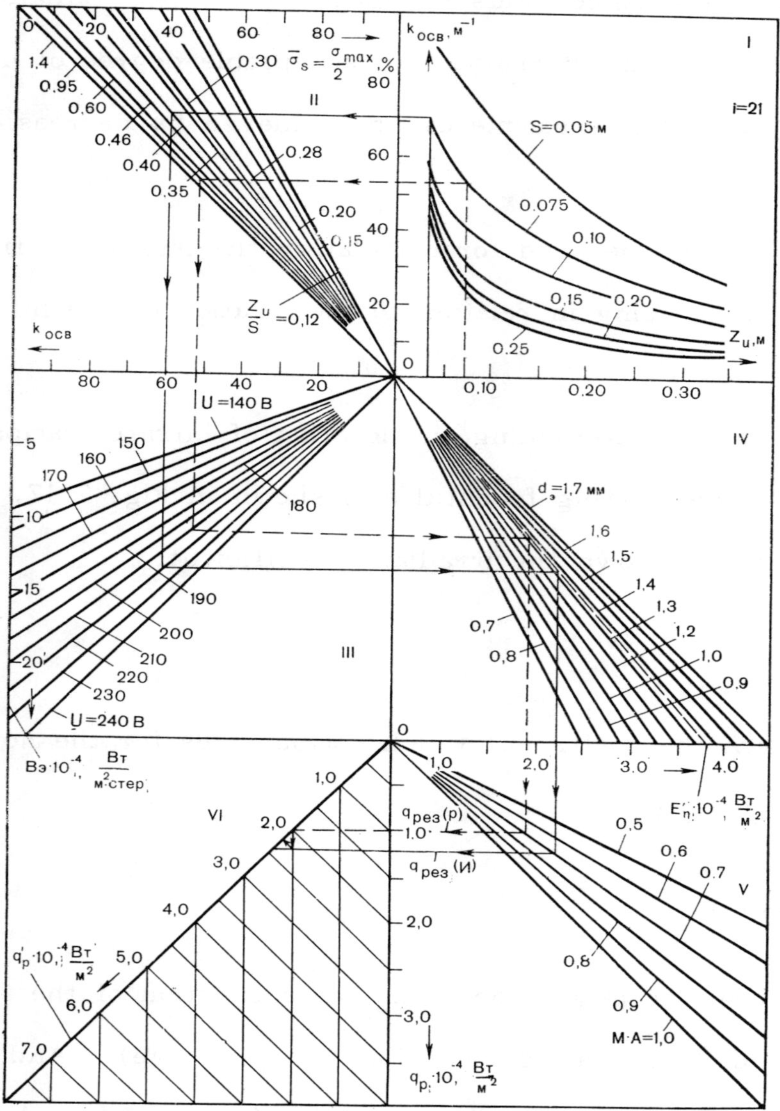

Figure 7.17 Nomogram for determining the location of linear infrared generators in an oven with a plane reflector [16, 68, 88]: (E_{net}= 2.2·10^4 W/m^2; bread crumbs irradiated by an assembly of 21 KG-1000 lamps at z = 0.08 m, h_r = 0.02m, U=220 v, and B_{ef} = 22·10^4 W·m^{-2}·sr^{-1}).

The net radiant flux incident upon the sample is defined as follows:

$$E_{net} = B_{ef}d_{ef}K[A/(1 - RR_r)]. \qquad (7.57)$$

The component of the E_{net} coming from the reflector is determined as:

$$E_{net}(R_r) = B_{ef}R_r d_{ef}K[A/(1 - RR_r)]. \qquad (7.58)$$

The net total radiant flux is shown graphically ($q_{net} \equiv E_{net}$) in quadrant VI (Fig. 7.17):

$$E_{net} = E + E(R_r). \qquad (7.59)$$

The nomogram can be used to solve a reverse problem; namely the problem of determining the geometrical parameters for the optimal location of the lamps at a given radiation flux by using quadrant I and on the basis of the type of infrared source chosen (i.e., d_{ef}) and its operating conditions $B_{ef} = f(U)$.

Since during multiple reflection the spectral composition of the radiation flux changes the nomogram can be used only for determining E_r approximately on the basis of averaged A, R, and R_r. For more precise calculations of E_r it is necessary to know the dependence of $E_{r\lambda}$ on the radiation wavelength and then integrate it with respect to the entire spectrum of the infrared source.

7.6 Equation of heat balance for infrared drying installations

Calculations of the processes involved in the infrared treatment of foodstuffs (how much energy is needed, how much moisture is evaporated, and how long it will take to complete the treatment) should be based on a general differential heat-transfer equation which takes into account not only conductive heat transfer, but also all the sources of heat due to phase transformations, internal mass exchange, and the absorption of radiation deep inside the sample. The solution of such an equation is fraught with considerable difficulties. We shall consider a simplified form of a differential equation.

It should be borne in mind that a part of the radiant energy absorbed by foodstuffs exposed to infrared radiation is consumed in biochemical reactions taking place in the foodstuffs.

$$dF = dF_s + dF_e + dF_{rx} + dF_d; \qquad (7.60)$$

where, F - radiant energy absorbed by the sample; F_s - radiant energy consumed in heating the sample; F_e - radiant energy consumed during the evaporation of moisture; F_{rx} - radiant energy consumed in biochemical reactions taking place in the sample; F_d - radiant energy given off by the heated sample and dispersed in the surrounding space.

The amount of energy absorbed by the sample should be determined with account taken of the thermal radiational characteristics of all the components participating in the heat exchange, including the loss of radiant energy by the sample. The magnitude of F_s can be determined without compensation for the losses of radiant energy by the sample; in this case there will be no need to introduce the term for the temperature of the sample raised to the fourth power. Then Eq. 7.60 can be rewritten:

$$dE_{net}S_o = dF_o + dF_e + dF_{rx} + dF_r; \qquad (7.61)$$

where, E_{net} - net radiation flux on the surface of the sample (Eqs. 7.11, 7.14); S_o - surface area of the irradiated sample.

With the aid of Eqs. 7.11 and 7.14 the following relationship is obtained which describes the heat balance for a layer exposed to infrared radiation having mass m, heat capacity C_s, and surface S:

$$E_{net}S_o d\tau = V_s C_s \rho_s dt_s(\tau) + q_m rSd\tau + \alpha[t_s(\tau) - t_c(\tau)]Sd\tau. \qquad (7.62)$$

The solution of this equation gives the time required for the heating of the sample.

In a particular case where E_A and q_m are constant [26] the solution of

Eq. 7.62 yields a known relationship [29] describing the duration of the heating process:

$$\tau=(1/D_1)\ln\big|[B_1 + D_1(t_{s2} - t_c)]/[B_1 + D_1(t_{s1} - t_c)]\big\};\qquad(7.63)$$

where, $B_1=(E_A - q_m r k_f)/C_s\rho_s k_f$, $D_1=-\alpha_{net}\tau/C_s\rho_s$. $\qquad$ (7.64)

In these equations, which differ from other known relationships [19], the term E_A and thus also B_1 are determined with account taken of the spectral thermal radiational and emissive characteristics of every component taking part in the heat exchange.

7.7 Estimating the efficiency of infrared generators

A single criterion Π has been worked out for evaluating quantitatively the operating conditions of infrared generators [20, 89]. It takes into account the R, T, and A characteristics of the foodstuffs to be irradiated; the ε and E characteristics of the infrared generators; the R, T, and A of the air-vapor medium; the R_r of the reflector and the walls; the nonsymmetrical conditions for bilateral irradiation; the technological requirements at different stages of the thermal treatment (k); and the thicknes of the substance. The efficiency criterion Π meets the said requirements [20, 89]:

$$\Pi=Q_r k\Xi.\qquad(7.65)$$

The coefficient k characterizes the correspondence between the spectral composition of the radiant flux and the requirements of the technological process. For optically thin layers whose bulk must be heated through, $k=k_v$. And k_v is then determined with the aid of Eqs. 3.2 and 3.7 as follows:

$$k_v=T/A=[(1 + R_\infty)/(\psi + \exp(L\ell))].\qquad(7.66)$$

If there is an intensive heat emission from the surface:

$$k = k_v^{-1} = A/T. \qquad\qquad (7.67)$$

The coefficient Ξ can be determined from the equation:

$$\Xi = F/P = \eta_{ef}; \qquad\qquad (7.68)$$

where, F - power of the infrared radiator; η_{ef} - radiant efficiency input of the radiator.

In Fig. 7.18 are shown the dependence of the efficiency criterion Π and the coefficients Q_r, k, and Ξ on the temperature of the infrared generator. The efficiency criterion, however, does not take into account the biological, biochemical, physico-chemical, and other effects of infrared radiation on the substance being irradiated. Therefore, the method for determining the coefficient k should be improved, and the degree of significance of each coefficient in Eq. 7.65 should be considered separately.

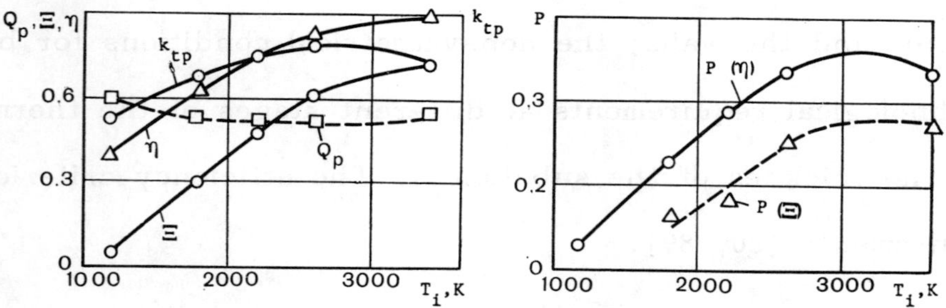

Figure 7.18   The efficiency criterion Π as a function of the temperature of the infrared source in the case of the irradiation of a bread-dough system [88, 89].

These factors are taken into account in the following relationship [39, 61]:

$$\xi = E_{net}^{*}{}^{\alpha} n_T^{\beta} r_\lambda^{\gamma} \eta_{ef}^{\delta}; \qquad\qquad (7.69)$$

where, E_{net}^{*} - dimensionless flux density of radiation absorbed by the sample and determined similarly as Q_{net}; n_T - coefficient which takes into account the specific technological requirements; r_λ - coefficient which takes into account the specific effect of infrared radiation on the sample; α, β, γ, δ - degree of significance of the coefficients.

The coefficient n_T is determined as follows. Let us assume that owing to the technological requirements of the process of thermal treatment the layer of sample should be so heated that the absorbed energy is distributed according to the function $w_\lambda(x)$ (Fig. 7.19,a). In that case the following quantity averaged with respect to the thickness of the layer can be accepted as the coefficient $n_{T\lambda}$, which takes into account the correspondence of the $w_\lambda(x)$ function to the technological requirements:

$$n_{T\lambda}=(1/\ell)\int_0^\ell [w_\lambda(x) - \Delta w_\lambda(x)/w_\lambda(x)]dx; \tag{7.70}$$

From this equation it follows that when $w_\lambda(x)$ is absent the technological requirements are: $w_\lambda(x) = w_{net}(x)$, $\Delta w_\lambda(x) \equiv 0$, and $n_{T\lambda}=1$.

For the total radiation of an arbitrary spectral composition the expression for n_T is defined by the integration of Eq. 7.71 with respect to the entire spectrum of the incident flux E:

$$n_T=[\int_{\lambda_1}^{\lambda_2} E_\lambda n_{T\lambda} d\lambda][\int E_\lambda d\lambda]^{-1}. \tag{7.72}$$

Here, the range of wavelength λ_1 and λ_2 is determined according to Eqs. 4.114 and 4.115.

We shall now consider two examples of how the coefficient $n_{T\lambda}$ for foodstuffs exposed to infrared radiation is calculated. In such processes as desiccation and baking the entire bulk of a substance should be heated through. In this case the following technological requirements must be fulfilled (Fig. 7.19,b):

$$w(x)=const, \quad 0\leq x\leq\ell. \tag{7.73}$$

As has been shown in Chapters 2 and 4, such a distribution cannot be achieved. In the case of optically thin layers it is possible to obtain a $w(x)$ function which changes slightly with the depth in the case of bilateral irradiation (Fig. 7.4).

Under real conditions the $w(x)$ function is described by the sum of exponential relationships. In the case of diffuse irradiation the maximum values of $w(x)$ are observed at the boundaries of the layer ($x=0$); they should not exceed the assigned value of the heat source because of the limitation imposed, for example, by the thermal lability of the substance (in order to prevent the dextrinization of starch, scalding, the formation of crust, etc). In this case, under the conditions of bilateral irradiation, by using Eq. 2.26 we get:

$$n_{T\lambda}(V)=(E_{\lambda 1} + E_{\lambda 2})[1 - R_{\lambda \infty}\exp(-L_\lambda^* \tau_\ell)][1 - \exp(-L_\lambda^* \tau_\ell)]/$$

$$L_\lambda^* \tau_\ell \{E_{\lambda 1}[1 - R_{\lambda \infty}\exp(-2L_\lambda^* \tau_\ell)] + E_{\lambda 2}(1 - R_{\lambda \infty})\exp(-L_\lambda^* \tau_\ell)\}. \qquad (7.74)$$

For an optically infinitely thick layer subjected to irradiation on both sides (at $L_\lambda^* \tau_\ell > 10$, $\exp(-L_\lambda^* \tau_\ell \ll 1$) Eq. 7.74 is considerably simplified:

$$n_{T\lambda}(V)=(E_{\lambda 1} + E_{\lambda 2})/E_{\lambda 1}L_\lambda^* \tau_\ell = 1/L_\lambda^* \tau_\ell(1 + K); \qquad (7.75)$$

where, $K=E_{\lambda 2}/E_{\lambda 1}$ — coefficient which takes into account the nonsymmetrical irradiation conditions; E_λ — incident radiation flux determined with account taken of multiple reflection and the absorption by the air-water vapor medium, and of the location of each of the objects taking part in the heat transfer inside the infrared installation with respect to that of the others, and emissive characteristics. When $E_{\lambda 1}=E_{\lambda 2}$, Eqs. 7.74 and 7.75 become simplified:

$$n_{T\lambda}(V)=[2(1 - \exp(-L_\lambda^* \tau_\ell))]/[L_\lambda^* \tau_\ell(1 + \exp(-L_\lambda^* \tau_\ell))]; \qquad (7.76)$$

when $L_\lambda^* \tau_\ell > 10$, we have:

$$n_{T\lambda}(V)=2/L_\lambda^* \tau_\ell. \qquad (7.77)$$

In the case of unilateral irradiation of an optically infinitely thick layer (at $L_\lambda^* \tau_\ell > 10$) we have:

$$n_{T\lambda}(V) = 1/L_\lambda^* \tau_\ell. \qquad (7.78)$$

The last two equations show that the optical thickness $L_\lambda^* \tau_\ell \equiv L_\lambda \ell$ can serve as a criterion which takes into account the technological requirements in the case of a semi-infinite layer and when $L_\lambda \ell \geq 10$. It can be seen from Eqs. 7.73-7.78 that bilateral irradiation under symmetrical conditions yields the largest value of $n_{T\lambda}(V)$: at $L_\lambda \ell \geq 10$ the magnitude of $n_{T\lambda}$ is twice as large as that in the case of unilateral irradiation. The magnitude of $n_{T\lambda}$ is strongly affected by the layer $L_\lambda^* \tau_\ell$, which is a function of the optical characteristics and determines the thermal radiational characteristics R_λ and T_λ of the layer.

In the case of many foodstuffs their thermal processing requires intensive heating of the surface by low-temperature infrared radiators. In the process the absorbed energy is concentrated within a narrow surface layer $\Delta\ell$ (Fig. 7.19,c):

$$w_\lambda(x)= \begin{cases} w_{\lambda 1}(x), & 0 \leq x \leq \Delta\ell_1 \\ 0, & \Delta\ell_1 \leq x \leq \ell - \Delta\ell_2 \\ w_{\lambda 2}(x), & \ell - \Delta\ell_2 \leq x \leq \ell \end{cases} \qquad (7.79)$$

The coefficient $n_{T\lambda}(s)$, which takes into account the emission of the absorbed energy in the upper layers, in a general case of nonsymmetrical bilateral irradiation is defined as follows:

$$n_{T\lambda}(s) = 1 - 1/\ell \left\{ \int_0^{\Delta\ell_{k1}} [1-w_\lambda(x)w_{\lambda 1}(x)]dx + \int_{\Delta\ell_{k1}}^{i-\Delta\ell} [w_\lambda(x)/w_{\lambda m}(x)]dx + \right.$$

$$\left. + \int_{\ell-\Delta\ell_{k2}}^{\ell} [(w_{\lambda 2}(x)/w_{\lambda m}(x)) - w_\lambda(x)/w_{\lambda 2}(x)]dx \right\}. \qquad (7.80)$$

where, $w_{\lambda m}(x)$ – maximum value of either the $w_{\lambda 1}$ or $w_{\lambda 2}$ function. For unilateral irradiation this equation becomes simplified:

$$n_{T\lambda}(s)=1-(1/\ell)\left\{ \int_0^{\Delta\ell_{K1}} [1-w_\lambda(x)/w_{\lambda 1}(x)]dx + \int_{\Delta\ell_{k1}}^{\ell} (w_\lambda(x)/w_{\lambda 1}(x))dx \right\}. \qquad (7.81)$$

The approximate value of $n_{T\lambda}(s)$ in the case where $\Delta\ell_1$ and $\Delta\ell_2$ are unknown can be determined from the following relationship:

$$n_{T\lambda}(s) = 1 - n_{T\lambda}(V). \qquad (7.82)$$

The maximum value of the coefficient ($n_{T\lambda}=1$) indicates the best correlation between the technological requirements of the process, on the one hand, and the type of infrared generator chosen, the spectral composition of its radiation, and its operating conditions, on the other.

The coefficient r_λ is introduced in order to take into account the specific effect of infrared radiation on the substance being irradiated. This effect can be of a desirable or an undesirable nature. Thus, in some instances short-wave irradiation can cause overheating, while in such operations as the disinfection of grains such radiation is useful. It is impossible to choose the right infrared generators solely on the basis of the absorptance data; one must also consider the criterion r_λ. Short-wave visible and infrared radiation is often unsuitable because it destroys vitamins and has a detrimental effect on the color of the foodstuffs being irradiated.

The magnitude of r_λ is determined as follows. Let us assume that irradiation is undesirable in the range $\Delta\lambda_i (i=1,2,3,\ldots)$, but is desirable at $\Delta\lambda_j$ (Fig. 7.19,a-c). In this case an ideal infrared generator is one which has a discontinuous spectrum with $\Delta\lambda_j$ and which does not emit radiation in the $\Delta\lambda_i$ range.

A lack of correspondence between ΔR_λ of the spectral irradiance of the infrared source $R_\lambda(T)$ and the assigned R_1 is defined as:

$$\Delta R_\lambda = |R_{\lambda 1} - R_\lambda(T)|. \qquad (7.83)$$

The coefficient r'_λ, which takes into account the specific effect only of the spectral composition of the infrared radiation on the substance being irradiated, is defined as follows:

$$r_\lambda' = (1/\Delta\lambda) \int_{\lambda_{11}}^{\lambda_2} [(R_\lambda(T_s) - \Delta R_\lambda)/R_\lambda T] d\lambda =$$

$$= 1 - (1/\Delta\lambda) \int_{\lambda_1}^{\lambda_2} (\Delta R_\lambda/R_\lambda(T) d\lambda. \tag{7.84}$$

When $\Delta R_\lambda = 0$, the non-correspondence of the spectral composition is zero, and the coefficient $r_\lambda' = 1$. When $R_{\lambda 1} \leqq R_\lambda(T)$, the coefficient r_λ' can be determined from the following relationship:

$$r_\lambda' = (1/\Delta\lambda) \int_{\lambda_1}^{\lambda_2} (R_{\lambda 1}/R_\lambda(T)) d\lambda. \tag{7.85}$$

With the aid of the criterion r_λ, and by using the spectral thermal radiational characteristics A_λ, R_λ, T_λ, and $L_\lambda \tau_\ell$, we can also evaluate quantitatively the effect which the spectral composition has on the degree of influence of the infrared radiation. In this case the criterion r_λ can be determined according to the following equation:

$$r_\lambda = \sum_j [\int_{\lambda_1}^{\lambda_2} R_\lambda(T)A_\lambda d\lambda]/\int R_{\lambda 1}A_\lambda d\lambda] -$$

$$- \sum [\int_{\lambda_1}^{\lambda_2} R_\lambda(T)A_\lambda d\lambda/\int_{\lambda_1}^{\lambda_2} R_\lambda(T) d\lambda]. \tag{7.86}$$

By substituting T_λ, $L_\lambda^* \tau_\ell$, and R_λ for A_λ in Eq. 7.86 we can determine the share of the desirable (j) or undesirable (i) effect.

Figs. 7.20 and 7.21 show how the criterion ξ can be used for estimating the efficiency of infrared generators.

Thus, in designing new, technologically advanced industrial infrared installations, a thorough knowledge of the physical principles of infrared irradiation is required.

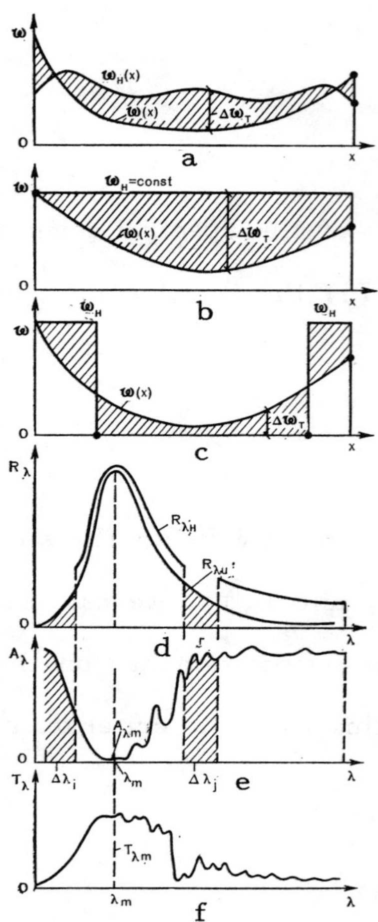

Figure 7.19 Diagram for determining coefficients k_T and r_λ [36, 61].

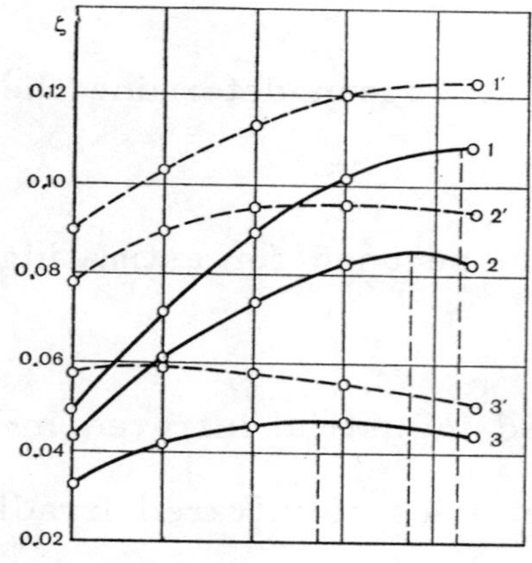

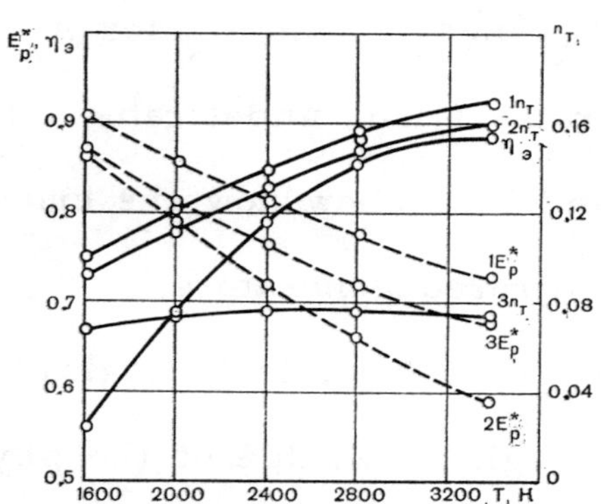

Figure 7.20 Dependence of $\overset{*}{E}_{net}$, n_T, and η_{ef} on the temperature of the infrared generator in the case of irradiation of different kinds of starch (ℓ=5.0 mm) (dashed lines represent the values obtained with no account being taken of η_{ef}) [61]: 1,1' - potato starch; 2,2' - cornstarch; 3,3' - rice starch.

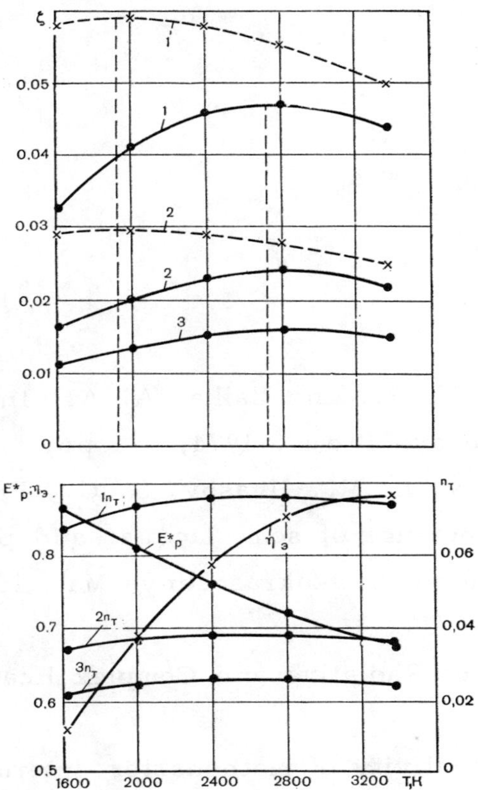

Figure 7.21 Dependence of E^*_{net}, n_T, and η_{ef} on the temperature of the infrared generator in the case of irradiation of layers of rice starch of different thicknesses (dashed lines represent the values of ξ obtained with no account being taken of η_{ef}) [61]: 1 – 5.0 mm; 2 – 10.0 mm; 3 – 15.0 mm.

CONCLUSION

The increasing use of infrared technology in the food industry calls for further investigations into the mechanism of heat and mass transfer processes during infrared irradiation of foodstuffs.

The physical principles of infrared irradiation of foodstuffs set forth in this book can serve as a basis for developing progressive food technologies and quality control methods, as well as for designing highly efficient industrial infrared installations.

1. Avramenko, V. V., Eselson, M. P., and Zaika, A. A. Infrared Spectra of Foodstuffs. Moscow: Pishchevaya Prom., 1974, 174 pp.

2. Afanaseva, G. D., Vinogradov, L. M., Ilyasov, S. G., Fridzon, M. B., and Tyurin, B. F. Spectral reflectance of some lacquer and paint coatings in the 0.25-15.0 μm spectral region. Lakokrasochnye Mat. i Prim., 1969, No. 4, p. 38.

3. Adrianov, V. N. Principles of Radiative and Complex Heat Transfer. Moscow: Energiya, 1972, 464 pp.

4. Aleksandrov, A. N. The sensitivity of methods for determining the water concentration of substances according to their infrared absorption spectra. Zhur. Priklad. Spektr., 1969, vol. 10, No. 5, pp.792-796.

5. Ambartsumyan, V. A. Problem of multiple scattering of light in a plane-parallel medium with internal reflection from the boundary surface. Uchenye Zap. LGU, C, Mat., 1964, No. 37, pp. 3-11.

6. Angersbakh, N. I. Thermal radiational convective drying of grapes by using solar energy. Summary of dissertation (Cand. Tech. Sc.). Moscow, 1986, 23 pp.

7. Angersbakh, A. K. Intensification of the process of thermal radiational convective drying of apples and quince. Summary of dissertation (Cand. Tech. Sc.). Moscow, 1987, 24 pp.

8. Andreeva N. P., Gorbacheva, M. P., and Mizonova, V. K. New types of quartz thermal radiators. Svetotekhnika, 1975, No. 9, pp. 12-13.

9. Ashkenazi, G. I. Color in Nature and in Engineering. Moscow: Energiya, 1974, 88 pp.

10. Ageenko, I. S., Ilyasov, S. G., Krasnikov, V. V., and Tyurev, E. P. Method for measuring the thermal radiational characteristics of foodstuffs subjected to infrared irradiation by a hemispherical flux. Ing.-Fiz. Zhur., 1984, vol. 46, No. 6, p. 952.

11. Bulyandra, A. F. Scientific and technical principles of selecting efficient conditions for drying processes and designing of drying equipment for the food industry. Summary of doctoral dissertation. Kiev, 1978, 48 pp.

12. Bramson, M. A. Infrared Radiation of Heated Substances. Moscow: Nauka, 1969, 224 pp.

13. Buger, P. Optical Treatise on the Gradation of Light. Moscow, Leningrad: Akad. Nauk SSSR, 1950, 173 pp.

14. Vavilov, S. I. Collected Works, vol. 1. Moscow: Akad. Nauk SSSR, 1954, 80 pp.

15. Voishvillo, N. A. The Gurevich-Kubelka-Munk theory for scattering layers with two reflecting boundaries. Optika i Spektroskopiya, 1974, vol. 37, No. 1, p. 136.

16. Galin, N. M. Thermal radiational drying of bread crumbs. Summary of dissertation (Cand. Tech. Sc.). Moscow, 1976, 21 pp.

17. Garcheva, T. V. Optical properties of flavoring substances in the Bulgarian food industry. Summary of dissertation (Cand. Tech. Sc.). Moscow, 1978, 29 pp.

18. Gershun, A. A. Selected Works on Photometry and Lighting Engineering. Moscow: Gosizdat, 1958, p. 548.

19. Ginzburg, A. S. Infrared Technology in the Food Industry. Moscow: Pisschevaya Prom., 1966, 408 pp.

20. Ginzburg, A. S. Theoretical and Technological Principles of Drying Foodstuffs. Moscow: Pishchevaya Prom., 1976, 528 pp.

21. Van de Hulst, H. C. Scattering of Light by Small Particles (in Russian translation). Moscow: Inostr. Lit., 1961, p. 49.

22. Grigorev, B. A. Pulse Heating by Radiation. Moscow: Nauka, 1974, Part I, 319 pp., Part II, 727 pp.

23. Gribkova, G. N. Optical properties of cereals and cereal products. Summary of dissertation (Cand. Tech. Sc.). Moscow, 1973, 25 pp.

24. Gurevich, M. M. Coefficients of reflection, transmission, and absorption of layers of different substances. Optiko-Mekh. Prom., 1976, No. 5, pp. 10-11.

25. Danilevich, A. F., Ilyasov, S. G., and Dolatsis, Ya. A. Thermal radiational properties of cooked and dried buckwheat groats. Trudy Lat. Selsk. Khoz. Akad., 1976, No. 88, pp. 17-19.

26. Dolatiis, Ya. A., Ilyasov, S. G., and Krasnikov, V. V. Effect of Infrared Radiation on Wood. Riga: Zinatne, 1973, 276 pp.

27. Dolin, L. S. Propagation of a narrow light beam in a medium with strongly anisotropic scattering. Izvestiya VUZ, Radiofizika, 1966, vol. 9, No. 1, pp. 61-71.

28. Dushchenko, V. P., Bulyandra, A. F., and Kucherik, I. M. Optical properties of some foodstuffs. Izvestiya VUZ, Pischchevaya Tekhnol., 1967, No. 4, pp. 152-155.

29. Dushchenko, V. P., Kucheruk, I. M., Berezhnoi A. F., and Petrova, R. S. Systematic investigation of the thermal radiational desiccation of capillary-porous substances. In: Heat and Mass Transfer, vol. VI, Part II. Kiev: Naukova Dumka, 1968, pp. 14-18.

30. Zhushman, A. I., Kovalenok, V. A., Syroedov, V. I., Ilyasov, S. G., and Tyurev, E. P. Optical and thermal radiational properties of some types of starch. Sakhar. Prom., 1975, No. 6, p.61.

31. Zagoruiko, A. L. and Korobeinikova, L. A. Method for determining the color index of crystalline sugar. Certificate of Invention No. 256666.

32. Zuev, V. E. Transparency of Atmosphere for Visible and Infrared Beams. Moscow: Soviet. Radio, 1966, 318 pp.

33. Ivanov, A. P. Optics of Scattering Media. Minsk: Nauka i Tekhnika, 1969, 592 pp.

34. Ivanov, V. V. Radiative transfer in multilayered optically thick atmosphere. Uchenye Zapiski LGU, Parts I and II, 1976, No. 385, pp. 3-23, 23-29.

35. Ivanov V. V. Invariance principle and internal radiation field in semi-infinite atmospheres. Astronom. Zhur., 1975, vol. 52, No. 3, pp. 217-226.

36. Ivens, B., Chu, C., and Churchill, S. Effect of anisotropic scattering on radiation transfer. Teploperedach, Ser. C, 1965, No. 3, pp. 69-76.

37. Ilich, G. K. Reflection coefficients of a semi-infinite layer in a scattering medium. Zhur. Priklad. Spektr., 1968, vol. 9, No. 2, pp. 294-297.

38. Ilich, G. K. and Ivanov, A. P. Diffuse reflection of optically thick layers in light-scattering media. Izv. Akad. Nauk SSSR, Fiz. Atm. i Okeana, 1965, vol. 1, No. 6, pp. 589-598.

39. Ilyasov, S. G. Theoretical principles of infrared irradiation of foodstuffs. Summary of doctoral dissertation. Moscow, 1977, 47 pp.

40. Ilyasov, S. G. and Krasnikov, V. V. Classification of foodstuffs according to their optical properties. Izv. VUZ, Pishchev. Tekhnol., 1970, No. 4, p. 21.

41. Ilyasov, S. G., Angersbakh, A. K., Angersbakh, N. I., and Kazimov, V. N.

Color index of dried fruits and grapes. Piscchevaya Prom., Ser. 6, No. 5, 1986, pp. 9-12.

42. Ilyasov, S. G., Angersbakh, N. I., and Angersbakh, A. K. Radiation energy transfer in scattering substances subjected to diffuse and directional irradiation at a certain angle. Inzhenerno-Fiz. Zhur., 1986, vol. 51, No. 4, pp. 633-638.

43. Ilyasov, S. G., Angersbakh, A. K., Angersbakh, N. I., and Soliev, A. Kh. Transfer of infrared radiation energy and heat transfer in thermal micro-biological processing of fruits. Agropishcheprom, 1988, No. 11, p.151.

44. Ilyasov, S. G. and Krasnikov, V. V. Methods for Determining the Optical and Thermal Radiational Properties of Foodstuffs. Moscow: Pishchevaya Prom., 1972, 175 pp.

45. Ilyasov, S. G. and Krasnikov, V. V. Radiation energy transfer in absorb-ing and scattering substances subjected to directional irradiation at a certain angle. Ing-Fiz. Zhur., 1972, vol. 23, No. 2, pp. 267-277.

46. Ilyasov, S. G., Lyamtsev, A. K., and Alekseev, B. V. Evaluation of in-frared radiation generators. Mekh. i Elektro. Sots. Selsk. Khoz., 1974, No. 10, pp. 21-22.

47. Ilyasov, S. G. and Krasnikov, V. V. Transfer of radiant energy in multi-layered scattering and absorbing substances. Ing.-Fiz. Zhur., 1974, vol. 27, No. 6, pp. 1053-1060.

48. Ilyasov, S. G., Krasnikov, V. V., and Tyurev, E. P. Methods for the simultaneous measurement of bihemispherical thermal radiational characteris-tics of scattering substances. Ing.-Fiz. Zhur., 1975, vol. 28, No. 6, pp. 1031-1039.

49. Ilyasov, S. G. and Krasnikov, V. V. Heat and mass transfer in capillary-porous substances subjected to infrared irradiation. In: Heat and Mass Transfer, vol. 5. Minsk: Inst. Teplo i Mass Obmena, 1976, pp. 165-176.

50. Ilyasov, S. G., Krasnikov, V. V., and Tyurev, E. P. Mrthods for investi-gating the spreading of the effective cross section in the case of direc-tional irradiation of absorbing and scattering substances. Ing.-Fiz. Zhur., 1977, vol. 32, No. 2, pp. 264-269.

51. Ilyasov, S. G., Krasnikov, V. V., and Sergeev, M. B. Transfer of radia-tion energy in substances with variable optical properties. Ing.-Fiz. Zhur., 1976, vol. 30, No. 3, pp. 389-396.

52. Ilyasov, S. G. and Angersbakh, A. K. Principles of heat and mass trans-fer in the case of infrared treatment of apples and quince. Pishch. i Pererab. Prom., 1987, No. 6, p.26.

53. Ilyasov, S. G., Angersbakh, N. I., Angersbakh, A. K., Kazimov, V. V., Tyurev, E. P., and Ageenko, I. S. Methods of producing dry fruits. Certificate of Inventions No. 1279574, USSR Byul. Izobr., 1986, No. 48, p. 9.

54. Ilyasov, S. G., Krasnikov, V. V., Fridzon, M. F., and Shlyakhov, V. I. Methods for measuring the spectral thermal radiational characteristics of light-scattering substances. Trudy TsAO, 1973, No. 114, 72 pp.

55. Kakalashvili, A. N., Bibilishvili, V. I., Barabadze, I. I., and Selyukov, N. G. Spectral optical characteristics of tea. In: Tea. Cultivation and Production. Tbilisi, 1971, No.4, pp. 30-36.

56. Kamenshchikova, S. V., Smirnov, A. M., and Mokhanchev, I. A. Spectrophotometrical methods for determining the quality of tobacco. Tabak, 1975, No. 3, pp. 36-40.

57. Kard, P. G. Theory concerning multilayered nonsymmetrical reflectors. Optika i Spektroskopiya, 1961, vol. 10, No. 3, pp. 384-389.

58. Keshanidi, Kh. L. Apparatus for determining cracking in wheat grains. Pishchevaya Tekhnol., 1971, No. 4, pp. 136-140.

59. Kizel, V. A. Reflection of Light. Moscow: Nauka, 1973, 351 pp.

60. Kislovsky, L. D. Optical properties of water and ice in infrared and radio-wave spectral range. Optika i Spektroskopiya, 1959, vol. 7, No. 3, pp. 311-320.

61. Kovalenok, V. A. Drying and infrared thermal processing of starch. Summary of dissertation (Cand. Tech. Sc.). Moscow, 1977, 26 pp.

62. Kozyrev, B. P. Vacuum radiation thermocouples. Izv. LETI, 1960, No. 44, pp. 3-21.

63. Kozyrev, B. P. and Vershinin, O. E. Determining the spectral coefficients of diffuse reflection of infrared radiation from blackened surfaces. Optika i Spektroskopiya, 1959, vol. 6, No. 4, p. 542.

64. Kondratev, K. Ya. Actinometry. Leningrad: Gidrometizdat, 1965, 691 pp.

65. Kovalev, A. E., VEKSHINA, A. A., Nikolaev, E. G., and Kairov, E. A. Method for calculating the distribution of the energy density of the radiation flux in the image spot of a hemispherical attachment for measuring diffuse reflection. Ing.-Fiz. Zhur., 1975, vol. 28, No. 5, pp. 838.

66. Kostenko, V. G., Abragam, D. R., Bakulina, L. F., and Vasiev, M. G. Photometric evaluation of the quality of potato starch. Sakhar. Prom., 1976, No. 7, pp. 67-69.

67. Korolev, A. F. Theoretical Optics. Moscow: Vyssh. Shkola, 1966, 555 pp.

68. Krasnikov, V. V., Plaksin, Yu. P., Galin, N. M.m and Ilyasov, S. G.
     Graphic analytical method for determining the parameters which indicate the
     location of linear infrared generators in multilayrede installations. In: Equip-
     ment for the Food Industry. Moscow: TsNIITEI, 1976, No. 3, pp. 16-23.

69. Krasnikov, V. V., Plaksin, Yu. M., and Galin, N. M. Radiant heat transfer
     in multilayered band-type infrared installations. Khleb. i Konditer. Prom.,
     1976, No. 2, pp. 20-22.

70. Kropotkin, M. A., Kutilina, N. B.,and Sheveleva, T. Yu. Apparatus for
     measuring the transmission and reflection coefficients of light-scattering sub-
     stances in the spectral region from 1 to 20 μm. Izv. LETI, 1975, No. 161,
     pp. 110-114.

71. Kropotkin, M. A. Comparison of the focusing properties of hemispherical and
     semi-ellipsoidal mirrors. Izv. LETI, 1974, No. 142, pp. 72-74.

72. Lav, T and Grosh, R. Radiant heat transfer in absorbing, emitting and
     scattering media. Teplopredacha, 1965, vol. 87, No. 2, p.64.

73. Levitin, I. B. On the termindogy, characteristics, and classification of in-
     frared generators. Svetotekh., 1976, No. 5, pp. 11-13.

74. Latyev, L. M., Petrov, V. A., Chekhovsky, V. Ya., and Shestakov, E. N.
     Emissive properties of solid substances (Ed: A. E. Sheindlin). Moscow:
     Energiya, 1974, 472 pp.

75. Lebedev, P. D. Drying with infrared beams. Moscow: Gosenergoizdat.,
     1955, 232 pp.

76. Lisovenko, A. T., and Mikhelev, A. A. Penetration of infrared radiation
     into some capillary-porous substances. In: The Food Industry: Bread-Baking,
     Pastry-Making, Yeast-Making. Moscow: Nauka, 1962, No. 3, pp. 21-33.

77. Lykov, A. V. Theory of Heat Conduction. Moscow: Gostekhizdat, 1967,
     600 pp.

78. Lykov, A. V. Methods for solving nonlinear equations describing transient
     heat conduction. Izv. Akad. Nauk SSSR, Energetika i Transport, No. 5,
     pp. 109-150.

79. Lykov, A. V. and Akerman, L. Ya. Drying bread into hard and crisp bis-
     cuits. Trudy MTIPP, 1952, No. 2, p.97.

80. Lykov, A. V. Theory of Desiccation. Moscow: Gosenergoizdat, 1968, 470 pp.

81. Markov, M. N. Detectors of Infrared Radiation. Moscow: Nauka, 1968, 167
     pp.

82. Mikhailov, M. D. Transient heat and mass transfer in unidimensional sub-
     stances. Minsk: ITMO, 1969, 184 pp.

83. Ostrovsky, L. V. and Martynov, V. N. Permeability of vegetables with respect to integral infrared radiation flux. Konserv. i Ovoshch. Prom., 1975, No. 9, pp. 34-36.

84. Ostrovsky, L. V. Method for the complex determination of integral optical characteristics of foodstuffs. Izv. Vuzov, Pishchevaya Tekhnol., 1975, No. 2, pp. 168-170.

85. Ozisik, M. N. Radiative Transfer and Interactions with Conduction and Convection. New York: John Wiley & Sons, 1976, pp. 616.

86. Panfilov, A. S. Reflecting properties of natural substances. Kosm. Issled., 1976, vol. 14, No. 2, pp. 316-318.

87. Petrol, V. A. and Stepanov, S. V. Effect of reflection at the boundaries and of the selectivity of the optical properies of a medium on the radiational-conductive heat transfer in a plane layer. Teplofizika Vys. Temp., 1976, vol. 14, No. 5, pp. 957-964.

88. Plaksin, Yu. M. Baking pastry in infrared ovens. Summary of dissertation (Cand. Tech. Sc.). Moscow, 1972, 24 pp.

89. Plaksin, Yu. M. and Syroedov, V. I. External heat transfer under the condition of pulsed energy supply and quantitative method for determining the efficiency of infrared generators. In: Heat and Mass Transfer, vol. 6. Minsk: ITMO, 1972, pp. 287-294.

90. Popov, Yu. A. Some principles governing heat transfer by means of radiation in scattering media. Teplofizika Vys. Temp., 1969, vol. 7, No. 1, pp. 97-104.

91. Prishivalenko, A. P. A simple graphic method for determining the optical constants based on two measurements of transmittance. Optika i Spektroskopiya, 1966, vol. 21, No. 1, pp. 124-125.

92. Prishivalenko, A. P. Reflection of Light from Absorbing Media. Minsk: BSSR Akad. Nauk, 1963, p. 302.

93. Prishivalenko, A. P. and Astafeva, L. G. Absorption, scattering and attenuation of light by water-coated atmospheric particles. Izv. Akad. Nauk SSSR, Fiz. Atm. i Okeana, 1974, vol. 10, No. 12, pp. 1322-1327.

94. Rabinovich, G. P. and Slobodkin, L. S. Thermal radiational and convective drying of paint and varnishes. Minsk: BSSR Akad. Nauk, 1966, 172 pp.

95. Rvachev, V. P. and Sakhnovsky, M. Yu. Determining the optical properties of substances having arbitrary scattering indicatrixes with the aid of an integral photometer. Zhur. Priklad. Spektr., 1966, vol. 4, No. 2, pp. 172-174.

96. Rozenberg, G. V. Some aspects of the problem of spectral analysis of

light-scattering media and the reflectance of colored polydispersed substances. Izv. Akad. Nauk SSSR, Fiz. Atm. i Okeana, 1974, vol. 10,
No. 12, pp. 1322-1327.

97. Rogov, I. A. and Gorbatov, A. V. Physical methods of processing foodstuffs. Moscow: Pishchevaya Prom., 1974, 584 pp.

98. Rogov, I. A. and Nekrutman, S. V., Superhigh frequency and infrared
heating of foodstuffs. Moscow: Pishchevaya Prom., 1976, 210 pp.

99. Rogov, I. A., Adamenko, V. Ya., Nekrutman, S. V., Ilyasov, S. G.,
Krasnikov, V. V., Kopylov, Yu. a., Tyurev. E. P., Zhukov, N. N., and
Vigderman, V. Sh. Electrophysical, Optical and Acustic Properties of Food-
stuffs (Ed: I. A. Rogov). Moscow: Leg. i Pishch. Prom., 1981, 286 pp.

100. Romanova, L. M. Solving the radiative transfer equation in the case where
the indicatrix differs sharply from the spherical one. I. Optika i Spektroskopiya, 1962, vol. 13, No. 3, pp. 429-435.

101. Rubtsov, N. A. Radiational and combined heat transfer. Summary of doctoral dissertation. Novosibirsk, 1970, 46 pp.

102. Rychkov, V. I. Infrared drying and heating. In: Lighting and Infrared
Engineering, vol. 3. Moscow: VINITI, 1973, pp. 196-247.

103. Ryzhkov, L. N. and Luchsheva, Z, A., Choosing the spectrum of an infrared source for heating and drying purposes. In: Trudy VNIIETO. Moscow
Energiya, 1970, No. 4, pp. 116-120.

104. Sadovsky, G. N. Assessing the quality of flour on the basis of its whitness.
Izv. Vuzov, Pishchevaya Prom., 1974 No. 3. pp. 142-144.

105. Samson, A. M. Resonance luminescence of a plane-parallel layer. Doklady
Akad. Nauk BSSR, 1958, vol. 2, No. 1, pp. 15-19.

106. Selyukov, N. G. Dependence of the optical properties of capillary-porous
substances on their moisture content. In: Heat and mass transfer, vol. 6.
Minsk: ITMO, 1972, pp. 274-283.

107. Selyukov, N. G. Optical properties of foodstuffs subjected to thermal radiational treatment. Summary of dissertation (Cand. Tech. Sc.). Moscow,
1968, 34 pp.

108. Skverchak, V. D. Baking of bread and flour products in infrared ovens.
Summary of dissertation (Cand. Tech. Sc.). Moscow, 1970, 25 pp.

109. Sidorov, S. N., Smolkin, M. N., and Nikiticheva, A. M. Integral and
spectral characteristics of powerful halogen incandescent lamps. Optiko-
Mekh. Prom., 1976, No. 2, pp. 79-80.

110. Slobodkin, L. S. and Sotnikov-Yuzhik, Yu. M. Methods for determining

the Thermal Radiational Characteristics of Polymer Coatings. Minsk: Nauka, i Tekhnol., 1977, 160 pp.

111. Sobolev, V. V. Scattering of Light in the Atmosphere of Planets. Moscow: Nauka, 1972, 336 pp.

112. Skverchuk, V. D., Plaksin, Yu. M., Syroedov, V. I., and Ginzburg, A. S. Analytical investigations of the locations of electrical infrared generators in thermal radiational installations. Elektr. Obrab. Mater., 1970, No. 6, p. 53.

113. Syroedov, V. I., Ilyasov, S. G., Zhushman, A. I., Kovalenok, V. A., and Lukin, N. D. Method for producing dextri n. Certificate of Invention No. 576884. USSR BUl. Izobr., 1978, p.21.

114. Stepanov, B. I., Chekalinskaya, Yu. I., and Girin, O. P. Methods for determining the optical constants of light-scattering media. Trudy Inst. Fiz. i. Mat. Akad Nauk BSSR, 1956, No. 2, pp. 152-175.

115. Titarchuk, L. G. Radiative transfer in spherical multilayred atmosphere of a planet. Kosmos. Issled., 1973, vol. 11, No. 1, pp. 130-139; Ibid., 1972, vol. 10, No. 6, pp. 905-915.

116. Toporets, A. S. Reflection of light by uneven surfaces and light-scattering media. Summary of doctoral dissertation. Leningrad, 1980, 52 pp.

117. Trofimova, L. S. Indicatrix of scattering of light by a polydispersed medium. In: Atmospheric Optics. Moscow: Nauka, 1974, pp. 210-214.

118. Tyurev, E. P. Thermal radiational characteristics of foodstuffs and methods of their investigation under various irradiation conditions. Summary of dissertation (Cand. Tech. Sc.). Moscow, 1976, 25 pp.

119. Fabrikant, V. A. On Bouguer's law. Izv. Akad. Nauk SSSR, Ser. Fiz., 1962, vol. 26, No. 1 pp. 61-66.

120. Khusainov, U. M. Drying Fruits and Grapes with the Aid of Stored Solar Energy. Moscow: Leg. i Pishch. Prom., 1983, 41 pp.

121. Fridzon, M. B., Pakhomova, L. A., Selyukov, N. G., and Gorbatova, E. S. Spectral characteristics of reflection of some substances used in meteorology. Meteorologiya i Gidrologiya, 1967, No. 9, p. 97.

122. Khrustalev, B. A. Methods for investigating the radiational properties of solid surfaces. In: Radiative Heat Transfer. Kaliningrad: KGU, 1974, pp. 3-51.

123. Khrustalev, B. A. Radiational properties of construction materials used in engineering design of heat exchangers. Summary of doctoral dissertation. Moscow, 1973, 55 pp.

124. Khvashchevskaya, Ya. S. Apparent violation of Bouguer's law in the infrared spectral region. Doklady Akad. Nauk SSSR, 1958, vol. 2, No. 3, pp. 100-103.

125. Tsirlin, Yu. A., Daii, A. R. Diffusion of light-scattering media. Optika i Spektroskopiya, 1962, vol. 12, No. 2, pp. 304-310.

126. Chandrasekhar, S. Transfer of Radiant Energy (in Russian translation). Moscow: Inostr. Lit., 1953, p. 180.

127. Chekalinskaya, Yu. I. Distribution of radiation inside a powder layer. Ind.-Fiz. Zhur., 1960, vol. 3, No. 7, pp. 43-50.

128. Shvets, V. A. and Kazansky, V. V. Alteration of the SF-4 spectrometer for obtaining the reflection spectra of powders. Prib. i Tekh. Eksp.,1966, No. 4, p. 213.

129. Shulgin, I. A. Plants and Sun. Leningrad: Gidrometeoizdat, 1973, pp. 251.

130. Yukhnevich, G. V. Infrared Spectroscopy of Water. Moscow: Nauka, 1973, 208 pp.

131. Yakovenko, V. A., Evdokimova, G. I., and Sedova, Z. A. Physico-chemical properties of potato starch. Izv. Vuzov Pishch. Prom., 1976, No. 4, pp. 34-36.

132. Abrams, M. and Viskanta, R. The effects of radiative heat transfer upon the melting and solidification of semitransparent crystals. J. Heat Transfer. Trans. ASME, 1974, vol. 96C No. 2, pp. 75-82.

133. Abdulkader, A. and Birkebak, R. C. Optical surface roughness and slopes measurements with a double beam spectrometer. Rev. Sci. Instrum., 1974, vol. 45, No. 11, pp. 1356-1360.

134. Adrianov, V. N. and Polyak, G. L. Differential methods for studying radiant heat transfer. Int. J. Heat and Mass Transfer, 1963, vol. 6, p. 355.

135. Allen, E. Basic equations used in computer color matching. II. Tristimulus match, two-constant theory. JOSA, 1974, vol. 64, No. 7, pp. 991-993.

136. Armaly, B. F. and Lam, T. T. Influence of refractive index on reflectance from a semi-infinite absorbing-scattering medium with collimated incident radiation. Int. J. Heat and Mass Transfer, 1975, vol. 18, pp. 893-900.

137. Beach, A. D. A simple device for the efficient coupling of a light emitting diode to an optical fiber. J. Physics. E. Sci. Instrum., 1975, vol. 8, No. 9, pp. 745-747.

138. Blandamer, M. I. and Fox, M. F. Spectroscopic properties of water. Comprehensive treatise, vol. 2. Water Crystalline Hydrates. Aqueous Solutions of Simple Nonelectrical Substances. New York: John Wiley & Sons, 1973, pp. 459-494.

139. Blevin, W. R. and Brawn, W. J. Large-area bolometers of evaporated gold. J. Sci. Instrum., 1965, vol. 42, No. 1, pp. 19-23.

140. Brandenberg, W. Focusing properties of hemispherical and ellipsoidal mirror reflectometers. JOSA, 1964, vol. 54, No. 10, pp. 1235-1237.

141. Brandenberg, W. and Neu, J. Unidirectional reflectance of imperfectly diffusive surfaces. JOSA, 1966, vol. 56, No. 1, p. 97.

142. Cess, R. D., Mighdoll, P., and Tiwari, S. N. Infrared radiative heat transfer in nongray gases. Int. J. Heat and Mass Transfer, 1967, vol. 10, p. 1521.

143. Chandrasekhar, S. Radiative Transfer. New York: Dover Publ., 1960, p. 101.

144. Chu, C, Churchill, S., and Pang, S. Electromagnetic Scattering (Ed: M. Karker). New York: Macmillan Co., 1963, p. 507.

145. Coblentz, W. The diffuse reflecting power of various substrates. Nat. Bur. Standards, 1913, vol. 9, p. 283.

146. Devellers, P., Lemaitra, A., Lescure, J. P., and Roger, J. Coloration des sucress tentative de measure absolue, 3 partie. Sucrerie Francaise, 1970, vol. 111, No. 1, pp. 35-43.

147. MacAdam, D. L. Uniform color scales. JOSA, 1974, vol. 64, No. 12, pp. 1691-1702.

148. Domoto, G. A. and Wang, W. C. Radiative transfer in homogeneous nongray gases with nonisotropic particle scattering. J. Heat Transfer. Trans. ASME, 1965, vol. c96, No. 3, pp. 385-390.

149. Dunn, T., Richmond, J., and Wiebelt, J. Ellipsoidal mirror reflectometer. I. Res. Nat. Bur. Stand., 1966, vol. 70C, No. 2, p. 75.

150. Duntley, S. Q. The optical properties of diffusing materials. JOSA, 1942, vol. 32, No. 2, pp. 61-70.

151. Eddington, A. The Internal Constitution of the Stars. Cambridge: Univ. Press, 1926, p. 71.

152. Geist, J. C. and Richmond, J. C. On the absorptance of cavity-type receivers. NBS Tech. Note, 1971, No. 575, pp. 1-38.

153. Goebel, D. G. Generalized integrating-sphere theory. Applied Optics, 1967, vol. 1, No. 1, pp. 125-128.

154. Grosjean, C. C. In: Electromagnetic Scattering (Ed: M. Kerker). New York: Macmillan and Co., 1963, pp. 485-506.

155. Gurevich, M. M. Über eine Rationalle Klassification der Lichtstreuenden Medien. Physik Zeitsch., 1930, vol. 31, p. 753.

156. Helmholtz, H.   Theorie der Wärme.   Leipzig: J. A. Barth Verlag, 1903, p. 158.

157. Hopf, E.   Mathematical Problems of Radiative Equilibrium.   Cambridge Univ. Press, 1934, p. 70.

158. Hottel, H. C., Sarofim, A. F., and Sze, D. K.   Radiative transfer in an absorbing-scattering medium.   Proc. 3-rd Int. Heat Transfer Conf.   Chicago, 1966, vol. 5, pp. 112-121.

159. Hughes, H. K.   Beer's law and the optimum transmittance in absorption measurements.   Applied Optics, 1963, vol. 2, No. 9, pp. 937-945.

160. Ilyasov, S. G. and Krasnikov, V. V. Radiation propagation in multilayered systems with variable optical properties.   Intern. J. Heat and Mass Transfer, 1975, vol. 18, No. 6, pp. 769-774.

161. Ilyasov, S. G. and Krasnikov, V. V.   Optical properties of food products and infrared radiation heat transfer in drying.   Int. Symp. Heat and Mass Transfer Problems.   Food Eng., Wageningen, Netherlands, 1972, October 24-27, F.8, pp. 1-24.

162. Jansen, J. E. and Torberg, R. H.   Measurement of spectral reflectance using an integrating hemisphere.   Measurement of thermal radiational properties of solids. Washington, D. C., NASA, 1963, p.169.

163. Judd, D. B.   Terms, definitions and symbols in reflectometry, JOSA, 1967, vol. 57, No. 4, p. 445.

164. Karvaly, B.   Some comments on theories concerning the connections between true (bulk) absorption and diffuse reflectance spectra of powdered solids. Part I.   On Bodo's theory.   Acta Physica et Chemica, 1970, vol. 16, Nos. 3-4, pp. 107-117.

165. Kortüm, G. and Delfs, H.   Reflexionspektrometrische Messung im infraroten Spektralgebiet.   Spectrochimica Acta, 1964, vol. 20, pp. 405-413.

166. Kronstein, M., Kraushaar, R. J., Deacle, R. E.   Sulfur as a standard of reflectance in the infrared.   JOSA, 1963, vol. 53, No. 4, pp. 458-465.

167. Kubelka, P. and Munk, F.   Ein Betrag zur Optik der Farbanstriche.   Z. für tech. Physik, 1931, vol. 12, p. 593.

168. Lathrop, A. L.   A Kubelka-Munk calculation chart for light-scattering materials.   Tappi, 1964, vol. 47, No. 12, pp. 789-791.

169. Lathrop, A. L.   Diffuse scattered radiation theories of Duntley and Kubelka-Munk.   JOSA, 1965, vol. 55, No. 9, pp. 1097-1104.

170. Lin, C. C. and Chan, S. H.   An analytical solution for a planar nongray medium in radiative equilibrium.   J. Heat Transfer, Trans. ASME, 1975, vol. C97, No. 1, pp. 26-32.

171. Look, D. C. and Love, T. J. Investigation of the effects of surface roughness upon reflectance. AIAA Paper, 1970, Nos. 70-820, pp. 1-11.

172. Luikov, A. V. Systems of differential equations of heat and mass transfer in capillary-porous bodies. (Review). Intern. J. Heat and Mass Transfer, 1975, vol. 18, No. 1, pp. 1-14.

173. Mcbean, D. Sun-drying of foods. Food Technol. Australia, 1975, vol. 27, No. 11, pp. 474-475.

174. Mark, C. The spherical harmonics method. II. Atomic energy of Canada Ltd. Res. Dev. CRT-338. Chalk River, Ontario, 1947.

175. Mie, G. Beitrage zur Optitrüber Medien speziell Kollodler Metallösangen. Ann. Phys., 1908, vol. 25, No. 2, p. 377.

176. Morris, J. C. Integrating sphere for the infrared. Appl. Optics, 1966, vol. 5, No. 6, pp. 1035-1037.

177. Neher, R. T. and Edwards, D. K. Far infrared reflectometer for imperfectly diffuse specimens. JOSA, 1965, vol. 4, No. 7, pp. 775-780.

178. Neher, R. T. Appl. Optics, 1965, vol. 4, No. 7, p. 775.

179. Palmer, K. F. and Williams, D. Optical properties of water in the near infrared. JOSA, 1974, vol. 64, No. 8, pp. 1107-1110.

180. Paschen, von F. Über die Verteilung der Energie im Spektrum des schwarzen Korpers bei neideren Temperaturen. Gesammtsitzungvom, 1899, Bd. 27, Apr., S. 405.

181. Perrelle, E. Moss, T., and Herbert, H. The measurement of absorptivity and reflectivity. Infrared Physics, 1963, vol. 3, p. 35.

182. Phillips, D. G. and Billmeyer, F. W. Predicting the reflectance and color of paint films by Kubelka-Munk analysis. IV. Kubelka-Munk scatering coefficient. J. Coat. Technol., 1976, vol. 48, No. 616, pp. 30-36.

183. Pfening, H. Zur Trocknung von Kartoffel-starke mit infraroten Strahlen. Die Starke, 1968, Band 20, No. 1, S. 16-26.

185. Putley, E. H. Solid state devices for infrared detection. J. Sc. Instrum., 1966, vol. 43, pp. 857-868.

186. Richmond, J. C. and Dunn, S. T. Ellipsoidal mirror relectometer incuding means for averaging the radiation reflected from the sample. USA Pat. No. 3504983, 1970.

187. Schadach, V. P. Bestimmung der räumlichen spektralen Strahlungseigenschaften (Reflexions- und Absorptionsvermögen) von Werkstoffen in Abhändigkeit von ihrer Temperatur. Inter. Z. für Elektrowärme, 1965, Band 23, No. 1, S. 8.

188. Schuster, A.   Radiation through a foggy atmosphere.   Astrophysical J.,
      1905, vol. 21, pp. 1-22.

189. Schwartzschild, K.   Über diffusion und absorption in der Sonnenatmosphäre.
      Sitzungsber, Königl. Preuss. Akad. Wiss., 1914, vol. 47, p. 1183.

190. Seveik, V. J.   Emissivity of beef.   Food Technol., 1962, vol. 16, pp. 124-
      126.

191. Sherrell, F. G. and Shahrokhi, F.   Determination of hemispherical emittance
      by measurements of infrared bihemispherical reflectance. ASAA Paper, 1970,
      No. 65, pp. 1-9.

192. Snider, D. M. and Viskanta, R.   Radiation-induced thermal stratification
      in surface layers of stagnant water.   J. Heat Transfer ASME, 1975, vol. C97,
      No. 1, pp. 33-39.

193. Sorensen, C. M., Mockler, R. C., and Sullivan, W. J.   Depolarized correla-
      tion function of light doubly scattered from a system of Brownian particles.
      Phys. Rev. A. Gen Phys., 1976, vol. 14, No. 4, pp. 1520-1532.

194. Sparrow, E. M. and Cess, R. D.   Radiation Heat Transfer.   Belmont: Brooks·
      Cole Publ., 1966, p. 37.

195. Stokes, G. G.   On the intensity of light reflected from or transmitted
      through a pilr of plates.   Proc. Roy. Soc., London, 1860-62, vol. 11, p.
      545.

196. Swift, J. R.   Moisture measurements in the food industry.   An infrared ab-
      sorption technique.   Food Technol. in Australia, 1971, vol. 23, No. 7, pp.
      352-353.

197. Taudon, S. P. and Gupta, J. P. A note on the Kubelka-Munk theory of
      diffuse reflectance.   Indian J. Pure Appl. Phys., 1970, vol. 8, April, pp.
      231-232.

198. Taylor, A.   The measurement of diffuse reflection factors and new absolute
      reflectometer, JOSA, 1920, vol. 4, p. 9.

199. Tien, L. C. and Churchill, S. W.   An evaluation of approximations for ra-
      diant transport with isotropic scattering.   Che. Eng. Prog., Symposium Ser.,
      Heat Transfer, 1965, vol. 61, No. 59, pp. 155-161.

200. Torrance, K. E. and Sparrow, E. M.   Theory for off-specular reflection
      from roughened surfaces, JOSA, 1967, vol. 57, No. 9, pp. 1105-1114.

201. Traugott, S. C.   Radiation through a plane-parallel absorbing medium be-
      tween directional surfaces.   AIAA J., 1971, vol. 9, No. 3, pp. 500-507.

202. Tung, P. L. and Kan, N. K. A.   Calculation of the reflectance of a light
      diffuser with nonuniform absorption. JOSA, 1970, vol. 60, No. 9, pp. 1252-
      1256.

203. Viskanta, R. Effectiveness of a layer of an absorbing-scattering gas in schielding a surface from incident thermal radiation. J. of the Franklin Inst., 1965, vol. 280, No. 6, pp. 483-492.

204. Vrugt, J. W. Optical properties of thin powder layers. Philips Res. Repts., 1965, vol. 20, pp. 23-33.

205. Wendland, W. W. and Hecht, H. G. Reflectance Spectroscopy. New York: Int. Publ., 1966, p. 150.

206. Weston, K. C., Howe, J. J., and Greene, M. J. Approximate temperature distribution for a diffuse, highly reflecting material. AIAA Journal, 1972, vol. 10, No. 9, pp. 1252-1254.

207. Wiebelt, J. A. Ellipsoidal mirror reflectometer. J. Res. Nat. Bur. Stand., 1966, vol. 70, No. 2, pp. 75-88.

208. Wolfe, R. N., DePalma, J. J., and Saunders, S. B. Measurement of volume reflectance and volume transmittance of turbid media. JOSA, 1965, vol. 55, No. 8, pp. 956-962.

209. White, J. U. A new method for measuring diffuse reflectance in the infra-red region. JOSA, 1964, vol. 54, No. 11, pp. 1332-1337.

210. Ulbricht, R. Das Kugelphotometer. Munich-Berlin: R. Oldenburg Verlag, 1920, p. 99.

211. Umoff, N. A. Phys. Z., 1905, No. 6, p. 674.

212. Diffuse Reflectance Accessory SP375, SP890. Cambridge. Unicam Instruments Ltd., 1967.

213. Zeutner, R. C., Mac Gregor, R. K., and Pogson, J. T. Bidirectional reflectance characteristics of integrating sphere coatings. AIAA Paper, 1971, Nos. 71-77, pp. 1-10.

214. Zerlaut, G. and Krupnick, A. An integrating sphere reflectometer for the determination of absolute hemispherical spectral reflectance, AIAA J., 1966, vol. 4, No. 7, p. 1227.

215. Clark, C., Vinegar, R., and Hardy, J. Goniometric spectrometer for the measurement of diffuse reflectance and transmittance of skin in the infrared spectral region. JOSA, 1953, vol. 43, No. 11, p. 993.

216. Toporets, A. S. and Mazurenko, M. M. Goniospectrophotometric apparatus with high angular resolution. Optiko-Mech. Prom., 1964, No. 1, p.37.

217. Shibata, K. New concept in the measurement of translucent and opaque materials. Optical Instruments, 1966, No. 1, p.1.

218. Duncle, R. V. Spectral reflectance measurements. I. Symposium: Surface-effective spacecraft materials. Palo Alto, Calif., 1959. New York: John Wiley

& Sons, Inc., 1960, p. 117.

219. Reshina, I. I., Sakin, I. L., and Novak, I. I. A small attachment for the IKS-12 infrared spectrometer. Optiko-Mekh., Prom., 1961, No. 11, p. 18.

220. Norris, K. Sc. Instr., 1954, No. 31, p. 284.

221. Birbek, R., Ekkert, E. Effect of roughened metallic surface on the angular distribution of reflected monochromatic radiation. Teploperedacha, 1965, No. 1, p. 102.

222. An attachment for measuring the reflection of IPO-12. Instructions on its use. Leningrad: LOMO, 1963.

223. Bennett, H. E., Bennett, J. M., and Ashley, E. J. Infrared reflectance of evaporated aluminum films. J. Opt. Soc. Amer., 1962, vol. 52, p. 1245.

224. Kingslake, R. Appl. Optics and OPt. Eng. , III, 1965.

225. Deribere, M. Practical Use of Infrared Rays. Moscow, Leningrad: Gos-energoizdat, 1959, p. 100.

226. Weinberg, T. I., Voishvillo, N. A., and Makhlona, G. A. Neutral diffuse-reflecting glass. Optiko-Mekh. Prom., 1975, No. 1, p.28.

227. Melnikova, I. S. Drying of Pasta. Moscow: Pishchepromizdat, 1954.